Gap Junction Channels and Hemichannels

Edited by **Donglin Bai and Juan C. Sáez**

CRC Press
Taylor & Francis Group
Boca Raton London New York

CRC Press is an imprint of the
Taylor & Francis Group, an **informa** business

CRC Press
Taylor & Francis Group
6000 Broken Sound Parkway NW, Suite 300
Boca Raton, FL 33487-2742

First issued in paperback 2021

© 2017 by Taylor & Francis Group, LLC
CRC Press is an imprint of Taylor & Francis Group, an Informa business

No claim to original U.S. Government works

ISBN-13: 978-1-4987-3862-0 (hbk)
ISBN-13: 978-0-367-65844-1 (pbk)

Library of Congress Cataloging-in-Publication Data

Names: Bai, Donglin, editor. | Saéz, Juan Carlos, editor.
Title: Gap junction channels and hemichannels / edited by Donglin Bai and Juan C. Saéz.
Other titles: Methods in signal transduction.
Description: Boca Raton : CRC Press, Taylor & Francis Group, 2017. | Series: Methods in signal transduction series | Includes bibliographical references and index.
Identifiers: LCCN 2016004060 | ISBN 9781498738620 (hardcover : alk. paper)
Subjects: | MESH: Gap Junctions | Connexins--genetics | Synaptic Transmission--physiology
Classification: LCC QH603.C4 | NLM QU 350 | DDC 571.6--dc23
LC record available at http://lccn.loc.gov/2016004060

Visit the Taylor & Francis Web site at
http://www.taylorandfrancis.com

and the CRC Press Web site at
http://www.crcpress.com

Gap Junction Channels and Hemichannels

METHODS IN SIGNAL TRANSDUCTION SERIES

Joseph Eichberg, Jr. and Michael X. Zhu
Series Editors

Published Titles

Gap Junction Channels and Hemichannels, Donglin Bai and Juan C. Sáez
Cyclic Nucleotide Signaling, Xiaodong Cheng
TRP Channels, Michael Xi Zhu
Lipid-Mediated Signaling, Eric J. Murphy and Thad A. Rosenberger
Signaling by Toll-Like Receptors, Gregory W. Konat
Signal Transduction in the Retina, Steven J. Fliesler and Oleg G. Kisselev
Analysis of Growth Factor Signaling in Embryos, Malcolm Whitman and
    Amy K. Sater
Calcium Signaling, Second Edition, James W. Putney, Jr.
G Protein–Coupled Receptors: Structure, Function, and Ligand Screening,
    Tatsuya Haga and Shigeki Takeda
G Protein–Coupled Receptors, Tatsuya Haga and Gabriel Berstein
Signaling Through Cell Adhesion Molecules, Jun-Lin Guan
G Proteins: Techniques of Analysis, David R. Manning
Lipid Second Messengers, Suzanne G. Laychock and Ronald P. Rubin

Contents

Series Preface

The concept of signal transduction is now long established as a central tenet of biological sciences. Since the inception of the field close to 50 years ago, the number and the varieties of signal transduction pathways, cascades, and networks have steadily increased and now constitute what is often regarded as a bewildering array of mechanisms by which cells sense and respond to extracellular and intracellular environmental stimuli. It is not an exaggeration to state that virtually every cell function is dependent on the detection, amplification, and integration of these signals. Moreover, there is increasing appreciation that in many disease states, aspects of signal transduction are critically perturbed.

Our knowledge of how information is conveyed and processed through these cellular molecular circuits and biochemical switches has increased enormously in scope and complexity since this series was initiated 15 years ago. Such advances would not have been possible without the supplementation of older technologies, drawn chiefly from cell and molecular biology, biochemistry, physiology, and pharmacology, with newer methods that make use of sophisticated genetic approaches as well as structural biology, imaging, bioinformatics, and systems biology analysis.

The overall theme of this series continues to be the presentation of the wealth of up-to-date research methods applied to the many facets of signal transduction. Each volume is assembled by one or more editors who are preeminent in their specialty. In turn, the guiding principle for editors is to recruit chapter authors who will describe procedures and protocols with which they are intimately familiar in a reader-friendly format. The intent is to assure that each volume will be of maximum practical value to a broad audience, including students and researchers just entering this area, as well as seasoned investigators.

It is hoped that the information contained in the books of this series will constitute a useful resource to the life sciences research community well into the future.

Joseph Eichberg
Michael Xi Zhu
Series Editors

Preface

Gap junction channels are a group of intercellular channels ubiquitously expressed in tissues and organs to synchronize many physiological processes. Hemichannels were discovered about two decades ago (a hemichannel is half of a gap junction channel) as a relevant cell membrane pathway for the release and uptake of ions and small molecules, many of which are autocrine and paracrine signal molecules. Mutations in 14 out of 21 genes encoding gap junction protein subunits, called connexins, have been found to be associated with inherited human disorders and diseases, including hearing loss, skin diseases, peripheral and central neuropathic disorders, lens cataracts, cardiac arrhythmias, and developmental disorders. The prevalence of connexin 26 (Cx26) gene mutation-linked congenital sensorineural hearing loss is so high that complete DNA sequencing of the coding region of *GJB2* (the gene encoding Cx26) has become routine for newborns or during pregnancy. Because of the important physiological role of connexins and the linkage of the connexin gene mutations with many diseases, the field of connexin research has recently exploded and has become one of the most active areas of research. Numerous novel approaches and techniques have been developed, but there is no single book dedicated on the unique techniques and protocols for the research on these large pore channels. The last book on gap junction methods and protocols was published 15 years ago, which cannot meet the current need of researchers and trainees in the fields. To fill this gap, we recruited expert researchers in the field to share their state-of-the-art techniques, protocols, and thoughts/reviews on various approaches commonly used in the studies of gap junction channels and hemichannels.

Chapters 1 to 6 in this book are focused on the protocols on, approaches to, and reviews of gap junction channels, and Chapters 7 to 12 are on connexin hemichannels. We hope that this book will be a very useful reference book for graduate students, postdoctoral fellows, and researchers with an interest in gap junction channels and hemichannels. It should be noted that we have covered many common approaches in studying these channels, but some areas/approaches may not be discussed in detail in this book because of the difficulties in finding appropriate contributors.

We would like to thank all the contributors for taking their valuable time to put their chapters together to share their detailed experimental protocols and their views on different approaches. Without their willingness to help, consistent efforts in developing and revising their chapters, and their contribution in helping to reviewing some chapters, we would not be able to complete this book.

A video that relates to Chapter 3 is available from the CRC Press Web site. Interested readers should click on the Downloads/Updates tab at the following URL: https://www.crcpress.com/Gap-Junction-Channels-and-Hemichannels/Bai-Sez/9781498738620.

MATLAB® is a registered trademark of The MathWorks, Inc. For product information, please contact:

The MathWorks, Inc.
3 Apple Hill Drive
Natick, MA 01760-2098 USA
Tel: 508-647-7000
Fax: 508-647-7001
E-mail: info@mathworks.com
Web: www.mathworks.com

Cover Image: The cover image of this book was generously provided by Dr. Sandra A. Murray from Department of Cell Biology, University of Pittsburgh.

Editors

Dr. Donglin Bai received his PhD from the University of Cambridge, United Kingdom, in 1994. He then moved to Canada and worked on neurotransmitter receptors at the Loeb Research Institute, Ottawa, in the Department of Physiology of the University of Toronto, and at the Samuel Lunenfeld Research Institute, Toronto, as a postdoctoral fellow. In 2002 he was recruited as an assistant professor in the Department of Physiology and Pharmacology of the University of Western Ontario, London, Ontario, Canada. He is currently a tenured associate professor at the same university. His current research interests are the physiology of gap junction channels, including gap junction channel docking, single channel conductance, and gating properties. He is also interested in revealing how gap junction gene mutations are linked to human diseases (cardiac arrhythmias, hypomyelination, cataracts, deafness, and skin diseases) and developing strategies for rescuing the mutants. He received grant support from the following Canadian funding agencies: Canadian Institutes of Health Research, Canada Research Chairs, Heart and Stroke Foundation of Canada, Natural Sciences and Engineering Research Council, Canada Foundation for Innovation, and Early Researcher Award.

Dr. Juan C. Sáez received his PhD in neuroscience from Albert Einstein College of Medicine (AECOM), New York, New York, in 1986. He stayed for one year as an instructor in the Department of Neuroscience of AECOM, and then, he became an assistant professor in the same department. In 1993, he joined the Physiology Department of the Pontificia Universidad Católica de Chile, Santiago, Chile, where he has been a professor since 2003. His current research interest is understanding the regulation and the function of connexin- and pannexin-based channels in different cell types, including the cells of the nervous system, the immune system, and the gastrointestinal system and, more recently, on skeletal muscles. He has also characterized biophysical features of the mentioned channels. Recently, he has also used protocols for identifying highly selective inhibitors of connexin hemichannels without an effect on gap junction channels and with potent anti-inflammatory activity to treat chronic diseases. He has been continuously funded through the National Institutes of Health and different foundations of the Chilean government.

Contributors

Guillermo A. Altenberg
Department of Cell Physiology and
Molecular Biophysics and Center for
Membrane Protein Research
Texas Tech University Health Sciences
Center
Lubbock, Texas

Donglin Bai
Department of Physiology and
Pharmacology
University of Western Ontario
London, Ontario, Canada

Roger Cachope
Dominick P. Purpura Department of
Neuroscience
Albert Einstein College of Medicine
New York, New York

John A. Cameron
Department of Physiology and
Pharmacology
University of Western Ontario
London, Ontario, Canada

Juan Manuel Valdez Capuccino
Department of Pharmacology,
Physiology and Neuroscience
New Jersey Medical School
Rutgers University
Newark, New Jersey

Ricardo Ceriani
Centro Interdisciplinario de
Neurociencia de Valparaíso
Facultad de Ciencias
Universidad de Valparaíso
Valparaíso, Chile

Jorge E. Contreras
Department of Pharmacology,
Physiology and Neuroscience
New Jersey Medical School
Rutgers University
Newark, New Jersey

Sebastian Curti
Laboratorio de Neurofisiología Celular
Departamento de Fisiología
Facultad de Medicina
Universidad de la República
Montevideo, Uruguay

Maarten De Smet
Department of Basic Medical Sciences–
Physiology Group
Faculty of Medicine and Health Sciences
Ghent University
Ghent, Belgium

José F. Ek Vitorín
Department of Physiology
University of Arizona
Tucson, Arizona

Eliseo A. Eugenin
Public Health Research Institute and
Department of Microbiology,
Biochemistry and Molecular Genetics
Rutgers New Jersey Medical School
Rutgers, the State University of
New Jersey
Newark, New Jersey

Matthias M. Falk
Department of Biological Sciences
Lehigh University
Bethlehem, Pennsylvania

Charles G. Fisher
Department of Biological Sciences
Lehigh University
Bethlehem, Pennsylvania

Isaac E. García
Centro Interdisciplinario de
    Neurociencia de Valparaíso
Facultad de Ciencias
Universidad de Valparaíso
Valparaíso, Chile

Colin R. Green
Department of Ophthalmology and
    New Zealand National Eye Centre
Faculty of Medical and Health Sciences
University of Auckland
Auckland, New Zealand

Oscar Jara
Centro Interdisciplinario de
    Neurociencia de Valparaíso
Facultad de Ciencias
Universidad de Valparaíso
Valparaíso, Chile

Rachael M. Kells Andrews
Department of Biological Sciences
Lehigh University
Bethlehem, Pennsylvania

Yeri Kim
Department of Ophthalmology and
    New Zealand National Eye Centre
Faculty of Medical and Health Sciences
University of Auckland
Auckland, New Zealand

Tia J. Kowal
Department of Biological Sciences
Lehigh University
Bethlehem, Pennsylvania

Mohamed Kreir
Nanion Technologies GmbH
Munich, Germany

Luc Leybaert
Department of Basic Medical Sciences–
    Physiology Group
Faculty of Medicine and Health
    Sciences
Ghent University
Ghent, Belgium

Alessio Lissoni
Department of Basic Medical Sciences–
    Physiology Group
Faculty of Medicine and Health Sciences
Ghent University
Ghent, Belgium

Jaime Maripillán
Centro Interdisciplinario de
    Neurociencia de Valparaíso
Facultad de Ciencias
Universidad de Valparaíso
Valparaíso, Chile

Agustín D. Martínez
Centro Interdisciplinario de
    Neurociencia de Valparaíso
Facultad de Ciencias
Universidad de Valparaíso
Valparaíso, Chile

Paula Mujica
Centro Interdisciplinario de
    Neurociencia de Valparaíso
Facultad de Ciencias
Universidad de Valparaíso
Valparaíso, Chile

Sandra A. Murray
Department of Cell Biology
University of Pittsburgh, School of
    Medicine
Pittsburgh, Pennsylvania

Alberto E. Pereda
Dominick P. Purpura Department of
    Neuroscience
Albert Einstein College of Medicine
New York, New York

T. I. Shakespeare
Department of Biology
Fort Valley State University
Fort Valley, Georgia

Karin R. Sipido
Division of Experimental Cardiology
Department of Cardiovascular Diseases
Katholieke Universiteit Leuven
Leuven, Belgium

David C. Spray
Dominick P. Purpura Department of
    Neuroscience
Albert Einstein College of Medicine
New York, New York

Randy F. Stout, Jr.
Dominick P. Purpura Department of
    Neuroscience
Albert Einstein College of Medicine
New York, New York

Courtney A. Veilleux
Public Health Research Institute
    and Department of Microbiology,
    Biochemistry and Molecular
    Genetics
Rutgers New Jersey Medical School
Rutgers, the State University of
    New Jersey
Newark, New Jersey

Nan Wang
Department of Basic Medical Sciences–
    Physiology Group
Faculty of Medicine and Health
    Sciences
Ghent University
Ghent, Belgium

1 Immunofluorescence: Applications for Analysis of Connexin Distribution and Trafficking

Sandra A. Murray and T. I. Shakespeare

CONTENTS

1.1   SUMMARY

Immunofluorescence methods and techniques for visualizing the subcellular location of gap junction proteins (connexins) and their associated proteins will be presented in this chapter. Specifically, a step-by-step guide for localizing connexin in cell cultures and tissues with immunofluorescence will be presented. The concepts, rationale, and advantages of using particular protocols will be discussed. The methods of preparing and processing cells and tissues for immunofluorescence will be provided. This chapter will serve as a guide for researchers interested in the morphological analysis of gap junctions.

1.2   INTRODUCTION

Gap junctions, which are membrane channels composed of proteins called connexins (Goodenough et al. 1996), permit intercellular communication of regulatory molecules between contacting cells (Decker et al. 1978; Munari-Silem et al. 1995; Oyoyo et al. 1997; Shah and Murray 2001). Information on the assembly of connexins into functional channels and the removal of these channels from the plasma membrane (reviewed in the schematic seen in Figure 1.1) is critical to understanding gap junction turnover, the regulation of cell–cell communication, and the possible role of connexins as anchors for scaffold molecules. To study gap junction assembly and turnover, it is necessary to be able to identify the localization of gap junction proteins at the cell surface and within the cytoplasm (as seen in Figure 1.2). While biochemical methods are powerful tools for detecting the presence, abundance, and types of connexin found in cell populations, immunofluorescence provides information on the specific compartments and cellular locations of connexin and associated proteins needed to study gap junction trafficking and turnover.

Immunofluorescence is a relatively straightforward and inexpensive method for detecting the presence and the subcellular location of connexins within cells. Variations in fluorescent stain intensity and location can be used to gain insight into connexin function and cell–cell communication in cell populations. By using a specific primary antibody that binds to connexin and a secondary antibody that has been conjugated to a fluorophore, connexins can be visualized with a fluorescence microscope (Oyoyo et al. 1997). Colocalization procedures for detecting the

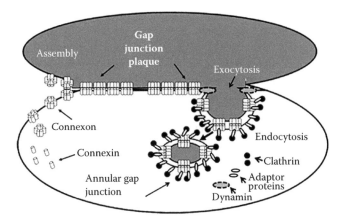

FIGURE 1.1 Schematic of gap junction plaque assembly and endoexocytosis. The delivery of connexons to the plasma membrane and the docking of the six-unit connexons between two adjacent cells result in the formation of a gap junction plaque. Endoexocytosis of a small region (or the entire plaque) results in the formation of cytoplasmic annular gap junction vesicles. Clathrin, clathrin adaptor proteins, and dynamin have been demonstrated to play a pivotal role in the formation of the annular gap junctions. (Murray, S.A. et al.: ACTH and adrenocortical gap junctions. *Microscopy Research and Technique*. 2003. 61. 240–246. Copyright Wiley-VCH Verlag GmbH & Co. KGaA. Reproduced with permission.)

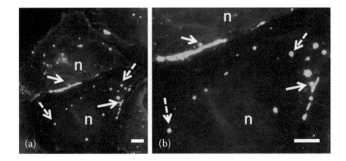

FIGURE 1.2 **(See color insert.)** Immunofluorescence of SW-13 human adrenal cortical tumor cells (a, b). Gap junction plaques (solid arrows) at the cell surface and annular gap junctions (dashed arrows) within the cytoplasm can be seen in cells expressing endogenous Cx43. The image in (a) has been enlarged in (b) to better demonstrate budlike areas of the gap junction plaque that are consistent with endoexocytosis (solid arrows). In these images, rabbit polyclonal anti-Cx43, diluted 1:100 in PBS, was applied to cells grown on coverslips for 1 hour at 37°C. After the cells were washed in PBS, they were incubated in a secondary antibody, goat antirabbit Alexa 488-tagged fluorophore, diluted 1:1000 in PBS, for 1 hour. In this cell population, filamentous actin was stained (red) with rhodamine phalloidin, which allowed the cell borders to be better distinguished. The cells were treated with 80 μM dynasore (Sigma-Aldrich), a dynamin inhibitor, for 1 hour, to inhibit the release of annular gap junctions from the gap junction plaque and thus allow the imaging of buds (solid arrows in (a)) on the gap junction plaque. The image was taken with a Nikon Microphot FXA fluorescence phase microscope. Bars = 10 μm, n = nucleus.

cellular location of more than one protein can be used to observe and evaluate connexin-associated proteins and to predict the role of these proteins in controlling connexin trafficking to and from the plasma membrane as well as in connexin degradation (Nickel et al. 2008). Since the development of immunofluorescence by Kaplan et al. (1950), this technique has evolved over the years into a standard laboratory method for localizing proteins in cells and tissues (Odell and Cook 2013). There are now many different protocols available, all based on the use of an antibody specific to the molecule of interest and a fluorescent dye to visualize antibody binding (Odell and Cook 2013; Platt and Michael 1983; Wouterlood et al. 1987). If the "primary" antibody is tagged with a fluorescent dye, the method is termed *direct immunofluorescence* (Kupper and Storz 1986; Odell and Cook 2013). However, as seen in Figure 1.3a, if the fluorescent tag is on a second antibody, which is directed against the primary antibody, the method is termed *indirect immunofluorescence.* In the case of connexin localization, indirect immunofluorescence is used mainly because the commercial anticonnexin primary antibodies and the corresponding fluorescent-tagged secondary antibodies are readily available. Further, indirect staining methods have the potential to allow two or more of the fluorescent-tagged secondary antibodies to bind to the primary, thus increasing the intensity of the signal. This signal amplification is particularly valuable in situations where protein expression is low. An additional advantage of using indirect immunofluorescence to localize connexin is the ease of colocalizing other proteins within the same sample. This is particularly useful, for example, in demonstrating the association of connexin with its trafficking proteins.

In the next section we list some of the materials and reagents needed for connexin detection, is provided and general procedures for indirect immunofluorescence of connexin localization are also given. The details of sample preparation and connexin visualization may need to be altered to meet the needs of the investigator and to optimize the detection of connexin in various cell and tissue types being analyzed.

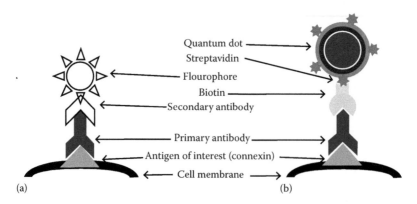

Quantum dot

Streptavidin

Flourophore

Biotin

Secondary antibody

Primary antibody

Antigen of interest (connexin)

Cell membrane

(a)   (b)

FIGURE 1.3 Schematic illustrations of (a) indirect immunofluorescence with primary/secondary antibody and (b) quantum dot streptavidin-biotin conjugate methods.

1.3 MATERIALS AND REAGENTS

1.3.1 FIXATIVES

The selection of the appropriate fixative is a key to good immunofluorescence results (Larsson 1988). A good fixative should prevent the loss of cellular structures without destroying the antigen-binding sites needed to bind the primary antibody. The investigator will need to optimize fixation based on the cell or tissue types being studied, but methanol and acetone at −20°C are the two most typically used solvents to fix cells for connexin localization. In addition, chemically cross-linking compounds, such as aldehydes, formaldehyde, and glutaraldehyde, are commonly used (Schutte et al. 1987). However, because glutaraldehyde can produce artificial staining patterns due to autofluorescence and often prevents antigen/antibody recognition, it is used only at a very low concentration, if at all, when performing immunofluorescence.

1.3.2 PERMEABILIZATION SOLUTIONS

Since connexin is expressed intracellularly, the cells will need to be permeabilized so that the anticonnexin antibody can penetrate the cell and gain access to connexin-binding sites. Permeabilization buffers such as 0.1–0.25% Triton X-100, 100 μM digitonin, or 0.5% saponin, all prepared in phosphate buffered saline (PBS), can be used to increase antibody penetration into the cell. In addition to these, there are many different solvents and detergents that can be used, but they all do considerable membrane damage. Depending on the cell type and the size of the molecule that needs to penetrate the cell membrane, one buffer may be more suitable than another. In choosing a permeabilization buffer, it is important to select a permeabilization buffer that does as little damage to the cells being used as possible. Triton X-100, at low concentrations, is one that is frequently used for immunofluorescence and has been shown to give good results for connexin localization. Saponin, however, is not as harsh as Triton X-100 or digitonin. Note that if acetone at −20°C is used to fix the cells, there is no need for permeabilization, since acetone fixation also increases the capacity for antibody penetration into the cell (Jamur and Oliver 2010). Acetone does less damage to the cell membrane than treatments with Triton, digitonin, or saponin. Further, there is no need to permeabilize if using frozen sections, rather than cells grown on coverslips, since the sectioning process allows the primary antibody to access intracellular connexin binding sites.

1.3.3 BLOCKING SOLUTIONS

To block nonspecific binding sites, the cells can be incubated in 1% bovine serum albumin (BSA) in PBS for 30 minutes at room temperature. As an alternative, PBS containing 10% serum from the host species of the secondary antibody can be used if available. In addition, blocking solutions can be purchased for use. For example, the investigator may find that Blocking One Histo (Nacalai Tesque, San Diego, California) is better at reducing nonspecific background staining than BSA or normal serum.

1.3.4 PRIMARY ANTIBODIES

Monoclonal mouse and affinity purified polyclonal rabbit antibodies (immuno-globulin G) against connexins can be purchased (Zymed Laboratory Inc., San Francisco, California; Sigma-Aldrich, St. Louis, Missouri; Abcam, Cambridge, Massachusetts; Santa Cruz Biotechnology, Dallas, Texas; and Life Technologies, Grand Island, New York). While there are many connexin family members (Kumar and Gilula 1992), in this chapter the protocols have been optimized for localizing connexin 43 (Cx43). The polyclonal antibodies against Cx43 were prepared against synthetic peptides corresponding to the carboxyl terminus of the Cx43 molecule (residues 370–381) (Yeager and Gilula 1992). Details of the preparation and characterization of the connexin antibodies have been previously described (Kumar and Gilula 1992), and their use in immunofluorescence localization of connexin has been demonstrated (Davis et al. 2002; Nickel et al. 2008; Oyoyo et al. 1997).

1.3.5 SECONDARY ANTIBODIES

Secondary antibodies are available from a number of commercial sources (Life Technologies, Grand Island, New York; Abcam, Cambridge, U.K.; and AbD Serotec, a Bio-Rad company, Berkley, California). A secondary antibody that is compatible with the primary antibody must be used. For example, if the primary connexin antibody is raised in a rabbit, the secondary should be antirabbit. Secondary antibodies can be purchased with tags to fluoresce at different wavelengths, depending on the desired color (e.g., red vs. green) or the display filters on the microscope. It is important to store the secondary antibody in the dark or the intensity of fluorescent staining will diminish and be lost over time.

Historically, fluorescein isothiocyanate (FITC), tetramethylrhodamine isothio-cyanate), and rhodamine were the most commonly used fluorophores (Enestrom 1986; Kupper and Storz 1986; Mossberg and Ericsson 1990). With the advances in immunofluorescence methods, newer molecules that are brighter and have greater stabilities when exposed to light have been introduced. Alexa Fluor® and cyanine-based dyes, such as Cy3™ (Life Technologies), are two groups of fluorophores that have gained popularity. Most recently, quantum dots (Qdot) (Invitrogen, Carlsbad, California), which are water-soluble semiconductor nanoparticles that are both fluorescent and electron dense, have been used to specifically label and colocal-ize connexin with other proteins (Giepmans et al. 2005; Nickel et al. 2008). The use of quantum dots, with various sizes and emission wavelengths (Qdot 655 [red], 565 [green], 525 [blue], 585 [yellow], 605 [orange]), can allow double or even triple labeling within cells at the light microscopic level of resolution (Giepmans et al. 2005). Importantly, since quantum dots are electron dense, they can be used at the trans-mission electron microscopic (TEM) level of resolution to correlate immunofluores-cence staining at the light level of resolution with specific organelles, identified by their ultrastructure (Giepmans et al. 2005; Nisman et al. 2004). For example, double-membrane (pentalaminar) annular gap junction vesicles (Jordan et al. 2001; Larsen and Hai 1978; Larsen et al. 1979; Murray et al. 1981) can be distinguished from gap junction structures, such as Cx43 biosynthetic pathway-derived single membrane secretory vesicles, with quantum dot immunoelectron microscopic techniques.

1.3.6 NUCLEAR STAINS

Fluorescent deoxyribonucleic acid (DNA) stains, such as Hoechst 33342 (2'-(4-hydroxyphenyl)-5-(4-methyl-1-piperazinyl)-2,5'-bi-1H-benzimidazole; membrane permeable) and DAPI (4',6-diamidino-2- phenylindole; less membrane permeable), can be prepared and used to localize the nucleus (Otto and Tsou 1985). To prepare the Hoechst stock solution, dissolve Hoechst (Sigma-Aldrich) in water at 1 mg/100 mL. The Hoechst powder should be stored in the dark at −20°C, and the stock solution should be stored at 4°C also in the dark. Wrap foil around the stock solution and powder containers to protect the Hoechst from light. Before use, dilute the Hoechst stock solution to the desired concentration (0.1–1 µg/mL) with water and then apply the solution to the tissue on the glass slides or to the cells on the glass coverslips for 30 seconds to 1 minute. In addition to diluting the powder to make the Hoechst solution, a 1 µg/mL Hoechst solution can be purchased from Life Technologies and applied directly to the cells or tissue sections.

For DAPI staining of the nucleus, DAPI (Sigma-Aldrich Co.) should be dissolved in ultrapure water to 1 mg/mL (stock solution). The stock solution is stable for several months if stored at −20°C with foil wrapped around the tube to protect the solution from light. The stock solution should be diluted just before use to 1:1000 in ultrapure water or PBS (1 µg/mL DAPI) and filtered to remove dye aggregates that can result in artifactual punctate staining. Hoechst stain can be used to stain live cells, as it is cell permeable, while DAPI is preferred when cells are fixed and permeabilized. Both fluoresce blue when viewed with a DAPI filter cube (Otto and Tsou 1985).

1.3.7 MEMBRANE STAINS

Membrane staining is needed in some studies to help define and distinguish the cell borders. This is particularly true when trying to differentiate between surface gap junctions, which sometimes appear punctate, and cytoplasmic annular gap junction structures. There are numerous membrane dyes commercially available for immunofluorescence methods. Long-chain dialkylcarbocyanines, such as DiI and DiO (Life Technologies), are lipophilic tracers that have been extensively used as a counterstain of the cell membrane when localizing connexin.

1.3.8 ACTIN STAIN

In addition to membrane staining, actin is sometimes used to distinguish the cell outlines and to study the relationships of connexin to the cytoskeleton. A number of actin-selective fluorescent probes are commercially available, including Alexa Fluor 568 phalloidin (red), FITC-phalloidin (green), or Alexa Fluor 488 phalloidin (green), or rhodamine-phalloidin (red) (Life Technologies).

1.4 EQUIPMENT

1.4.1 Humidified Chamber

To keep the samples from drying out during the various immunofluorescence incu-bation steps, a humidified chamber can be purchased or constructed. One method for constructing a humidified chamber is to line the bottom of a container with wet paper towels that have been soaked with sterile water. A platform should be placed within the chamber as a support for coverslips and slides during the staining process. The platform will ensure that the coverslips and slides remain above the level of the wet paper towels. One can construct a platform, for example, by placing a square plastic culture dish lid in the chamber to support the coverslips or the slides. The humidified chamber should be either light-tight or wrapped in foil to protect the light-sensitive fluorophores.

1.4.2 Microscopes

A fluorescence microscope equipped with filter sets for detecting fluorescent probes and with a ×20, a ×40, and a ×60 oil objective is needed. For the TEM comparison to light microscopic analysis of connexin structures with quantum dot techniques, an electron microscope is needed.

1.5 LOCALIZATION OF CONNEXIN IN CELL CULTURES

1.5.1 Acid Cleaning Coverslips for Growing Cells

For the best viewing and resolution, cells should be seeded onto glass cover-slips rather than plastic coverslips (which exhibit auto fluorescence). We are currently using coverslips with thickness measuring from 0.13 to 0.17 mm. For most microscopes, the best resolution will be obtained with coverslips that are 0.17 mm thick. The use of circular coverslips will allow them to be placed into 24-well plates if desired.

Dust and dirt on the coverslips could possibly result in artifacts and nonspecific staining. To eliminate these possibilities and to increase cell adhesion, the coverslips should be precleaned in acid and coated to increase the capacity of the cells to adhere to the coverslips. Some cell types, however, are capable of adhering to uncoated glass coverslips without the need for coating. The following is a method for acid cleaning and coating coverslips:

1. Spread the coverslips out on the bottom of a glass beaker and add 1 M HCl. Incubate the coverslips in the acid solution for 24 hours. (This should be done under a fume hood and by taking necessary precautions for working with acids.)
2. Carefully decant the acid solution, and rinse the coverslips by adding sterile water to the beaker and decanting five times.
3. Rinse the coverslips with 95% ethanol three times.

4. If needed, coat the coverslips by placing them in a rack (Electron Microscopy Sciences [EMS], Hatfield, Pennsylvania) and submerging the rack in a 0.1 mg/mL solution of gelatin or poly-L-lysine and allowing them to sit in the solution for 5 minutes to 1 hour.
5. Remove the rack of coverslips from the solution, and allow the coverslips to air-dry completely (or dry in a 37°C oven).
6. Put the coverslips into culture plates, and sterilize them by allowing them to sit under ultraviolet light in the laminar flow fume hood for 4 or more hours.
7. To monitor for possible pathogen growth and to test the sterility of the coverslips before seeding cells, add culture medium to a tissue culture plate containing one of the coverslips and incubate at 37°C for 1 day. The medium should remain clear and free of turbidity. Observe the medium in the tissue culture plate on an inverted microscope to check for the absence of microorganisms, such as bacteria, yeast, or fungi.

1.5.2 Cell Culture Protocol

The protocols, the specific culture medium, and the culture conditions vary according to the cell type being studied. Here, the most common method of maintaining cells is presented:

1. Place the acid-washed coverslips into culture plates.
2. Seed cells at a density of 5×10^5 cells per cm^2 in a 35 mm plate. The seeding density, however, will be cell type dependent and determined by the desired level of confluency needed for a given protocol.
3. Grow the cells overnight at 37°C in a 5% CO_2 atmosphere. (Growth time and atmosphere may need to be altered depending on the cell type. For example, Sertoli cells thrive at 32°C in a 5% CO_2 atmosphere [Shupe et al. 2011]. Also, some cells may need either more or less time to adhere to the coverslips.)

1.5.3 Cell Fixation Protocol

1. Remove the culture medium from the cells by aspirating.
2. To fix, add 3% formaldehyde, in PBS at pH 7.4, to the culture dish for 10–20 minutes at room temperature.
3. Remove the fixative.
4. Wash the culture dish, which contains the cells on coverslips, three times with PBS.
5. Fix with 100% acetone (chilled at −20°C) for 5 minutes by adding acetone to a glass petri dish and placing the dish in the freezer at −20°C. Note that after acetone fixation, you will not need to permeabilize the cells further. Proceed to the blocking solution step, as described in Section 1.5.5. If you decide to fix with methanol or skip the acetone fixation, you will need to permeabilize the cells.

1.5.4 CELL PERMEABILIZATION

1. To permeabilize, add a permeabilization buffer (either 0.1–0.25% Triton X-100, 100 μM digitonin, or 0.5% saponin, all prepared in PBS) to the culture dishes that contain the cells on coverslips. This will increase antibody penetration and thus staining. If Triton X-100 is used, generally the cells will incubated in the buffer for 15 minutes at 4°C. However, note that the time, the temperature, and the concentration of the permeabilization buffer, as well as the choice of a buffer, may have to be adjusted to optimize staining depending on the cell type.
2. Remove the permeabilization buffer, and wash three times with PBS.

1.5.5 BLOCKING

To block nonspecific binding sites, the cells on the coverslips should be incubated in a blocking buffer (1% BSA in PBS or in PBS containing 10% serum from the host species from which the secondary antibodies were derived) for 1 hour at room temperature before incubating in the primary antibody.

1.5.6 IMMUNOLABELING OF CONNEXIN PROTEINS: CELLS ON COVERSLIPS

1. Prepare a 1:100 dilution of anticonnexin antibody in 1× PBS and, to conserve the antibody, place a drop (~50 μL) of the solution on a sheet of Parafilm (Sigma-Aldrich) that has been spread on the support platform.
2. Use forceps to hold the edge of the coverslips, and place them cell side down on top of the drop of the antibody solution.
3. Incubate the cells on coverslips in the anticonnexin antibody in the humidified chamber for 1 hour at 37°C or overnight at 4°C. (In the example in Figure 1.2, the rabbit anticonnexin antibody [Life Technologies] was diluted 1:100 in PBS.) Try not to let the cells, in this or any of the following steps, dry out. In addition, between steps, allow excess PBS to drain from the coverslips, then use a piece of filter paper to remove excess PBS from the coverslips and from the forceps used to handle the coverslips. This will ensure that the incubation solutions are not further diluted.
4. At the end of the incubation period, lift the coverslips from the drop of anticonnexin antibody solution by holding an edge of the coverslip with forceps, and rinse the coverslips three to five times by dipping them in a beaker filled with PBS.
5. Place the coverslips cell side down on a drop of the secondary antibody solution (~50 μL, 1:200 dilution) that has been placed on a clean Parafilm sheet, and incubate the coverslips for 1 hour at room temperature in a dark humidified chamber. (In Figure 1.2, goat antirabbit Alexa 488-tagged fluorophore, diluted 1:1000 in PBS, was used and gap junction structures appear green. If, however, red staining is desired, use Alexa secondary 594-tagged fluorophore. The anticonnexin primary antibody should be used in a 1:100 dilution in either scenario).

6. At the end of the incubation, rinse the coverslips thoroughly in PBS as described in step 4.

If double labeling will not be performed, proceed to Section 1.5.8, 1.5.9 or 1.5.10. To evaluate the location of two (or more) different proteins, proceed to Section 1.5.7, step 7.

1.5.7 DOUBLE LABELING OF CELLS ON COVERSLIPS (OPTIONAL)

For labeling connexin and another protein in the same cell, the use of a second primary antibody to probe for the second protein of interest is needed. To prevent cross-reactivity, the second primary antibody should be raised in a species different from that for the first primary antibody. Double labeling of connexin and a different protein can be done simultaneously such that the two different primary and secondary antibodies are added into solution together. However, in some protocols, attempting to perform simultaneous staining may interfere with the binding of one of the two antibodies and thus protein localization. With the sequential approach, it will be easier to optimize staining by using longer times or temperatures of incubation for one of the two antibodies. Therefore, it is suggested that sequential incubation be used when working out a new labeling protocol or when staining a new cell type.

For sequential incubation after staining for connexin (as described in steps 1–6 in the preceding section), we continue as follows:

7. Incubate the cells with the second blocking solution for 30 minutes and then rinse.
8. Incubate the cells with the second primary antibody (the primary antibody against the protein of interest) in the humidified chamber for 1 hour at 37°C or overnight at 4°C.
9. At the end of the incubation period, lift the coverslips from the drop of the second primary antibody solution by holding an edge of the coverslips with forceps, and then rinse the coverslips by dipping them three to five times in a beaker filled with PBS.
10. Incubate the cells in the second secondary antibody for 1 hour at 37°C in the humidified chamber.
11. Lift the coverslips from the secondary antibody solution by holding an edge of the coverslips with forceps, and then rinse the coverslips by dipping them three to five times in a beaker filled with PBS.

For simultaneous incubations, mix the two different primary antibodies together and incubate the cells as in steps 1–4 of Section 1.5.6. Then mix the two different secondary antibodies together, and incubate the cells by following steps 4–6 of Section 1.5.6.

1.5.8 ACTIN STAINING (OPTIONAL)

To stain for actin, the coverslips should be incubated for 30 minutes, cell side down, in a drop of Alexa Fluor 568 phalloidin or rhodamine phalloidin (for red staining [Life Technologies]). The phalloidin dye should be diluted 1:40 with 1% BSA/PBS/0.1%

azide to reduce background staining. The incubation should be carried out in the humidified chamber. Wash the coverslips three to five times with PBS, and proceed to Section 1.5.9 or 1.5.10. Note that one must consider the colors of the fluorophores being used. If the connexin is stained to appear green, then a counterstain should be selected that will give another color (red) for the second protein of interest.

1.5.9   Nuclear Labeling of Cells on Coverslips (Optional)

To label the nucleus, incubate the coverslips cell side down for 30 seconds to 2 minutes in drops of DNA binding dye (~50 µL), such as Hoechst or DAPI (0.1–2 µg/mL), that have been placed on a sheet of Parafilm. Lift the coverslips from the nuclear labeling dye solution with forceps, and rinse the coverslips three times by dipping them in a beaker filled with PBS.

1.5.10   Mounting the Coverslips on Slides

1. Place the coverslips, cell side down, onto glass microscope slides on which a drop (~50 µL) of Fluoromount-G™ antiquench mounting medium (SouthernBiotech, Birmingham, Alabama) has been added.
2. Remove excess Fluoromount-G solution by blotting with a sheet of filter paper and applying gentle pressure.
3. Apply clear fingernail polish to the edges of the coverslips to seal and prevent the cells from drying. Allow the nail polish to air-dry.
4. Once the nail polish dries, the immunolabeled cells are ready for microscopic viewing. At this point, the slides can be placed in a slide box and stored at between −20°C and 4°C. To prevent photobleaching, exposure to prolonged ambient and microscopic light should be minimized.

1.6   IMMUNOLABELING OF CONNEXIN PROTEINS: TISSUE SECTIONS

In addition to localizing connexin in cells in culture, connexin can be analyzed in tissues. There are a number of protocols for preparing tissue slices for immunofluorescence (Peters 2010). Here, a quick-freeze method of freezing tissue will be described.

1.6.1   Quick-Freeze Method

After the tissue is removed from the animal, it should be submerged in a quick-freezing cyroprotectant medium, such as Tissue-Tek® OCT (EMS) in a freezing mold. Be careful to make sure that no air bubbles are in the Tissue-Tek OCT. The mold should then be placed on a block of dry ice until the Tissue-Tek OCT freezes, and then stored in a freezer or liquid nitrogen until the frozen sections are cut on a cryostat (Leica Biosystems, Buffalo Grove, Illinois). The sections can then be mounted onto commercially available positively charged (Globe Scientific Inc., Paramus, New Jersey) or gelatin-coated microscope slides to enhance the adhesion of the negatively charged tissue to the slides.

1.6.2   Tissue Staining

1. Allow the tissue section on the glass slides to thaw. To minimize the amount of solutions that will be needed, use a PAP pen (Ted Pella, Redding, California) to draw around the area of the section and thus create a confined area for subsequent staining.

2. Cover the tissue sections with blocking buffer (1% BSA in PBS or in PBS containing 10% serum from the host species of the secondary antibody) by adding the solution into the area marked by the PAP pen. Incubate in the blocking solution for 1 hour at 37°C in a humidified chamber. Note that, although it may be possible to skip the blocking buffer for cells in culture, the blocking buffer is needed for blocking nonspecific binding sites when staining tissue. The best blocking buffer is PBS containing 10% serum from the host species of the secondary antibody.

3. At the end of the incubation period, rinse the slide with the tissue section three to five times in PBS.

4. To localize connexin, incubate the tissue section for 1 hour at room temperature or overnight at 4°C in the anticonnexin antibody by adding ~50 μL of the antibody solution to the designated area marked by the PAP pen. Ensure that the entire tissue section is covered. Note that the volume of antibody needed to cover the sample may vary depending on tissue section size. Use enough antibody solution to ensure that the entire sample is covered. To prevent drying, place the slides in a humidified chamber during the incubation period.

5. Remove the anticonnexin antibody and rinse the slide three to five times in PBS.

6. Incubate the tissue slices in the secondary antibody (Alexa Fluor 488 [green] or Alexa Fluor 568 [red] [Life Technologies; diluted 1:1000 in PBS]) for 1 hour at 37°C. To minimize drying, place the slide in a humidified chamber to prevent fading of the fluorophore during the incubation period.

7. Rinse the sections three to five times by dipping them in a beaker filled with PBS.

1.6.3   Double Labeling of Tissue Sections (Optional)

For double labeling, use a second primary antibody to probe for the second protein of interest and a matched second secondary antibody and proceed with the protocol as described for double labeling cells on coverslips.

1.6.4   Nuclear Labeling of Tissue Sections (Optional)

To label the nucleus, cover the tissue section with ~50 μL of a DNA-binding dye such as Hoechst or DAPI (0.1–2 μg/mL) for 1 minute, remove the dye solution, and rinse three to five times in PBS buffer.

1.6.5 Mounting Coverslips over Tissue Sections

1. Cover the tissue section with ~50 µL of the Fluoromount-G antiquench mounting medium (SouthernBiotech), and apply a coverslip over the tissue section, trying not to trap air bubbles between the tissue and the coverslip.
2. Remove excess Fluoromount-G solution by placing a sheet of filter paper and applying gentle pressure.
3. Apply fingernail polish to seal the edges of the coverslips and prevent the cells from drying.
4. Once the nail polish is dry, the tissue section is ready for microscopic viewing and analysis. The slides can be placed in a slide box and stored between −20°C and 4°C. To prevent photobleaching, exposure to prolonged ambient and microscopic light should be minimized.

1.7 LOCALIZING CONNEXINS WITH QUANTUM DOTS (QDOT): LIGHT AND TRANSMISSION ELECTRON MICROSCOPY

Quantum dots (water-soluble, semiconductor, nanocrystal particles) purchased from Invitrogen can be used for labeling of connexins. In addition, because dots of different sizes fluoresce in different colors (for example, the larger Qd 655 are red, while the smaller Qd 565 are green), it is possible to use Qdot techniques to label and observe different proteins with both immunofluorescence methods at the light microscopic level of resolution and immunoelectron methods at the TEM level (as seen in Figure 1.3b and 1.4). In addition, it is possible to compare and confirm staining of structures seen at the light microscopic level with the added power of resolution provided by TEM. In this section, we will describe protocols for localizing connexin in cells on coverslips by immunofluorescence techniques with Qdots. Further, since the Qdots can be used to correlate light microscopic staining with structures seen with TEM (Deerinck et al. 2007; Giepmans et al. 2005), we have included protocols for Qdot immunoelectron microscopy of cells on coverslips. Note that, although only cells on coverslips are discussed here, similar Qdot immunoelectron microscopy protocols could be used for localizing connexin in tissue sections (Deerinck et al. 2007; Giepmans et al. 2005).

1.7.1 Immunolabeling Connexins

1. Grow the cells to approximately 80% confluency on coverslips.
2. Rinse the coverslips with the wash buffer, tris(hydroxymethyl)aminomethane (Tris)-buffered saline (TBS). To prepare TBS, mix 10 mM Tris with 130 mM sodium chloride (NaCl) and adjust the pH to 7.4.
3. Fix the cells with a solution of 4% paraformaldehyde in TBS for light microscopy or 2% paraformaldehyde and 0.1% glutaraldehyde in TBS for electron microscopy.
4. Rinse the coverslips three times with TBS chilled to 4°C.
5. Permeabilize the cells with 0.1% Triton X-100 in blocking solution (either 3% goat serum and 20 mM glycine in TBS or 1% BSA in TBS) for 15 minutes at 4°C.

6. Rinse the coverslips three times with TBS chilled to 4°C.

7. Incubate the cells for 30 minutes in the blocking solution (3% goat serum, 20 mM glycine in TBS) at 4°C for 5 min.

8. Incubate the cells overnight at 4°C in the connexin primary antibody. For example, in Figure 1.4, the cells were incubated in anti-Cx43 antibody raised in rabbit (Life Technologies), diluted 1:100 in the blocking buffer.

9. Rinse the coverslips three to five times with chilled TBS.

10. Incubate the cells on the coverslips for 10 minutes in the incubation buffer supplied with the Qdot streptavidin conjugate kit supplied by Life Technologies at 37°C in a humidified chamber.

11. Remove the buffer and incubate the cells on coverslips for 1–3 hours at 4°C with biotin-conjugated antibody (Life Technologies) diluted 1:100 in the incubation buffer that is supplied with the Qdot streptavidin conjugate kit.

12. Rinse the coverslips three times with chilled TBS.

13. Incubate the cells overnight at 4°C in the solution of Qdot conjugated to streptavidin (Invitrogen) overnight. As seen in Figure 1.4, the cells were incubated in Qdot 655 conjugated to streptavidin, diluted 1:25 in the incubation buffer supplied with the Qdot streptavidin conjugate kit.

15. Rinse the coverslips three times with PBS in preparation for processing them for either light (continue with steps 16–19 in the next section) or TEM (skip to steps 20–33).

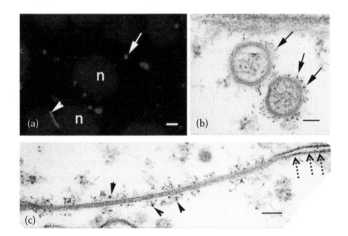

FIGURE 1.4 (See color insert.) Immunolabeling Cx43 with (a) immunofluorescence and (b, c) quantum dot streptavidin-biotin conjugate methods. The arrowheads designate Cx43 associated with gap junction plaques and the solid arrows point at Cx43 associated with annular gap junction vesicles. With TEM, the presence of quantum dots can be seen (arrows in b and c) and the morphology of the pentilaminar membrane typical of gap junctions can be appreciated in both (b) the annular vesicles and (c) the surface gap junction plaque. The nonjunctional membrane is devoid of quantum dots (dashed arrows in c). Bars = 10 µm (a) and 100 nm (b, c) n = nucleus.

1.7.2 Light Microscopy

16. For light microscopy, place the coverslips cell side down onto glass slides on which a drop of Fluoromount-G antiquench mounting medium (SouthernBiotech) has been added.

17. Remove excess Fluoromount-G solution by blotting with a sheet of filter paper and applying gentle pressure.

18. Apply fingernail polish to the edges of the coverslips to seal and prevent the cells from drying.

19. Once the nail polish is dry, the cells are ready for microscopic viewing. Note that it is best to evaluate and optimize Qdot staining at the light microscopic level before proceeding with the Qdot immunoelectron microscopic protocols. Quantum dots are semiconductor nanoparticles whose fluorescence is dependent on the dot size. Therefore, in the fluorescent microscope, Qdot 655, for example, can be detected (red) with the rhodamine excitation cube, while Qdot 565 will be detected (green) with the fluorescein (FITC) cube. The investigator should evaluate the sample with the fluorescent microscope, making sure to use the appropriate excitation cube (Qdot 655 [red], 565 [green], 525 [blue], 585 [yellow], 605 [orange]) for the Qdot size being used.

1.7.3 Transmission Electron Microscopy

20. For TEM, postfix the cells on coverslips in 2.5% glutaraldehyde in PBS, for 1 hour at room temperature.

21. Rinse three times with PBS.

22. Postfix the cells on coverslips with 1% osmium tetroxide with 1% potassium ferricyanide in water (alternatively 0.05 M cacodylate buffer at pH 7.2 can be used) for 30 minutes at 4°C. Note that it is best not to dilute the osmium in PBS, since PBS can result in artifactual precipitants.

23. Rinse the cells five times in ice-cold PBS.

24. Dehydrate the cells in ascending grades of ethanol (20%, 30%, 50%, 70%, and 90%) on ice, for 5–10 minutes each time, and then in three changes of absolute alcohol, 15 minutes each change, at room temperature. The absolute alcohol should be either freshly opened or desiccated by adding molecular sieves (Fisher Scientific, Grand Island, New York) to the bottle, to ensure that it remains dry.

25. Infiltrate the cells on coverslips by adding three changes of Epon resin, 1 hour for each time.

26. Fill a Beem capsule (BEEM Inc,, West Chester, Pennsylvania) with fresh Epon resin, and invert the capsules over the cell monolayer on the coverslips.

27. Polymerize the Epon at 37°C overnight and then at 60°C for 48 hours.

28. Detach the Beem capsules and the underlying cells from the coverslips. To do this, alternately dip the Beem capsule, with the attached coverslip, into a benchtop Dewar (Fisher Scientific, Pittsburgh, Pennsylvania) that contains liquid nitrogen and then into a beaker of boiling double-distilled water until

the Beem capsule detaches from the coverslip (two to three times), leaving the cells in the Epon.

29. Trim the block, cut ultrathin sections (65 nm thick), and mount them on TEM copper grids.
30. Stain the samples by inverting the grids onto a drop of 2% uranyl acetate in methanol for 1 hour at 4°C.
31. Rinse the grids with water.
32. Place the grid on a drop of Reynolds lead citrate for 7 minutes. The lead citrate staining should be performed in a chamber that contains NaOH pellets to prevent lead carbonate precipitation.
33. Rinse the grids with water and store them. Once dry, they are ready to be viewed in an electron microscope.

1.7.4 Important Considerations

- When working with coverslips, it is important to make sure that the coverslips do not become inverted. If this happens, you will not see staining, since the cells will not have been incubated in the various solutions.
- If the antibodies being used are not compatible, there will be no staining. Check before starting to make sure, for example, that if the anticonnexin antibody was raised in a rabbit, than a secondary antirabbit antibody is used, or if the primary was raised in a mouse, than the secondary antibody should be antimouse.
- If the staining is faint and there is a desire to increase the intensity of staining, incubate the cells or tissue sections for a longer time in the primary antibody. In addition, increasing the primary and/or secondary antibody concentration can increase the staining intensity; however this could, if the concentration is too high, result in artifactual staining.
- When optimizing staining or troubleshooting a staining problem, use a cell population that is known (positive control) and one that is not known (negative control) to express the connexin being stained. These controls will provide information about the staining procedure and allow the user to evaluate the conditions of the primary and secondary antibodies.

ACKNOWLEDGMENTS

We gratefully acknowledge Simon Watkins, Margaret Bisher, Cheryl Bell, Beth Nickel, and the Center of Biological Imaging at the University of Pittsburgh for their technical support. This research was supported by National Science Foundation (NSF) grant numbers MCB-1023144 and MCB-1408986.

REFERENCES

Davis, K.T., Prentice, N., Gay, V.L., Murray, S.A. 2002. Gap junction proteins and cell–cell communication in the three functional zones of the adrenal gland. *J Endocrinol* 173:13–21.

Decker, R.S., Donta, S.T., Larsen, W.J., Murray, S.A. 1978. Gap junctions and ACTH sensitivity in Y-1 adrenal tumor cells. *J Supramol Struct* 9:497–507.

Deerinck, T.J., Giepmans, B.N., Smarr, B.L., Martone, M.E., Ellisman, M.H. 2007. Light and electron microscopic localization of multiple proteins using quantum dots. *Methods Mol Biol* 374:43–53.

Enestrom, S. 1986. Stabilization of fluorochromes after immunofluorescent staining of tissue and embedding in Epon. *Stain Technol* 61:65–69.

Giepmans, B.N., Deerinck, T., Smarr, B.J.Y., Ellisman, M. 2005. Correlated light and electron microscopic imaging of multiple endogenous proteins using quantum dots. *Nat Methods* 2:743–749.

Goodenough, D.A., Goliger, J.A., Paul, D.L. 1996. Connexins, connexons, and intercellular communication (review). *Ann Rev Biochem* 65:475–502.

Jamur, M.C., Oliver, C. 2010. Permeabilization of cell membranes. *Methods Mol Biol* 588:63–66.

Jordan, K., Chodock, R., Hand, A.R., Laird, D.W. 2001. The origin of annular junctions: A mechanism of gap junction internalization. *J Cell Sci* 114:763–773.

Kaplan, M.E., Coons, A.H., Deane, H.W. 1950. Localization of antigen in tissue cells: Cellular distribution of pneumococcal polysaccharides types II and III in the mouse. *J Exp Med* 91:15–30.

Kumar, N.M., Gilula, N.B. 1992. Molecular biology and genetics of gap junction channels. *Semin Cell Biol* 3:3–16.

Kupper, H., Storz, H. 1986. Double staining technique using a combination of indirect and direct immunofluorescence with monoclonal antibodies. *Acta Histochem* 78:185–188.

Larsen, W.J. 1977. Structural diversity of gap junctions, A review. *Tissue Cell* 9:373–394.

Larsen, W.J., Hai, N. 1978. Origin and fate of cytoplasmic gap junctional vesicles in rabbit granulosa cells. *Tissue Cell* 10:585–598.

Larsen, W.J., Tung, H.N., Murray, S.A., Swenson, C.A. 1979. Evidence for the participation of actin microfilaments and bristle coats in the internalization of gap junction membrane. *J Cell Biol* 83:576–587.

Larsson, L. 1988. *Immunocytochemistry: Theory and Practice*. Boca Raton: CRC.

Mossberg, K., Ericsson, M. 1990. Detection of doubly stained fluorescent specimens using confocal microscopy. *J Microsc* 158:215–224.

Munari-Silem, Y., Lebrethon, M.C., Morand, I., Rousset, B., Saez, J.M. 1995. Gap junction-mediated cell-to-cell communication in bovine and human adrenal cells: A process whereby cells increase their responsiveness to physiological corticotropin concentrations. *J Clin Invest* 95:1429–1439.

Murray, S.A., Larsen, W.J., Trout, J., Donta, S.T. 1981. Gap junction assembly and endocytosis correlated with patterns of growth in a cultured adrenocortical tumor cell (SW-13). *Cancer Res* 41:4063–4074.

Murray, S.A., Davis, K., Gay, V. 2003. ACTH and adrenocortical gap junctions. *Microsc Res Tech* 61:240–246.

Nickel, B.M., DeFranco, B.H., Gay, V.L., Murray, S.A. 2008. Clathrin and Cx43 gap junction plaque endoexocytosis. *Biochem Biophys Res Commun* 374:679–682.

Nisman, R., Dellaire, G., Ren, Y., Li, R., Bazett-Jone, D.P. 2004. Application of quantum dots as probes for correlative fluorescence, conventional, and energy-filtered transmission electron microscopy. *J Histochem Cytochem* 52:13–18.

Peters, S. 2010. *A Practical Guide to Frozen Section Technique*. Newark. Springer.

Odell, I.D., Cook, D. 2013. Immunofluorescence techniques. *J Invest Dermatol* 133:1–4.

Otto, F., Tsou, K. 1985. A comparative study of DAPI, DIPI, and Hoechst 33258 and 33342 as chromosomal DNA stains. *Stain Technol* 60:7–11.

Oyoyo, U.A., Shah, U.S., Murray, S.A. 1997. The role of alpha1 (connexin-43) gap junction expression in adrenal cortical cell function. *Endocrinology* 138:5385–5397.

Platt, J.L., Michael, A.F. 1983. Retardation of fading and enhancement of intensity of immunofluorescence by *p*-phenylenediamine. *J Histochem Cytochem* 31:840–842.

Schutte, B., Reynders, M.M., Bosman, F.T., Blijham, G.H. 1987. Effect of tissue fixation on
   anti-bromodeoxyuridine immunohistochemistry. *J Histochem Cytochem* 35:1343–1345.

Shah, U.S., Murray, S.A. 2001. Bimodal inhibition of connexin 43 gap junctions decreases
   ACTH-induced steroidogenesis and increases bovine adrenal cell population growth.
   J Endocrinol 171:199–208.

Shupe, J., Cheng, J., Puri, P., Kostereva, N., Walker, W.H. 2011. Regulation of Sertoli-germ
   cell adhesion and sperm release by FSH and nonclassical testosterone signaling. *Mol
   Endocrinol* 25:238–252.

Wouterlood, F.G., Bol, J.G., Steinbusch, H.W. 1987. Double-label immunocytochemistry:
   Combination of anterograde neuroanatomical tracing with *Phaseolus vulgaris* leucoag-
   glutinin and enzyme immunocytochemistry of target neurons. *J Histochem Cytochem*
   35:817–823.

Yeager, M., Gilula, N.B. 1992. Membrane topology and quaternary structure of cardiac gap
   junction ion channels. *J Mol Biol* 223:929–948.

2 Imaging Gap Junctions in Living Cells

Matthias M. Falk, Charles G. Fisher,
Rachael M. Kells Andrews, and Tia J. Kowal

CONTENTS

2.1   INTRODUCTION

Research over the past years has shown that gap junctions, comprised of connexins, are highly dynamic structures. Connexins are synthesized, oligomerized into connexons, trafficked to the plasma membrane, docked into double-membrane spanning gap junction channels, assembled into gap junction plaques, internalized, and degraded. This dynamic nature apparently ensures proper gap junction function by regulating the level of direct cell-to-cell communication and of physical cell–cell adhesion. Gap junction dynamics correlates with the short half-life of connexins that has been determined to range from 1–5 hours in vivo and in situ (Berthoud et al. 2004). The dynamics of protein structures such as gap junction channels are best studied and demonstrated in live cells. To investigate the individual steps of the life cycle of connexins and gap junctions over time, the time-lapse imaging of fluorescently tagged connexins appears to be the most suitable method, as it provides spatiotemporal resolution in real time. Imaging fixed cells with either fluorescently tagged connexins or untagged connexins stained with a fluorophore is useful for colocalization studies of connexins with associated proteins. Additionally, it can offer hints at the dynamics of the gap junctions when fixing and analyzing multiple dishes at successive time points. However, fixed cells cannot offer real-time analysis of the dynamics of the gap junctions. Biochemical studies of gap junctions, which usually involve western blotting, can give a broad picture of the total connexin population in cells, but cannot offer information on the dynamics of individual gap junction plaques. Hence, a combination of approaches including biochemical and imaging analyses appears most desirable and is likely to produce the most convincing results. In the following sections, we provide an overview of live-cell imaging techniques as they have been applied over the past two decades in the authors' laboratory to investigate the biosynthesis, the structure, and the function of gap junctions. We briefly discuss microscope requirements, fluorescent probes, and imaging systems. Then we discuss how to image specific stages of the gap junction life cycle such as trafficking, endocytosis, and degradation. Next, we describe how to make observations of the gap junction plaque structure and dynamics, the colocalization of gap junctions with their binding partners, and how to conduct quantitative fluorescence analysis of these observations. We then explain the techniques to allow for the imaging of gap junctions in living cells for several days without significant detrimental side effects. Finally, we point out pitfalls and typical mistakes that can occur when conducting live-cell observations on gap junctions and give hints and tips on how to avoid them.

2.2   LIVE-CELL MICROSCOPE EQUIPMENT

Imaging living cells over time requires forethought, planning, and special equipment to maintain the cells at physiological conditions. Time-lapse imaging of gap junctions can be accomplished either through consistent manual imaging by the operator or with an automated live-cell video microscope. In general, an automated system is preferable to a manual system because most time-lapse sessions can last for hours and require the user to open and close illumination shutters, rotate filters, correct focal drift, and potentially track several positions all while maintaining identical time intervals for the entire recording session. Another major advantage of automated systems is that the cells are exposed to excitation light only during exposures, which reduces negative, light-induced effects (see Section 2.2.6). Of course, automation comes with an extra price tag, easily selling for tens of thousands of dollars.

2.2.1   CELL CULTURE CONDITIONS

Imaging metabolically driven processes in live mammalian cells requires maintaining normal physiological conditions over time. These include maintaining the cells at 37°C, 100% humidity, a pH of about 7.5, and a steady unlimited nutrient supply. Conditions such as these are normally provided in a standard laboratory cell culture incubator. However, outside of the incubator conditions can rapidly change due to lower ambient room temperature and lower CO_2 concentration of air. Indeed, most dynamic processes in mammalian cells come to a halt at ≤20°C (Saraste et al. 1986; Fuller et al. 1985; Saraste and Kuismanen 1984; Matlin and Simons 1983; Rotundo and Fambrough 1980), so at the very least, a temperature equivocal to body temperature needs to be maintained. Moreover, the pH of the culture medium, which is normally adjusted to pH 7.5 at 5% CO_2 atmosphere, can rise within minutes above pH 8, creating conditions that are clearly not physiologically ideal and can adversely affect cellular processes, including dynamics. If live-cell image recordings are designed to last only a few minutes, the issues mentioned earlier may not pose severe problems since the temperature can be maintained by placing a small heater or a hair dryer within close range of the microscope stage. The culture medium may also be amended with 4-(2-hydroxyethyl)-1-piperazineethanesulfonic (HEPES) acid buffer (10–25 mM), which maintains the physiological pH of the medium when outside of the incubator. Any culture medium that is normally supplemented with phenol red as a pH indicator typically generates background fluorescence that is especially pronounced in the red fluorescence channel and may need to be replaced by a culture medium without a pH indicator (commercially available for standard medium formulations). Finally, oxygen free radicals, which are toxic to cells, are typically generated by the intense illumination energy in the culture medium (occurring on the microscope stage over time) and may need to be neutralized by the addition of OxyFluor™ (Oxyrase, Inc.) to the culture medium.

2.2.2 MICROSCOPE SETUPS

Live-cell imaging is typically performed on inverted microscopes, meaning the objectives are mounted below the microscope stage (Figure 2.1). This arrangement provides the most convenient setup, as it allows for the imaging of cells attached to

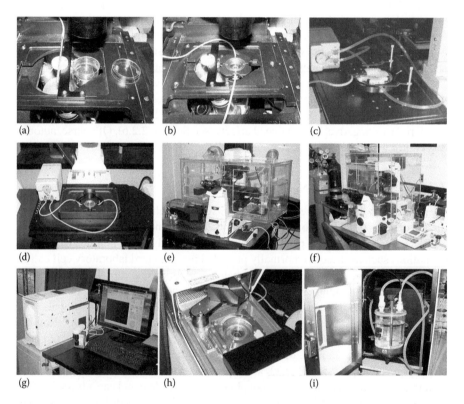

FIGURE 2.1 (See color insert.) Live-cell microscope equipment. Live cell chambers: (a) MatTek glass bottom dish, (b) POCmini, and (c) Focht Chamber System 2. (c, d) Peristaltic pumps used to circulate or exchange medium over time. Environmental incubators (e) partially or (f) entirely encasing microscopes. (g) Nikon's BioStation, a combined microscope/cell culture incubator, featuring (h) a temperature- and 5% CO_2-controlled imaging cell chamber, and (i) a gas bubbler to maintain 100% atmosphere humidity.

the bottom surface of a dish placed on the stage while keeping the cells immersed in the medium. However, conventional upright microscope systems can also be used for live-cell imaging if they are equipped with special commercially available dip-in objectives. The front lens on these objectives is specially sealed so that the tip can be submersed in the culture medium without damaging it due to corrosion. These objectives are water-immersion objectives that offer an intermediate quality (typical NA*: 0.8–1.2) to air (typical NA: ≤0.75) and oil immersion objectives (typical NA: ~1.4) (note that the NA of an objective directly dictates image resolution). For example, we have used 40× and 60× Nikon Fluor dip-in water-immersion objectives on an upright Nikon Eclipse E800 fluorescence microscope with satisfactory results.

2.2.3   LIVE-CELL CHAMBERS

Standard cell/tissue culture (TC) polystyrene plastic dishes are not suitable for high-quality oil immersion objectives, as TC plastic is of poor optical quality compared to glass and is too thick to focus on cells in the dish. Specially designed cell culture dishes in which a portion of the bottom is replaced with a high-quality, 0.17 mm thick borosilicate glass coverslip that is coated with extracellular matrix proteins for cell adhesion (collagen, fibronectin, etc.) are commercially available (MatTek Corporation as well as other manufacturers) and are a convenient solution for many live-cell applications (Figure 2.1a). For longer time-lapse recordings, specially designed live-cell chambers such as the POCmini-2 (POC Cell Cultivation Systems, PeCon GmbH) and the Focht Chamber System 2 (FCS2®; Biptechs) are commercially available (Figure 2.1b and c) and offer better control of environmental conditions, including medium exchange either by gravity or a peristaltic pump (Figure 2.1b through d).

2.2.4   ENVIRONMENTAL INCUBATORS

For longer time-lapse recordings that require maintaining 37°C for hours, the construction of environmentally controlled incubators that encase the microscope either partially or entirely is necessary (Figure 2.1e and f). These incubators not only keep the cells alive and metabolically active but also help with preventing focal drift (the gradual tendency of specimens to move out of focus over time), an issue that especially occurs when different parts of the microscope are maintained at different temperatures (ambient vs. 37°C). Of course, environmental incubators reduce the access to the working parts of the microscope; thus, multiple doors on all sides are desirable to ensure adequate access (see Figure 2.1e and f).

---

* NA = numerical aperture. The numerical aperture of a microscope objective is a measure of its ability to gather light and resolve specimen details. It is defined by NA = $n \sin\alpha$ (n represents the refractive index of the medium between the objective front lens and the specimen, and α is the one-half angular aperture of the objective).

2.2.5 COMBINED MICROSCOPE/CELL CULTURE INCUBATORS

More recently, live-cell microscope systems that combine a cell culture incubator, a fluorescent microscope, and an air-cooled charge-coupled device (CCD) camera all in one convenient unit, such as Nikon's BioStation IM-Q (Figure 2.1g), have become commercially available at an affordable price. They optimize the environmental conditions and hence eliminate many of the detrimental issues that are prohibitory to extended live-cell recordings. These systems contain a heated live-cell chamber (Figure 2.1h) and are connected to a CO_2 tank. The consistency of the atmosphere is maintained by a gas mixer, which maintains 5% CO_2. The gas mixture is bubbled through a concealed water-filled jar (Figure 2.1i) before being pumped into the chamber, thus maintaining 100% humidity to prevent unintentional evaporation and concentration of the culture medium. These systems typically allow live-cell recordings for multiple days (with an image acquired every minute) even with short wavelength (blue light) illumination as shown for gap junctions in Figure 2.2a and b. The image resolution/quality is not typically as high as with high-end microscope systems described earlier. This is because the imaging is performed using lower numerical aperture, long-distance 20× and 40× objectives that can also be used at 10×, 20×, ×40, and ×80, depending on the configuration, by swinging additional collector lenses into the light path. However, as evident from image sequences shown in Figure 2.2a and b, the imaging quality is clearly sufficient to resolve the dynamics of the gap junctions in living cells, allowing one to follow their entire life cycle (from plaque assembly to degradation) over time. Two fluorescent wavelengths can be chosen and recorded in addition to phase contrast white light illumination. Additionally, multiple points of interest in the dish can be programmed in sequence, allowing for the simultaneous imaging of several locations (gap junctions) over time. As video microscopes are typically used for longer durations, it is critical to ensure that the microscope and the live-cell chamber function properly to maintain a favorable environment or else the cells will likely die during imaging, or irrelevant recordings of cells may be generated that do not reflect physiological conditions.

2.2.6 POTENTIALLY TOXIC EFFECTS OF EXCITATION LIGHT

A general issue associated with live cell imaging is phototoxicity, which is especially pronounced if high-energy short wavelength light (ultraviolet [UV], blue) is used for excitation (as is needed for cyan fluorescent protein [CFP] and green fluorescent protein [GFP]-excitation, see Section 2.4.2). Extensive illumination with bright light (either continuous or administered in repeated pulses) must be kept to a minimum as the cells typically react negatively by rounding up as shown in the images in Figure 2.2c and d. The use of neutral density filters, bright high-quantum yield fluorophores, the reduction of laser power, and simply restricting to a minimum time that one looks at a specimen before beginning a time-lapse recording will help to reduce phototoxicity.

Another issue related to illumination that is generally encountered with live-cell imaging is known as *unintentional photobleaching*, which is the loss of emitted light by the permanent destruction of the fluorophores. Some fluorescent probes used in biological applications photobleach quicker (e.g., blue fluorescent protein, CFP,

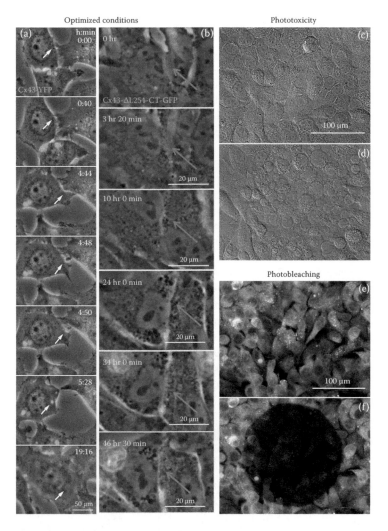

FIGURE 2.2   (See color insert.) Optimized imaging conditions and negative effects of excitation light. Optimizing cell culture and imaging conditions allow the cells to reside physiologically active for many hours as shown on the selected still images of the extended time-lapse recordings of the cells (over 20 hours in (a) [Reproduced from Fong, J. T., R. M. Kells, A. M. Gumpert, J. Y. Marzillier, M. W. Davidson, and M. M. Falk, *Autophagy*, 8, 794–811, 2012. With permission.] and over 46 hours in (b) [Reproduced from Fong, J. T., R. M. Kells, and M. M. Falk, *Mol Biol Cell*, 24, 2834–48, 2013. With permission.]) expressing wild type Cx43-eYFP (in (a)) and an internalization-deficient mutant (Cx43-ΔL254-CT-eGFP) (Fong, Kells, and Falk 2013.). Note the formation, internalization, and degradation of a gap junction plaque in (a), while the large internalization-deficient gap junction plaque in (b) stays in the membrane for the entire time of the recording. Phototoxicity resulting in rounded up cells (d) after exposing them to strong, short (blue) wavelength excitation light for several minutes. (c) The same area imaged before prolonged exposure. The loss of fluorescence intensity is due to unintentional fluorophore photobleaching resulting from prolonged exposure. The same area (e) before and (f) after exposure with the field diaphragm partially closed during the exposure time.

Lysotracker Red, Rhodamine) than others (e.g., eGFP, Alexa488, mCherry, mApple). An example is shown in Figure 2.2e and f. The cells stained with a fluorescently labeled antibody that decorates the plasma membrane were imaged right after selecting a region (Figure 2.2e), and again a few minutes later (Figure 2.2f). To demonstrate the significant loss of fluorescence emission, the field diaphragm was partially closed to restrict illumination to only the central portion of the imaging field. The diaphragm was then fully opened just before taking the second image. Again, avoiding unnecessary illumination in any form and selecting less sensitive and more stable probes may help to reduce this issue.

2.3 IMAGING TECHNIQUES

2.3.1 WIDE-FIELD MICROSCOPY

Wide-field and confocal fluorescence microscopy are standard imaging techniques which are routinely used in live-cell imaging, including gap junction research. Both techniques have their advantages and disadvantages and produce images of different appearance (see Figure 2.3a through c). This makes one technique preferable over the other depending on the research goal. More specialized imaging techniques such as total internal reflection fluorescence (TIRF) microscopy, a technique that only detects fluorescent signals in the immediate vicinity of the coverslip surface (up to a few hundred nanometers from the surface, see Section 2.7.2); and the now commercially available but costly superresolution and two-photon confocal microscope systems are not required for the investigation of connexins and gap junctions in living cells and are not further discussed here.

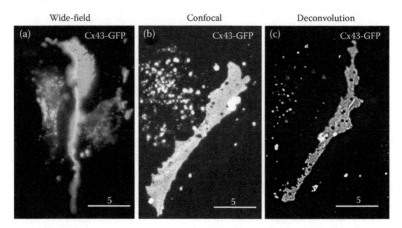

FIGURE 2.3   (See color insert.) Imaging techniques and image resolution. Gap junction plaques assembled from Cx43-GFP were imaged with (a, c) wide-field and (b) confocal microscopy. The Z image stack acquired in (c) was deconvolved postacquisition and a volume view was generated using mathematical image deconvolution algorithms. High primary image resolution achieved by high-quality, high NA oil immersion objectives is desirable and is required if the structural details of the gap junction plaques are to be resolved as shown for homomeric gap junction plaques assembled from

(Continued)

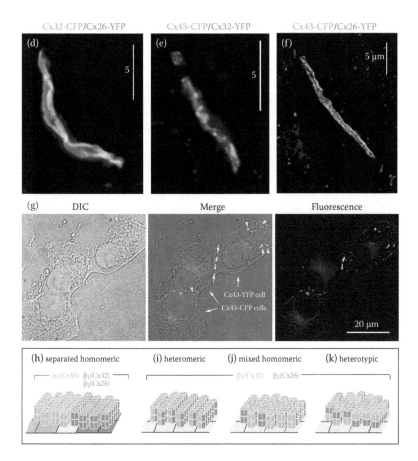

FIGURE 2.3 (CONTINUED)   (See color insert.) Cx43-eCFP and Cx32-eYFP in (e), or Cx43-eCFP and Cx32-eYFP in (f) (Reproduced from Falk, M. M., and U. Lauf: High resolution, fluorescence deconvolution microscopy and tagging with the autofluorescent tracers CFP, GFP, and YFP to study the structural composition of gap junctions in living cells. *Microsc Res Tech*. 2001. 52. 251–62. Copyright Wiley-VCH Verlag GmbH & Co. KGaA. Reproduced with permission.) Individual connexins and gap junction channels (docked connexons) are too small to be resolved by light microscopy and appear yellow (the resulting color of merged green and red fluorescence) shown for a gap junction plaque assembled from Cx32-eCFP and Cx26-eYFP in (d) (Falk 2000a) and for heterotypic gap junctions assembled from Cx43-eCFP and Cx43-eYFP connexons shown in (g). (Reproduced from Piehl, M., C. Lehmann, A. Gumpert, J. P. Denizot, D. Segretain, and M. M. Falk, *Mol Biol Cell*, 18, 337–47, 2007. With permission.) The scheme in (h) through (k) depicts the resulting image color of different types of gap junction plaques.

As living cells are sensitive to the repeated intense light pulses required for imaging wide-field microscopes are often more compatible with live-cell imaging compared to confocal point scanners. As the whole field of view is imaged in a single exposure, the less intense light typically provided by a mercury-arc or a xenon lamp (now also strong light-emitting diode illumination systems) is sufficient.

2.3.2 CONFOCAL FLUORESCENCE MICROSCOPY (POINT SCANNERS AND SWEPT FIELD/SPINNING DISK)

Confocal microscopes that use pinholes to achieve confocality (the thickness of the focal plane that is imaged) collect only a single, 1–2 μm thin plane at a time. These microscopes need to systematically scan the field of view to acquire an image, a process that generally takes much longer and requires more intense illumination solely provided by a laser light source (Figure 2.3b). Commercially available spinning disk and swept-field confocal systems image multiple focal points simultaneously and are thus faster. This reduces excitation light intensity and time, yet also reduces confocality. Another advantage of wide-field systems beside the lower price tag is that the entire depth of view is captured and thus objects such as vesicles trafficking along microtubules (see Section 2.7.2), which move out of the focal plane, become blurry, but remain visible. In a confocal system, these vesicles become "lost" (are no longer captured) when moving out of the focal plane, which could lead to the false interpretation that they fused with the plasma membrane or another subcellular structure. Out-of-focus fluorescence blur can be reduced or eliminated by applying mathematical deconvolution algorithms that are built into most advanced imaging software packages commercially available today (compare Figure 2.3a and c). However, the technique requires collecting Z stacks of equally spaced images (collected at ≤0.2 μm distance) that again add imaging and exposure time, increase unintentional photobleaching, and increase file size. It also may generate image distortion, especially if faster moving objects are to be resolved. In our lab, the imaging of gap junctions and the postacquisition deconvolution were performed on a DeltaVision 283 imaging system (Applied Precision Inc., Issaquah, Washington). The ability to intentionally photobleach, photoconvert, and photoactivate the chromophores in defined image regions and the ability to convincingly detect the colocalization of the chromophores make confocal microscopy the preferred imaging system for these applications. A comparison of both imaging techniques mentioning a number of relevant pros and cons is given in Figure 2.4.

2.3.3 IMAGING GAP JUNCTIONS WITH DIFFERENT IMAGING TECHNIQUES

In conventional light microscope systems (not superresolution), even at the highest magnification, the image resolution as defined by Rayleigh's law* depends on the wavelength of the light used for the excitation and is restricted to about 200 nm (e.g., 400 nm wavelength blue light). Thus, the fluorophores that are spaced less than this minimal distance normally cannot be resolved, as shown, for example, in Figure 2.3d and g for gap junction plaques assembled from heteromeric connexons (assembled from coexpressed Cx32-CFP [pseudocolored green] and Cx26-yellow fluorescent protein (YFP) [pseudocolored red]), and heterotypic gap junctions assembled between cocultured, Cx43-CFP (green) and Cx43-YFP (red) expressing

---

* $d = 1.22\lambda n/(2NA)$; δ = smallest resolvable distance between two points; λ = wavelength of the excitation light in nm; n = refractive index (= 1.0 in air; 1.33 in water; 1.52 in immersion oil); NA = numerical aperture (an index indicating how much light an objective can collect; high NA = high resolution).

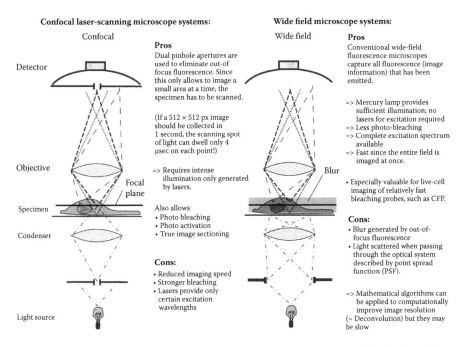

Confocal laser-scanning microscope systems:

Confocal

Detector

Objective

Focal plane

Specimen

Condenser

Light source

Pros

Dual pinhole apertures are used to eliminate out-of-focus fluorescence. Since this only allows to image a small area at a time, the specimen has to be scanned.

(If a 512 × 512 px image should be collected in 1 second, the scanning spot of light can dwell only 4 µsec on each point!)

=> Requires intense illumination only generated by lasers.

Also allows
• Photo bleaching
• Photo activation
• True image sectioning

Cons:

• Reduced imaging speed
• Stronger bleaching
• Lasers provide only certain excitation wavelengths

Wide field microscope systems:

Wide field

Blur

Pros

Conventional wide-field fluorescence microscopes capture all fluorescence (image information) that has been emitted.

=> Mercury lamp provides sufficient illumination; no lasers for excitation required
=> Less photo-bleaching
=> Complete excitation spectrum available
=> Fast since the entire field is imaged at once.

• Especially valuable for live-cell imaging of relatively fast bleaching probes, such as CFP.

Cons:

• Blur generated by out-of-focus fluorescence
• Light scattered when passing through the optical system described by point spread function (PSF).

=> Mathematical algorithms can be applied to computationally improve image resolution (= Deconvolution) but they may be slow

FIGURE 2.4 Pros and cons of wide-field and confocal microscopes. Both techniques have their positive and negative features that make both systems desirable depending on the goal that one aims to achieve.

cells, respectively. In both cases, the gap junction plaques (depicted with arrows in Figure 2.3g) and the internalized annular gap junction vesicles (depicted with arrowheads) appear yellow in the merged images, the resulting color of the superimposed green and red fluorescent signals (see Section 2.10). Signal separation is simpler if the fluorophores are well separated as in the gap junction plaque assembled from Cx43-CFP and Cx32-YFP shown in Figure 2.3e in which homomeric, homotypic channels consisting of either connexin type formed large domains (red or green). If the domains are small, as in the gap junction plaque assembled from Cx43-CFP and Cx26-YFP shown in Figure 2.3f, high primary image resolution (high quality, color-corrected [apochromat] 60× or 100×, high NA), oil immersion objectives, and CCD cameras with sufficiently large photo chips ($\geq$1024 × 512 px) are required. Domains smaller than the resolution limit (or two types of homomeric/homotypic gap junction channels placed next to each other akin to the field on a checkerboard) will not be resolved and will appear yellow. A schematic of different types of gap junction channels and the resulting merged image color is shown in Figure 2.3h through k. Note that a yellow stripe may appear on images of the plaques where red and green domains touch (as in Figure 2.3h) when both colors are detected by the same pixels (the ones located along the green/red fluorescence border line) on the CCD camera photo chip. Also note that the mixed homomeric (as depicted in Figure 2.3j) and heterotypic channels (as depicted in Figure 2.3k) will also appear yellow.

2.4 FLUORESCENT PROBES

The imaging of gap junctions in live cells requires fluorescent tagged connexins for detection in the microscope. Cloning the cDNA of a fluorescent protein or other probe in the frame with the connexin cDNA sequence and expressing the resulting fusion protein in cells can achieve this. It should be noted that antibody staining in general is not suitable for live-cell imaging because the use of antibodies requires that the cells be fixed and their membranes be permeabilized (unless microinjected into cells; not a trivial process and only a relatively small number of cells can be labeled at a time).

2.4.1 Green Fluorescent Protein and Derivatives

GFP and its derivatives today are the most commonly used fluorescent probes. Its discovery and development into a user-friendly live-cell fluorescence probe* was honored with the awarding of the Noble Prize in Chemistry in 2008 to Osamura Shimomura, Martin Chalfie, and Roger Y. Tsien. GFP's ability to autofluoresce (emit green fluorescent light when excited with blue light) and its inert, noninvasive properties resulting in a cytoplasmic expression in mammalian cells (Figure 2.5f) have significantly contributed to the enormous popularity of this probe. At present, over 53,000 papers come up in MedLine when *green fluorescent protein* is searched for. Not surprisingly, this remarkable probe has revolutionized and revitalized modern cell biology. Interestingly, a whole toolbox of monomeric fluorescent proteins emitting from blue to far red are now available (Shaner et al. 2004, 2007) allowing for the simultaneous detection of several proteins of interest in living cells. For the purposes of live-cell imaging, cells can be either transiently transfected with a fluorescent connexin prior to imaging or cell lines that stably express fluorescent connexins can be used.

2.4.2 Tagging Connexins with Green Fluorescent Protein and Other Fluorescent Proteins

Tagging connexins on their C-terminus (designated Cx-GFP) is normally well tolerated. C-terminal GFP-tagged connexins traffic normally (see Section 2.7) and assemble into typical gap junctions (Figure 2.5a) that are functional in respect to dye transfer (Falk 2000a; Jordan et al. 1999) and exhibit only minor alterations in electrophysiological channel characteristics (Bukauskas et al. 2000). Indeed, enough space is available in a connexon for six GFP tags placed on the C-terminus to simply extend a gap junction channel into the cytoplasms (Figure 2.5b) (Falk 2000a). The length of the linker that connects the connexin and the fluorescent

---

* GFP was isolated from the southwest Pacific jellyfish *Aquorea victoria*. In an effort to enhance the proteins as a fluorescent live-cell compatible probe, codon usage was optimized for mammalian applications, dimerization tendency was removed, and brightness and stability were enhanced (designated eGFP). All fluorescent protein tags that have been used in this chapter were purchased from Promega Inc., and are designated eCFP, eGFP, eYFP (yellow fluorescent protein), etc.

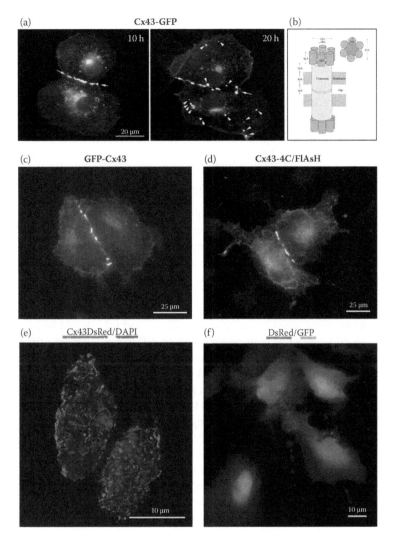

FIGURE 2.5   (See color insert.) Fluorescent probes. (a) Gap junctions assembled from Cx43 tagged with eGFP on the C-terminus imaged 10 and 20 hours posttransfection. (b) A schematic of a gap junction channel assembled solely from GFP-tagged connexins drawn to scale depicting that enough space is available in a connexon to spatially accommodate six C-terminally located GFP tags. (Reproduced from Falk, M. M, *J Cell Sci*, 113, 4109–20, 2000. With permission.) (c) Cx43 tagged with eGFP on the N-terminus also allows the nassembly of gap junction plaques, but these channels are not functional (Laird et al. 2001). (d) Gap junctions assembled from tetracysteine-tagged Cx43 and stained with FlAsH reagent (Gaietta et al. 2002). Cx43 tagged with the original red fluorescent protein, (e) DsRed; an obligate tetrameric protein, does not assemble into gap junctions and accumulates in the ER and the Golgi region while (f) DsRed expressed alone behaves inert and highlights the cytoplasm similar to GFP. (Reprinted from *FEBS Lett*, 498, Lauf, U., P. Lopez, and M. M. Falk, Expression of fluorescently tagged connexins: A novel approach to rescue function of oligomeric DsRed-tagged proteins, 11–5, Copyright (2001), with permission from Elsevier.)

protein sequence and its amino acid composition (providing more flexibility or rigidity) is likely to influence how well the fusion protein behaves. We have previously constructed a seven amino acid-long linker (including two central proline residues between the C-terminus and the GFP [Ala-Asp-*Pro-Pro*-Val-Ala-Thr]) with good results (Falk 2000a). It should be noted that fluorescent tags similar in size and conformation to GFP, such as CFP, YFP, mApple (and other color derivatives), or Dendra2 (see Section 2.8.2) and mEOS are similarly tolerated when placed on the C-terminus of the connexins (see Figures 2.2a, 2.3d through g, 2.7c,e, 2.9e through g, 2.10f,g and 2.11d).

2.4.3 POTENTIAL ISSUES RELATED TO SIZE AND LOCATION OF THE TAG

Despite its popularity, a few drawbacks of GFP and other fluorescent color variants need to be considered in terms of gap junction research. First, GFP is a 238 amino acid, 27 kDa protein that folds into a can-shaped 4.2×2.4 nm structure. Several connexins (including Cx23, 25, and 26) have a lower molecular weight compared to GFP, and all others have a comparable molecular weight or are little more than two times of the molecular weight (Cx62 being the largest known connexin). Thus, tagging connexins with GFP and GFP derivatives may potentially have a negative impact on the normal behavior of connexins, and this should be considered. Indeed, we do know of at least one significant difference between untagged and C-terminal-tagged connexins. Gap junction plaques assembled from endogenously or exogenously expressed untagged Cx43 in general are much smaller than plaques with a GFP tag on the Cx43 C-terminus (see Figures 2.6d and 2.12a compared to images of plaques assembled from Cx43-GFP). The tag is known to prevent the Cx43-GFP fusion protein from interacting with the membrane scaffolding protein, zonula occludens 1 (ZO-1), a regulatory protein believed to control plaque size (Hunter et al. 2005; Rhett and Gourdie 2012). The last C-terminal amino acid residues of Cx43 (and of other connexins) bind into a pocket of the ZO-1 PDZ2-domain (Chen et al. 2008). Thus, any addition to the connexin C-terminus will most likely prevent a Cx/ZO-1 interaction (reviewed by Thévenin et al. 2013). This change in gap junction behavior must be considered when performing relevant experiments.

Unexpectedly, placing GFP on the N-terminus of the connexins (designated GFP-Cx) also allows the assembly of gap junction plaques (Figure 2.5c); however, these channels are not functional (Laird et al. 2001), likely because the N-terminal domains of the connexin fusion protein subunits are misfolded. As we know from the crystal structure of a Cx26 gap junction channel, the N-terminal domains of the connexin proteins are located inside the channel and form the channel vestibule (Maeda et al. 2009; reviewed by Thévenin et al. 2013). As the diameter of the channel pore of a connexon only measures ~1.5 nm, not enough space for even a single GFP tag is available inside a gap junction channel, likely forcing the connexin N-terminus and the attached GFP tag outward into the cytoplasm (discussed by Thévenin et al. 2013). We know that the N-termini of the connexins harbor a flexible hinge (Kalmatsky et al. 2012, 2009; Purnick et al. 2000) that may facilitate such a conformational rearrangement. Albeit impaired channel function, N-terminal-tagged connexins may be helpful tools in addressing certain questions.

2.4.4 OTHER FLUORESCENT PROBES: HALOTAG AND TETRACYSTEINE TAGS

More recently, a novel protein-labeling technology based on a modified bacterial haloalkane dehalogenase, HaloTag (Los et al. 2008) (commercially available from Promega), has been developed that may provide additional versatility to the fluorescent protein–connexin toolbox. However, at 34 kDa, the HaloTag protein is even larger than the GFP. Additionally, and opposed to the GFP, the N-and C-termini are not next to each other in the HaloTag protein structure but located on opposite ends (Los et al. 2008), potentially creating additional issues if the HaloTag is to be placed inside the connexin sequence. Very recently, the first paper describing a HaloTag-tagged connexin, Cx36 has been published (Wang et al. 2015).

Another, much smaller class of tags are the biarsenical tetracysteine (4C) peptide tags (Griffin et al. 1998) that can allow real-time tracking of connexins and gap junctions in living cells (Figure 2.5d) (Boassa et al. 2010; Gaietta et al. 2002). A big advantage of these tags is their small size (only 6–12 amino acids) compared to fluorescent proteins, and they have been successfully placed inside the Cx43 C-terminal tail (Boassa et al. 2010). However, 4C tags are nonfluorescent, and the cells need to be stained for the tagged fusion proteins to become detectable. Biarsenical labeling reagents, FlAsH (green), and ReAsH (red), which bind to the uniquely arranged cysteine residues of the 4C tags, are commercially available as vector and staining kits (Life Technologies/Invitrogen). A comprehensive comparison of fluorescent protein and 4C tags has been published by Falk (2002).

2.4.5 PHOTOACTIVATABLE FLUORESCENT PROTEINS

Photoactivatable fluorescent tags that can be switched on and off by pulsing them with different distinct wavelengths of light have also been developed (reviewed by Lippincott-Schwartz and Patterson 2009; Shcherbakova et al. 2014). However, they are only useful for dynamic studies if the location of the protein of interest (connexin) is known, as the tagged proteins are not initially visible.

Taken together, fluorescent tags have made the investigation of connexins and gap junctions in living cells possible. In general, fluorescent tags are well tolerated when placed on the C-terminus of Cx43 and other connexins; however, not all connexins have been investigated as fluorescent protein fusion constructs. Due to the size and the potential aberrant effects of fluorescent tags, it is always advisable to conduct follow-up studies of untagged connexins using anticonnexin-specific antibodies. Indeed, the first available red fluorescent protein, DsRed, now known to be an obligate tetrameric protein, completely prevented gap junction formation and caused Cx43-DsRed and other DsRed-tagged connexins to aggregate in the endoplasmic reticulum (ER) (Lauf et al. 2001) (Figure 2.5e), even though DsRed expressed by itself behaved inertly similarly to GFP (Figure 2.5f).

2.5 GAP JUNCTION/CONNEXIN LOCALIZATION: IS THE FLUORESCENT PUNCTUM A GAP JUNCTION?

2.5.1 Cytoplasmic and Plasma Membrane Markers, Combined Differential Interference Contrast/Phase Contrast and Fluorescence Illumination

When gap junctions are abundantly expressed they typically appear as puncta and lines in the plasma membranes outlining the periphery of cells (as, e.g., in Figures 2.6d,f, 2.12a, and c). However, when the gap junctions are scarce, small, and not clearly appearing as dashed lines (as, e.g., in Figure 2.9a), it may be difficult to differentiate the gap junction plaques from the cytoplasmic connexin-containing structures (secretory vesicles, internalized annular gap junctions, inclusion bodies created by overexpression, etc.). Staining the plasma membrane or the cytoplasm with a second fluorescent stain may resolve this concern. In Figure 2.6, several examples in live

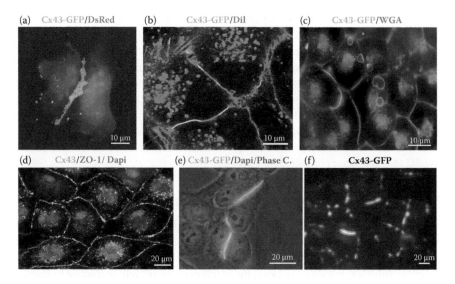

FIGURE 2.6 (**See color insert.**) Identification of gap junction plaques. To differentiate gap junctions from intracellular connexin-containing structures, it may be helpful to stain the cytoplasm by coexpressing an additional untagged fluorescent protein (in this case, DsRed), by staining the plasma membrane with membrane specific dyes such as (b) DiI (Reproduced from Falk, M. M., S. M. Baker, A. M. Gumpert, D. Segretain, and R. W. Buckheit III, *Mol Biol Cell*, 20, 3342–52, 2009. With permission.) or (c) WGA (With kind permission from **Springer Science+Business Media**: *J Membr Biol*, Degradation of endocytosed gap junctions by autophagosomal and endo-/lysosomal pathways: A perspective, 245, 2012, 465–476, Falk, M. M., J. T. Fong, R. M. Kells, M. C. O'Laughlin, T. J. Kowal, and A. F. Thevenin.), by staining in addition a membrane-localized protein, such as ZO-1 as in (d) (From Nimlamool, W., R. M. Kells Andrews, and M. M. Falk, *Mol Biol Cell*, 26, 2015. With permission.), or by acquiring and merging (e) phase contrast/DIC white light and fluorescent images. (Reproduced from Falk, M. M, *J Cell Sci*, 113, 4109–20, 2000. With permission.) Note that all the approaches shown with the exception of (d) are compatible with living cells.

(2.6a through c) and fixed cells (2.6d through f) are shown. The coexpression of a second fluorescent protein (not tagged to another protein) will label the cytoplasm (Figure 2.6a). Staining the cells for a few minutes with 1,1′-Dioctadecyl-3,3,3′,3′-tetramethylindocarbocyanine perchlorate (DiI) (a red lipid probe), fluorescent-labeled wheat germ agglutinin (WGA), or another commercially available lipid/plasma membrane marker before imaging will directly label the plasma membranes (Figure 2.6b and c). This can also be achieved in fixed cells by indirectly labeling the plasma membrane by costaining a plasma membrane-localized protein such as ZO-1, cadherin, catenin, etc. (Figure 2.6d). Finally, acquiring and merging phase contrast/differential interference contrast (DIC) white light images with fluorescent connexin images will not only demonstrate the plasma membrane localization, but also identify the location of other connexin-positive structures (see Figures 2.6e, and 2.9a through c). This can be done both in living as well as in fixed cells, and even the cell nuclei can be stained in the living cells if the DAPI label is replaced with a membrane-permeable chromatin dye such as Hoechst 33342.

2.6   GAP JUNCTION PLAQUE DYNAMICS

2.6.1   PLAQUE STRUCTURE, FUSION AND SPLITTING, SPATIAL MOVEMENT OF GAP JUNCTIONS, AND CONNEXIN-FREE JUNCTIONAL MEMBRANE DOMAINS

Time-lapse recordings of gap junctions can provide detailed information about plaque structure and dynamics when images are acquired at high primary magnification/resolution. Gap junctions can be oriented in two principal directions when viewed under a microscope depending on the cell morphology. If the lateral membranes of cells in a monolayer are oriented perpendicular to the image plane (e.g., found in general in polarized cells such as Madin–Darby canine kidney [MDCK] cells), the gap junction plaques will appear as lines and puncta providing an edge-on view (Figure 2.7a, top). If the cells are (partially) growing on top of each other (often found in HeLa and other non-contact inhibited cancer cells), the lateral membranes may be oriented more horizontal to the image plane, allowing a surface view (en face) of a gap junction plaque (Figure 2.7a, bottom). All examples of gap junctions shown in Figure 2.7b through f are en face views. Figure 2.7b depicts the structural dynamics of a gap junction plaque over time (55 minutes). Note the undulating edges, the deep invaginations, the more or less dark spherical connexin-free junctional membrane domains within the fluorescent plaque, and the drastic rearrangement of the two-dimensional shape of the plaque over time. Figure 2.7c depicts the fusion of two gap junction plaques by laterally moving in the plasma membranes. Note how the plaques move closer and closer together and suddenly fuse over their entire length. Figure 2.7d depicts the splitting of a gap junction plaque. In contrast to plaque fusion, the plaque portion that splits away slowly separates from the main plaque region with the connecting region becoming thinner and thinner before finally separating (comparable to an overstretched rubber band). Figure 2.7e depicts the fusion of plaque domains assembled from Cx43-CFP-labeled channels (green) present in a plaque that otherwise consists of Cx26-YFP-labeled channels (orange red; also

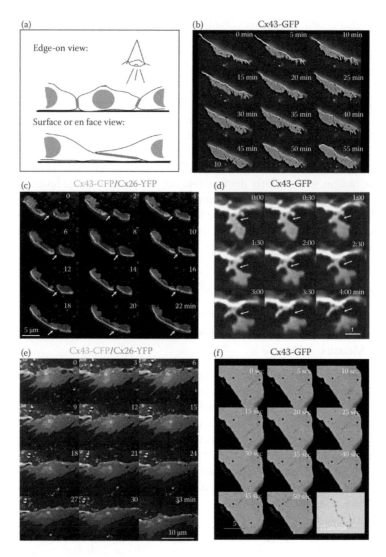

FIGURE 2.7 (See color insert.) Gap junction plaque dynamics. (a) Edge-on and en face surface views of gap junctions visible in cells depicted schematically. (b) Gap junction plaques are quite dynamic assemblies that structurally rearrange significantly over time. The gap junction plaques can fuse (depicted with arrows in c), split (depicted with arrows in d); (e) the domains (in this case, assembled from Cx43-eCFP [green]) in a gap junction otherwise consisting of Cx26-eYFP (orange/red) can split and fuse (Reproduced from Lopez, P., D. Balicki, L. K. Buehler, M. M. Falk, and S. C. Chen, *Cell Commun Adhes*, 8, 237–42, 2001. With permission.), and (f) connexin-free junctional domains (spherically shaped domains not containing connexins or gap junction channels and thus appearing dark) can move rapidly and saltatory throughout the gap junction plaques (Falk et al. 2009). While the structural rearrangements in (b) through (e) are relatively slow (in the minute range), the connexin-free junctional membrane domain dynamics (in f) can be fast (in the second range) requiring much shorter imaging intervals.

see the domain dynamics in the plaques shown in Figure 2.7c). These structurally dynamic rearrangements of the gap junctions are relatively slow in motion and can be detected by acquiring an image every 0.5–5 minutes. In contrast, the dynamics of connexin-free junctional membrane domains (not containing gap junction channels and appearing as dark, relatively spherical domains within the gap junction plaques [Falk et al. 2009] can rapidly move, requiring faster image acquisition [1–10-second intervals]). The tracking of one of these domains, moving 15 μm in 1 minute, is depicted in Figure 2.7f. In general, if the cells tolerate the extra excitation light and the probe is not rapidly bleaching, taking too many images is better than spacing them too far apart as rapid movements may become difficult or impossible to track with confidence (see, e.g., Section 2.7). Cutting out redundant images (as done in the collages shown in Figures 2.2a, 2.7b through f, 2.9f,g, 2.10a through g, 2.11b, and e, and in time-lapse movies) is always possible. Taken together, these recordings of gap junctions demonstrate that the channels within the gap junctions remain mobile resulting in a change of morphology of the plaques over time, to fuse and split, and to move throughout the plaque due to the fluidly arranged lipids that surround each gap junction channel (see Falk et al. 2009).

2.7 TRAFFICKING/SECRETORY PATHWAY

If Cx43-GFP fluorescence is imaged as soon as the fluorescence becomes detectable in cells (approximately 4–6 hours after transient transfection), a fluorescent haze highlighting the cytoplasm is detectable (suggesting ER localization) that then accumulates in the perinuclear regions (suggesting Golgi localization, labeled with arrows) before the gap junctions become visible (Figure 2.8a, 4 and 5 hours). Next, small gap junction plaques appear, which grow in size over time (Figure 2.8a, 5–20 hours, labeled with arrowhead). As described earlier, C-terminal-tagged connexins that are unable to interact with ZO-1 may grow very large and may occupy the entire lateral membrane space (as in Figure 2.8a, 40 hours). At this time, large, bright-fluorescent, spherical structures appear in the cytoplasm of some cells, suggestive of internalized gap junctions (Figure 2.8a, 20–80 hours; also compare Figure 2.5b, 20 hours). In a transient transfection, as shown in Figure 2.8a, new protein biosynthesis ceases after 24–48 hours due to the degradation of the cDNA, and no new Cx43 protein is biosynthesized at this time. As the gap junctions turn over, fewer and fewer remain (Figure 2.8a, 60 hours), and finally only internalized gap junctions and gap junction degradation products remain visible (Figure 2.8a, 80 hours).

2.7.1 COLOCALIZATION WITH RELEVANT COMPARTMENT MARKERS

As is typical for membrane proteins, Cx43 has been shown to cotranslationally insert into the ER membranes and to traffic via the Golgi apparatus to the plasma membrane (Koval 2006; Falk and Gilula 1998; Falk et al. 1997, 1994; Laird 1996; Musil and Goodenough 1993). We demonstrated Cx43-GFP colocalization by staining fixed

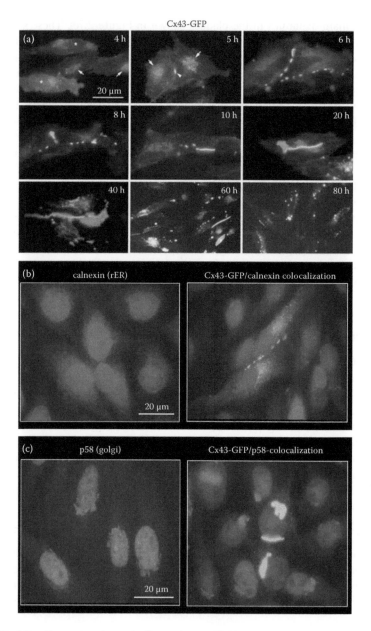

FIGURE 2.8 (See color insert.) Connexins trafficking along the secretory pathway: Colocalization with relevant compartment markers. (a) The stages of Cx43-GFP expression in transiently transfected HeLa cells including Cx43-eGFP ER and Golgi localizations (depicted with arrows in the images acquired earlier after transfection, 4–5 hours), formation of plasma membrane gap junctions (depicted with arrowhead, 5–40 hour), and cytoplasmic gap junction degradation vesicles (AGJs, 6–80 h). (b, c) Cx43-eGFP ER and Golgi localizations confirmed by costaining with ER (calnexin) and Golgi (p58) marker proteins. (Reproduced from Falk, M. M, *J Cell Sci*, 113, 4109–20, 2000. With permission.)

Cx43-GFP expressing cells with ER (calnexin) and Golgi (p58) marker proteins, respectively (Figure 2.8b and c). This can also be achieved in living cells by coexpressing an ER or a Golgi marker protein that is tagged with another fluorescent protein, or by staining the ER and the Golgi with compartment-specific membrane-permeable live-cell probes such as ER-Tracker™ Blue-White DPX or Texas Red-labeled BODIPY® ceramide, respectively (Cat. Nos. E12353 and D7540, Invitrogen/Molecular Probes).

2.7.2 TRAFFICKING ALONG MICROTUBULES

Cx43 connexons have been shown to traffic from the Golgi to the plasma membrane in secretory vesicles that migrate along microtubules driven by a kinesin motor protein (Fort et al. 2011; Shaw et al. 2007; Lauf et al. 2002). Secretory vesicles, including connexon-containing vesicles, move fast, up to a micrometer per second (Fort et al. 2011; Falk et al. 2009; Lauf et al. 2002) (Figure 2.9a). To convincingly capture these fast saltatory movements, it is best to collect images in a rapid sequence, at least every few seconds; otherwise, reliable connexon-containing vesicle tracking may not be possible. Tracking analyses can be done in different ways based on time-lapse image sequences; however, labeling micro-tubules (by, e.g., coexpressing YFP tubulin) together with connexins may provide the most convincing results (Figure 2.9e through g). We, for example, have merged images of all time points after color coding each image in Photoshop® (Lauf et al. 2002) (Figure 2.9a). This highlights the vesicle tracks, as well as the directionality based on color coding. Several transport vesicles containing Cx43 connexons transitioning from the Golgi to the plasma membrane can be seen in the merged image shown in Figure 2.9b and c.

The path and the position of vesicles at each time point can also be shown by connecting the vesicle locations with lines as shown in Figure 2.9e and f. As mentioned earlier, it is imperative to collect images in rapid sequence to allow for reliable tracking. Figure 2.9f shows the track of a Cx43-CFP-containing vesicle (green) that moves along YFP-labeled microtubules (red). Note that at several times, the vesicle changes direction by jumping onto different microtubules, resulting in a curved, zigzag path.

Successfully imaging the fusion event of the secretory vesicles with the plasma membrane as well as observing the delivery of the protein cargo into the plasma membrane also require capturing several images per second (Schmoranzer et al. 2000; Toomre et al. 2000). TIRF microscopes commercially available today are ideal platforms for this task. The microscope system available to us in 2002 was not able to capture the images in such a rapid sequence, yet we were able to indirectly demonstrate vesicle/plasma membrane fusion and Cx43 connexon delivery (Lauf et al. 2002). Figure 2.9g shows a sequence of selected images captured at 15-second intervals depicting a small area of the plasma membrane of a Cx43-CFP (pseudocolored red)/YFP tubulin (pseudocolored green) expressing cell. Four Cx43-CFP vesicles (labeled with arrows) are piled up at the end of a micro-tubule extending into a filopodial plasma membrane extension (0:45 minutes). Three minutes later (3:45 minutes), only three vesicles remain; at 4:30 minutes, two vesicles; at 4:45 minutes, one vesicle; and at 5:00 minutes, no vesicles remain. That the vesicles fused with the plasma membrane and did not move backward on the microtubule into the cytoplasm can be inferred from the fact that they are not visible on the microtubule closer to the minus end in any of the captured images. Indeed, two additional vesicles (on the same microtubule) do not move during the time of the recording, in this case, serving as convenient spatiotemporal markers.

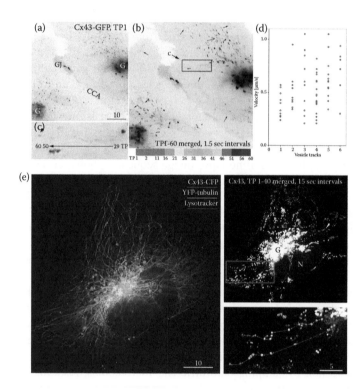

FIGURE 2.9 (See color insert.) Connexins trafficking along the secretory pathway: Trafficking along microtubules. To demonstrate the trafficking of connexons from the Golgi to plasma membrane, Cx43-eGFP-expressing HeLa cells were imaged at 1.5-second intervals. Time point 1 image (black and white inverted for contrast enhancement) is shown in (a). Each time point image was inverted and pseudocolored according to the color scheme shown in (b), and the colored images were merged resulting in vesicle tracks in which the directionality is indicated by the color coding. A selected track is shown in (c). The traveled distance measurements of six individual vesicles are shown in (d). (e) A HeLa cell expressing Cx43-eCFP and YFP-tubulin, and in addition stained with the live-cell compatible acidophilic stain, LysoTracker red (labeling lysosomes), was imaged every 15 seconds over time; the Cx43 channel images were merged, and the tracks of two Cx43-containing transport vesicles moving along microtubules were marked on the merged image insert. (Reproduced from Lauf, U., B. N. Giepmans, P. Lopez, S. Braconnot, S. C. Chen, and M. M. Falk, *Proc Natl Acad Sci U S A*, 99, 10446–51, 2002. With permission.)

(Continued)

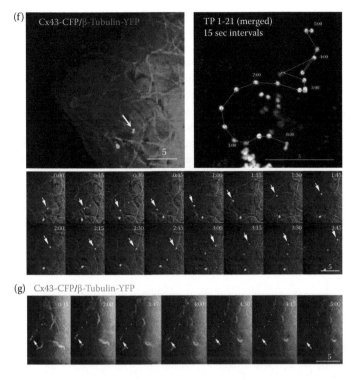

FIGURE 2.9 (CONTINUED) **(See color insert.)** Connexins trafficking along the secretory pathway: Trafficking along the microtubules. (f) A Cx43-eCFP-containing vesicle traveling along a curved microtubule (visualized by β-tubulin-YFP expression) and transitioning between different microtubules is depicted by outlining the traveled track with lines connecting the positions of the vesicle for each time point on the merged Cx43-eCFP channel image. The position of the vesicle at each time point is marked in the collage. (g) The delivery of Cx43-eCFP connexons packaged in transport vesicles into the plasma membrane inferred by the disappearance of the vesicles one after the other (marked with arrows) on the selected still images of a time-lapse recording of a Cx43-eCFP and β-tubulin-YFP-expressing HeLa cell. (Reproduced from Lauf, U., B. N. G. Giepmans, P. Lopez, S. Braconnot, S. C. Chen, and M. M. Falk, *Proc Natl Acad Sci U S A*, 99, 10446–51, 2002. With permission.)

2.8 PLASMA MEMBRANE DYNAMICS OF CONNEXONS AND GAP JUNCTIONS

2.8.1 Intentional Photobleaching: Fluorescence Recovery after Photobleaching and Fluorescence Loss in Photobleaching

Several microscopy-based techniques using living cells have been developed to investigate the dynamics of proteins in cells and can be applied to the study of connexins, connexons, and gap junctions. In fluorescence recovery after photobleaching (FRAP), the fluorescence signal emitted by a fluorescent protein in a defined area (square, bar, arbitrary shape) is intentionally permanently photobleached through the use of a strong laser (up to 100% power output). If the photobleached protein of interest is capable of diffusion (either throughout the cytoplasm if it is a soluble protein or in the membrane if it is a membrane protein) overtime, the bleached area will regain fluoresce because unbleached proteins will diffuse back into the bleached area. All biological fluorophores will eventually lose fluorescence when extensively excited (see Figure 2.2e and f), even probes such as GFP and YFP, which are quite resistant to photobleaching. Green-fluorescent probes in general were much easier to photobleach, as the argon ion lasers on conventional confocal microscopes (generating the blue 488 nm excitation line used to excite green fluorescent probes) have a much stronger energy output (approximately 50 mW) than the green-generating excitation line (543 nm, commonly used to excite red fluorescent probes) of helium/neon lasers (approximately 5 mW; the newest confocal microscope systems now have equally strong lasers). We have used this technique to demonstrate the diffusion of connexons in the plasma membrane (Figure 2.10a and b), to track the growth of the gap junctions, and to track the turnover of channels from the gap junctions (Falk et al. 2009,

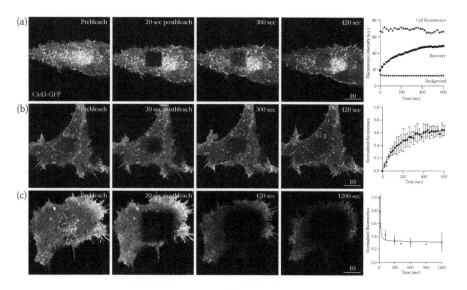

FIGURE 2.10 **(See color insert.)** Plasma membrane dynamics of connexons and gap junctions. (a, b, d, e) FRAP, (c) FLIP, and (f, g) Dendra2-photoconversion can be used to assess dynamics of connexons and gap junctions. (Reproduced from Lauf, U., B. N. Giepmans, P. Lopez, S. Braconnot, S. C. Chen, and M. M. Falk, *Proc Natl Acad Sci U S A*, 99, 10446–51, 2002. With permission.) In (a) and (b), the fluorescence in a $10 \times 10\,\mu m$ square in Cx43-eGFP-expressing HeLa cells was permanently photobleached using the laser of a confocal microscope. Note how the fluorescence "haze" indicative of dispersed plasma membrane-localized connexons recovers over time in the bleached area, while the intracellular, bright punctate fluorescence (probably predominantly inclusion bodies due to overexpression) in a remains largely immobile. In (c), the square area was repeatedly photobleached in 5-minute intervals resulting in an almost complete loss of cellular fluorescence over a 20-minute period.

(Continued)

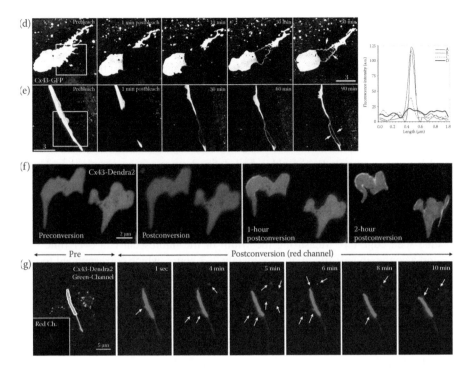

FIGURE 2.10 (CONTINUED) **(See color insert.)** Similarly, the fluorescence of the portions of Cx43-eGFP gap junction plaques (oriented en face in d, and edge on in e) was permanently photobleached. Note that the plaques recover a fluorescent rim of unbleached gap junction channels that grows wider over time, indicating that the newly synthesized channels are accrued to the outer edge of the gap junction plaques. (f) In a comparable approach, two gap junction plaques assembled from Cx43 tagged with Dendra2 were photoconverted from initially green to permanently red fluorescence using the 488 nm line of a confocal laser. Note that the plaques recover a rim of nonconverted green channels that grows wider over time, similar to the results obtained by FRAP experiments shown earlier. (g) Photoconversion also demonstrated that the vesicles (red, marked with arrows) were released from a photoconverted portion of an edge-on oriented gap junction plaque (outlined in white). (Reproduced from Falk, M. M., S. M. Baker, A. M. Gumpert, D. Segretain, and R. W. Buckheit III, *Mol Biol Cell*, 20, 3342–52, 2009. With permission.)

Lauf et al. 2002) (Figure 2.10d through g). Connexons/hemichannels delivered to the plasma membrane do not cluster into plaques comparable to docked gap junction channels. In confocal fluorescent images of Cx43-GFP expressing cells, the plasma membrane connexons appear as a fluorescent haze homogenously highlighting the plasma membrane (Figure 2.10a through 2.10c, panels labeled Prebleach). Note how the Cx43-GFP plasma membrane fluorescence in a photobleached square recovers as the connexons from surrounding unbleached membrane areas diffuse into the bleached area over time (Figure 2.10a and 2.10b). Interestingly, larger bright fluorescent Cx43-GFP-containing structures in the cytoplasm (internalized gap junctions and inclusion bodies) do not show a comparable dynamic behavior (Figure 2.10a).

In addition to FRAP, fluorescence loss in photobleaching (FLIP), a related technique, can be used to investigate connexon dynamics. In FLIP, the fluorescence in the defined area (e.g., a square) is photobleached repeatedly over time. If the proteins can laterally diffuse (such as connexons), the fluorescence will eventually be lost also outside of the bleached box, as the fluorophores diffuse into the boxed area and will be bleached in repeated bleach cycles. This technique demonstrates especially well the extent of dynamic movements either including the entire plasma membrane (as shown for connexons in Figure 2.10c) or being restricted to certain compartments (see Lauf et al. 2002).

When we permanently photobleached defined areas of gap junctions in Cx43-GFP expressing cells, the plaques recovered a fluorescent rim of unbleached channels clearly detectable within about 30 minutes. Fluorescent rims grew wider over time, indicating that the new channels were accrued along the edge of the Cx43-GFP gap junctions (Figure 2.10d depicts a plaque seen en face, Figure 2.10e depicts a plaque seen edge-on). Measuring the width of the fluorescent rim over time suggested a connexin half-life of approximately 2–3 hours, which corresponded to previously published data (Berthoud et al. 2004), to the Cx43 half-life calculated by Gaietta et al. (2002) who used successive FlAsH/ReAsH staining to demonstrate Cx43-based gap junction turnover, and to the experiments that used photoconvertible proteins (see Section 2.8.2) (Falk et al. 2009).

2.8.2 Photoconversion: Dendra2 and mEOS2

Photoconvertible fluorescent proteins such as Dendra2 and mEOS initially fluoresce green; however, they are photoconverted to a permanently red-emitting state when excited with moderate UV/blue light (Chudakov et al. 2007; Gurskaya et al. 2006). The photoconverted gap junctions assembled from Cx43-Dendra2 develop a progressive green outer rim of unconverted channels over time with similar kinetics to the photobleached gap junctions (Figure 2.10f). This technique also allowed us to directly visualize and quantify the vesicles generated by the internalization of the small plaque areas by photoconverting a portion of edge-on oriented gap junction plaques (Falk et al. 2009) (Figure 2.10g, released vesicles marked with arrows). Other more dynamic characteristics have been described for Cx43-based gap junction plaques (including an extremely fast, questionable turnover rate of only 2.7 minutes (Shaw et al. 2007), as it is not clear based on metabolic considerations how cells would maintain such a fast gap junction turnover rate), and for gap junctions assembled from other connexin types (Cx26) (Thomas et al. 2005). Fluorescent protein fusion tag and connexin type have also been reported to influence gap junction plaque stability (Stout et al. 2015).

When performing FRAP and FLIP experiments with connexins, it is important to keep the initial illumination to a minimum so as to not accidentally prebleach the fluorescent protein signal, and to make sure that the gap junctions do not move out of focus over time, as this may suggest falsified dynamic characteristics. This is especially important when using photoconvertible fluorescent proteins, as these easily convert unintentionally to red-emitting polypeptides if exposed to blue excitation light that is too strong. A detailed step-by-step protocol on how to successfully photoconvert Dendra2- and mEOS-tagged proteins including gap junctions has been published by us previously (Baker et al. 2010). See also the methods sections in the manuscripts of Lauf et al. (2002) and Falk et al. (2009) for details on how to perform plasma membrane connexon and gap junction FRAP and FLIP experiments.

2.9 INTERNALIZATION AND DEGRADATION OF GAP JUNCTIONS

2.9.1 IMAGING

Live-cell recordings of Cx43-GFP gap junctions demonstrated that the gap junction plaques and regions of plaques internalize as complete double-membrane spanning channels and do not split in half as is typical, for example, for adherens junctions (Ivanov et al. 2004a,b). The process generates cytoplasmic double-membrane gap junction vesicles (Falk et al. 2009; Piehl et al. 2007; Jordan et al. 1999), termed *annular gap junctions* (AGJs) or *connexosomes* (schemed in Figure 2.11a). Subsequent analyses have shown that clathrin as well as clathrin-endocytic machinery mediate this process (Xiao et al. 2014; Fong et al. 2013; Gumpert et al. 2008; Nickel et al. 2008; Piehl et al. 2007). Gap junctions assembled from other connexins have also been shown to internalize to form AGJs (Falk et al. 2014; Xiao et al. 2014; Johnson et al. 2013; Schlingmann et al. 2013). The process is quite impressive when captured in a time-lapse recording at high primary resolution with the image plane placed in the middle of a gap junction. In Figure 2.10b, the two gap junctions (a large and a small one, both labeled with arrows) are visible. Over time (89 minutes), the large gap junction successively bends toward the cytoplasm of the left cell, invaginates deeper and deeper and finally detaches forming a perfectly spherical AGJ in the cytoplasm of that cell, while the small gap junction resides undisturbed in the plasma membrane. The sphere is visible as a ring in a confocal and in a wide-field microscope when viewed at high primary magnification, as this generates only a thin focal plane (as seen in Figure 2.11b). However, Z sectioning will show the entire spherical morphology of the AGJs (Piehl et al. 2007). In addition, when combined with DIC white light illumination, the cell bodies become visible, demonstrating the translocation of the gap junction from its location in the plasma membrane into the cytoplasm of one of the connected cells (Figure 2.11c, depicted with arrowheads). Note that a new gap junction forms in the lateral membrane at a location proximal to that of the internalized gap junction (visible in the last image in Figure 2.11c, depicted with arrow).

2.9.2 COLOCALIZATION WITH RELEVANT COMPARTMENT MARKERS

The fate of endocytosed gap junctions and their degradation can be shown in living cells via colocalization by the coexpression of an endocytic marker protein tagged with a different fluorescent protein (as shown for LC3-GFP and Cx43-mApple as an example in Figure 2.11d), or by staining the subcellular compartments with specific live-cell probes (shown for proteolytic compartments stained with the acidophilic live-cell marker, LysoTracker™ Red; Invitrogen/Molecular Probes) (Figure 2.11e). Other comparable probes specific for other subcellular compartments are available from Molecular Probes/Invitrogen (ER, Golgi, mitochondria, etc.) If the cells are fixed, antibodies directed against a relevant marker protein can be used (shown for p62/sequestosome1 [SQSTM1] and endogenously expressed Cx43 in pulmonary artery endothelial cells as an example in Figure 2.12d).

When investigating connexin and gap junction degradation, it is important to not only use markers for lysosomes because these structures participate in several different cellular proteolytic pathways (endo-/lysosomal as well as autophagosomal

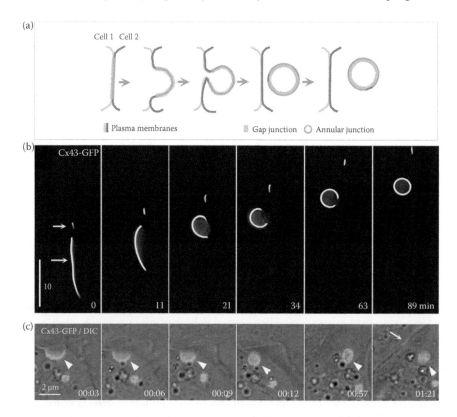

FIGURE 2.11 **(See color insert.)** Internalization and degradation of gap junctions. (a) Schematic depicting the internalization of gap junctions as complete double-membrane spanning channels. (Reproduced from Fong, J. T., R. M. Kells, A. M. Gumpert, J. Y. Marzillier, M. W. Davidson, and M. M. Falk, *Autophagy*, 8, 794–811, 2012. With permission.) The process leads to the formation of AGJ vesicles in the cytoplasm of one of the previously coupled cells. (b, c) The same process imaged by time-lapse microscopy in Cx43-eGFP-expressing HeLa cells (only the fluorescence is shown in b; the merged fluorescence and the DIC channels are shown in (c)). Note that the upper, small gap junction plaque in (b) (depicted with short arrow) does not internalize and remains in the plasma membrane; and that in (c), a new gap junction (depicted with arrow) is assembled at the location were the previous internalized gap junction plaque (depicted with arrowhead) was localized.

(Continued)

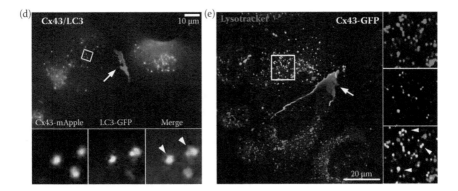

FIGURE 2.11 (CONTINUED) **(See color insert.)** Internalization and degradation of gap junctions. (d) Cx43-mApple-containing vesicles colocalize with LC3-GFP (a specific marker protein of autophagosomes) imaged in living cells, suggesting AGJ degradation via macroautophagy. (e) Cx43-eGFP-containing vesicles colocalize with LysoTracker Red-stained lysosomes imaged in living HeLa cells, further suggesting lysosomal-based degradation of AGJs. (Reproduced from Fong, J. T., R. M. Kells, A. M. Gumpert, J. Y. Marzillier, M. W. Davidson, and M. M. Falk, *Autophagy*, 8, 794–811, 2012. With permission.)

degradation). Indeed, AGJs have been shown as being degraded by macroautophagy, which converges with the lysosomal pathway (Bejarano et al. 2012; Fong et al. 2012; Lichtenstein et al. 2011; Hesketh et al. 2010). In addition, gap junctions have been presumed to also be degraded through the endo-/lysosomal pathway (Leithe et al. 2009; Qin et al. 2003; Laing et al. 1997), and misfolded connexin polypeptides can be degraded through the proteasomal pathway (Kelly et al. 2007; Laing and Beyer 1995). It is important to note that depending on the fluorescent probe that is used in degradation studies, the labeled degrading gap junctions may gradually lose their fluorescence because some GFP-types are sensitive to the low pH of the degradation compartments (see, e.g., the last still image in the time-lapse shown in Figure 2.2a).

2.10 FLUORESCENCE COLOCALIZATION

2.10.1 Tools and Techniques

Fluorescence colocalization analyses are typically performed on confocal microscopes.* Colocalization analysis is a powerful tool that can provide much information on the spatial interaction of connexins with binding partners throughout their life

---

* In contrast to a wide-field microscope, a conventional confocal microscope uses point illumination and a pinhole in an optically conjugate plane in front of the detector to eliminate out-of-focus emission light. As only the light very close to the focal plane is detected, the image's optical resolution, particularly in the sample depth direction (in Z), is much better than that of wide-field microscopes. However, as much of the light from the sample fluorescence is blocked at the pinhole and discarded, increased resolution is at the cost of decreased signal intensity, requiring light-intensive lasers for excitation and often long exposure times.

cycle (Interactome). In addition, temporal interactions of connexins with their binding partners may be resolved when colocalization analyses are performed in living cells. However, this technique also has the potential to produce significant false data when incorrectly performed. Colocalization is generally defined as two different fluorescently labeled proteins/molecules appearing in the same location on an image, which implies that they are interacting in some manner with each other. Colocalization will generate the resulting additive color by mixing both fluorescent labels in merged images (red colocalizing with green will result in yellow; blue colocalizing with green will generate cyan; red colocalizing with blue will give magenta; and colocalization of all three basic colors, red, green, and blue, will result in white* as in the examples shown in Figure 2.12a through d).

2.10.2 POTENTIALLY FALSE-POSITIVE RESULTS

To avoid errors, first it is important that the pinholes on the confocal microscope are small (≤1 airy units), so that the image sections are thin (≤ 1 µm); otherwise, confocality is lost (the thickness of the image plane increases), and the fluorescent labels localized in the same spot in X and Y, but in different depths in Z, may falsely colocalize (see Lauf et al. 2001 for examples). Second, the primary image magnification and the resolution should be optimal (60/63× NA 1.4 plan-Apochromat oil-immersion objectives). Immersion oils with a refractive index that matches that of glass coverslips and objective lenses are essential to reduce the image distortion. Third, it is important to not overexpose the image, which may occur when using too much laser power or too long of an exposure time or by using autoimage settings of the imaging software package. Indeed, as the excitation and emission spectra of most commonly used biological labels and fluorescent proteins overlap, it is likely that at a given wavelength (for example, to excite GFP), another fluorescent protein (CFP, YFP, RFP) is excited to some extent as well (see Figure 2.12e). It is important to check for this "bleed-through" into other emission channels (photomultipliers) and adjust the exposure level to a point where no bleed-through is detectable. Figure 2.12f shows a purposely false colocalization of Rhodamine-labeled phalloidin that decorates the actin filaments. Rhodamine is a red fluorescent dye that also emits somewhat green if overexposed. This bleed-through into the green channel was acquired, the image brightness was enhanced post-acquisition using Photoshop software, and the red emission channel and the enhanced green bleed-through image were merged, resulting in a complete, but false colocalization signal! Reducing the laser excitation power and the exposure times on wide-field microscopes (and not using autoexposure on these systems) can avoid such detrimental mistakes.

---

* Note that additive color mixing will generate different colors than subtractive mixing of pigment colors where the colors are adsorbed. Computer monitors use additive color mixing (RGB, for red, green, blue) to display color, while color printers use subtractive color mixing to generate colors (typically CMYK, for cyan, magenta, yellow, black). Publishers who print articles with color (including the publisher of this book series) often require images to be submitted in CMYK, not RGB color scheme. Colors may look somewhat different in CMYK format compared to RGB format and should be checked/adjusted before submission. Photoshop, for example, will allow one to interchange the color formats under "Image–Mode" settings.

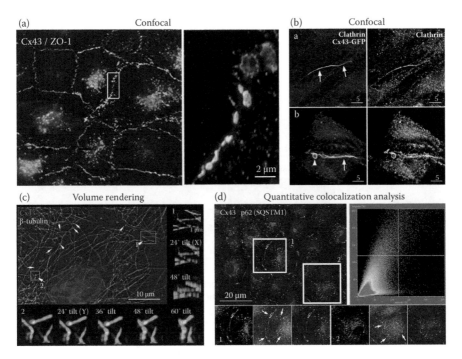

FIGURE 2.12 (**See color insert.**) Colocalization: Tools and techniques and potential false positives due to emission light bleed-through. (a) Cx43 gap junctions colocalizing with the scaffolding protein, ZO-1, generating a typical rim staining (Hunter et al. 2005), shown in fixed endogenously expressing primary PAECs. (Reprinted from *FEBS Lett*, 582, Baker, S. M., N. Kim, A. M. Gumpert, D. Segretain, and M. M. Falk, Acute internalization of gap junctions in vascular endothelial cells in response to inflammatory mediator-induced G-protein coupled receptor activation, 4039–46, Copyright (2008), with permission from Elsevier.) (b) Gap junctions and AGJs colocalizing with clathrin shown in fixed Cx43-eGFP-expressing HeLa cells. (Reproduced from Piehl, M., C. Lehmann, A. Gumpert, J. P. Denizot, D. Segretain, and M. M. Falk, *Mol Biol Cell*, 18, 337–47, 2007. With permission.) (c) Microtubules attached to Cx43 gap junctions (yellow dots in images of inset 2) shown in fixed and antibody-stained rat 1 cells. (Reprinted from *Curr Biol*, 11, Giepmans, B. N., I. Verlaan, T. Hengeveld, H. Janssen, J. Calafat, M. M. Falk, and W. H. Moolenaar, Gap junction protein connexin-43 interacts directly with microtubules, 1364–8, Copyright (2001), with permission from Elsevier.) To better demonstrate the colocalization, a stack of images in Z was acquired and a volume view was generated and rendered 48° horizontally (inset 1) and 60° sideways (inset 2), respectively. Note that at all angles, attachment points remain and do not spatially separate. (d) The colocalization of the Cx43 and the autophagic protein, p62/sequestosome1/SQSTM1 in PAECs, was demonstrated using an imaging software featuring quantitative colocalization analysis. (Reproduced from Fong, J. T., R. M. Kells, A. M. Gumpert, J. Y. Marzillier, M. W. Davidson, and M. M. Falk, *Autophagy*, 8, 794–811, 2012. With permission.) The fluorescence intensity of each pixel in both channels is analyzed and plotted. The blue signal in the shown scatter plot above the threshold (white lines, in this case, set high) indicates significant colocalization of both proteins.

(Continued)

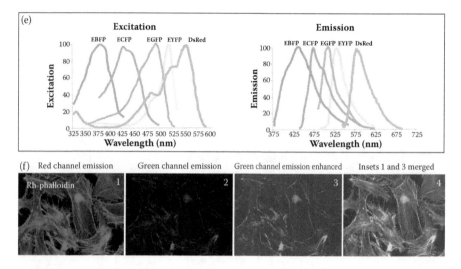

FIGURE 2.12 (CONTINUED)   **(See color insert.)** Colocalization: Tools and techniques and potential false positives due to emission light bleed-through. (e) Excitation and emission spectra of eGFP and color derivatives published by Promega Inc. Note the sometimes substantial overlap in excitation and emission signals requiring adequate excitation and emission bandpass filters, and correct exposure times. (f) A purposely falsified "colocalization" generated by enhancing the green bleed-through signal of the Rhodamine-labeled phalloidin (decorating actin filaments) and merging with the red emission signal to demonstrate that the images used for the colocalization analyses should be acquired at low exposure settings where bleed-through into the other emission channels is not occurring.

As colocalization is often restricted to small areas of the image, it can be more convincing to show additional higher secondary magnification images of selected, cropped areas (as in Figures 2.11d,e, 2.12a,c, and d). Furthermore, as monochrome often shows better than blue, green, or red on black (especially in print), it can be more convincing to show the single-color emission images in black and white and only the merged image in color (compare images in Figure 2.11d and e)*. To collect Z image stacks, reconstructing a volume view and tilting volume views sideways (as shown for microtubules attached to Cx43-based gap junctions in two different areas in Figure 2.12c) can further help to convincingly demonstrate a potential colocalization

---

* It is advantageous to use CCD cameras that are black and white, not color cameras, as they acquire images at 12- or even 16-bit image depth, compared to only 8 bits of RGB color images (the total bit depth a monitor can display is 24 bits). Also, they are more light sensitive compared to color CCD cameras. Black-and-white CCD cameras provide better images consisting of many more shades of gray (256, 4,096, and 65,536 shades of gray for an 8-, 12-, and 16-bit image, respectively) (Kimpe and Tuytschaever 2007). However, not all applications allow the display of 12- or 16-bit images and need to be converted into 8-bit images, unfortunately with an unavoidable loss of gray scales. This conversion can be done in Photoshop under "Image–Mode" settings. In addition, black-and-white images can be pseudocolored into any color, a feature that is useful to better show the most important object on an image (compare, for example, the images in Figure 2.9f and g in which Cx43-CFP and YFP tubulin are shown in either green or red, respectively) and to visualize the colocalization to red/green color-blind investigators/readers.

(Giepmans et al. 2001). Finally, a quantitative colocalization analysis can make a colocalization analyses even more compelling, especially if the maximum intensity threshold (the intensity of both signals on an image point) is set high. These can be performed with advanced imaging programs that measure the fluorescence intensities of each imaging point and show the values on scatter plots. This was performed with the image shown in Figure 2.12d, where the gap junctions and the AGJs colocalize with the autophagic marker protein, p62 sequestosome1, in endogenously Cx43-expressing primary pulmonary artery epithelial cells (PAECs) (Fong et al. 2012).

2.11 QUANTITATIVE FLUORESCENCE ANALYSES: A FEW EXAMPLES

Fluorescent images and image sequences of gap junctions not only contain a huge amount of qualitative and temporal information but also quantitative data as the fluorescence intensity correlates with the number of fluorophores. For example, quantitative fluorescence intensity analyses over time following photobleaching of the plasma membrane lipid (DiI-labeled) and Cx43-GFP demonstrate that the plasma membrane lipids are highly dynamic, both inside and outside of gap junctions, while the gap junction channels in the gap junction plaques are not. DiI fluorescence photobleached in defined areas inside and outside of gap junctions fully recovers to the prebleach levels within a few seconds, while the Cx43-GFP plaque areas do not (Figure 2.13a through c). These quantitative FRAP analyses were performed on a

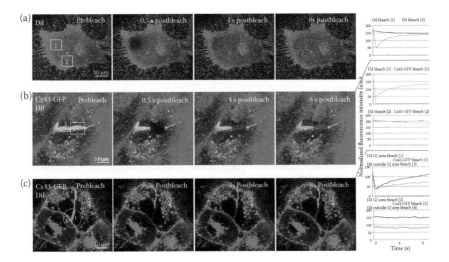

FIGURE 2.13 (**See color insert.**) Quantitative fluorescence analyses. Fluorescence images not only contain qualitative but also a wealth of quantitative data, as the fluorescence intensity correlates with the number of fluorophores. (a–c) Quantitative fluorescence intensity analyses of time-lapse recordings after permanently photobleaching the lipid (DiI) and the Cx43-GFP gap junction fluorescence in defined areas (boxed areas in (a) and (b) and white-outlined arbitrary areas in (c)). Note the rapid recovery of the lipid fluorescence inside and outside of the gap junctions, while the gap junction dynamics is significantly slower.

(Continued)

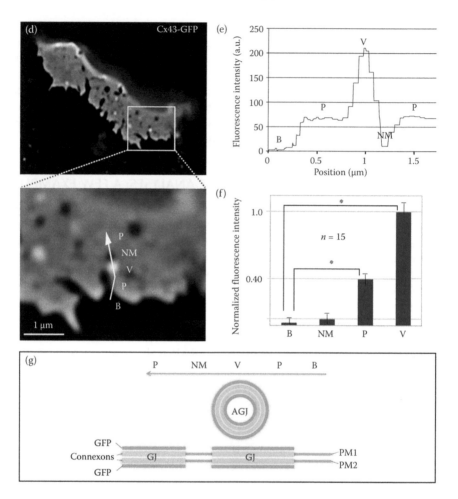

FIGURE 2.13 (CONTINUED) **(See color insert.)** Quantitative fluorescence analyses. (d–f) Quantitative fluorescence analysis performed along a line shown in the inset in d strategically placed to traverse a portion of a gap junction plaque P, the connexin-free junctional membrane domain inside NM and outside the plaque B, and a vesicle located in front of the plaque V. The fluorescence intensity along the line is shown in (e), and for 15 similar analyses in (f). (G) The schematic representation of the region shown in (d) with the layers of GFP highlighted that correspond to and correlate with the detected fluorescence intensities of areas B, NM, P, and V. (Also see the quantitative fluorescence colocalization analysis shown in Figure 2.12d.) (Reproduced from Falk, M. M., S. M. Baker, A. M. Gumpert, D. Segretain, and R. W. Buckheit III, *Mol Biol Cell*, 20, 3342–52, 2009. With permission.)

Zeiss LSM510 META laser-scanning confocal microscope running software package version 3.0 (Falk et al. 2009). Quantitative FRAP and FLIP experiments shown in Figure 2.10a through f were performed on a BioRad MRC-1024 laser-scanning confocal imaging system.

Furthermore, quantitative fluorescence image analyses also showed that gap junction channels internalized from gap junctions are endocytosed as complete double-membrane spanning channels and not as connexons (half-channels) (Falk et al. 2009). Figure 2.13d shows a still image of a time-lapse sequence of an en face-viewed gap junction acquired at high primary magnification. Several connexin-free junctional membrane domains within the gap junction (NM, appearing dark), bright fluorescent vesicles (V, AGJs generated by the endocytosis of gap junction plaque portions) in front of the plaque area (P), as well as connexin-free junctional membrane domain surrounding the gap junction plaque (B) are visible (schemed in Figure 2.13g). The fluorescence intensity along a line placed across these regions is shown in Figure 2.13e, and the average of 15 different plaques is shown in Figure 2.13f. Note that regions B and NM have comparable low background fluorescence. The fluorescence intensity in the plaque regions P emitted by two layers of GFP (on both sides of the docked connexons; compare Figure 2.13g) averages about 75 arbitrary fluorescence intensity units. The fluorescence intensity in the region of the AGJ in front of the plaque V is four times higher (averaging about 220 arbitrary units), indicating that four layers of GFP are present in the vesicle (two on the apical side and two on the basal side), totaling six layers of GFP when the vesicle is located in front of the gap junction plaque (as in the example shown in Figure 2.13d,e). These quantitative fluorescence analyses can be quite easily performed with advanced imaging software packages such as NIS Elements (Nikon), MetaVue and MetaMorph (Molecular Devices), Openlab (Improvision), ImageJ (National Institutes of Health, free shareware), etc.

2.12 DATA SIZE

2.12.1 CHALLENGES RELATED TO ACQUIRING LARGE TIME-LAPSE MOVIE FILES

As elaborated earlier, the best means of demonstrating the dynamic movements of gap junctions are time-lapse movie sequences. However, there are a few challenges with time-lapse movies that need to be considered. The raw data generated from time-lapse movies can easily be several hundred megabytes to gigabytes in size depending on the duration of the time lapse, the selection of image intervals, the number of channels acquired, and the number of points selected for study. A large external hard drive of at least several hundred gigabytes to a few terabytes should be available for the long-term storage of the generated data. It is best not to save the time-lapse images on an external drive as the data are generated, because the data transfer speed is often too slow for it to effectively work in real time and may slow down and unintentionally extend the set frame interval. It is better to store the captured frames on the imaging computer's hard drive and only transfer them to the external storage device after acquisition. It is advisable to erase the previous time-lapse movies from the hard drive from time to time, as they will gradually fill up the drive and slow down the computer system if the storage space becomes limited. Of course, a well-planned data filing system is essential if one wants to be able to reliably find and access earlier acquired image data sets.

2.12.2   COMPRESSION OF LARGE IMAGE FILES FOR SUBMISSION
AND PUBLICATION

For presentation, submission, and publication, movie sequences need to be trans-ferred into compatible formats such as AVI (for personal computers [PCs]), MOV (on Macintosh computers [Macs]), or MPEG which is accepted by Microsoft PowerPoint® and journal publishers. Compatibility issues between MOV and AVI formats unfortunately continue to exist between Macs and PCs, thus downloading a compatible media player that plays all formats (such as VLC or VideoLAN, a free shareware) is advisable. The presentation of movie files and the submission for pub-lication in general also require substantially reducing the file size. Most, if not all, the publishers limit the movie file size to 10 MB. It can be challenging to compress a 100 GB time-lapse recording into such a small size given that some time-lapse recordings may last for many hours. This, for example, was needed for the presen-tation of the over 20- and 50-hour-long recordings shown in Figure 2.2a and b. We had satisfactory success using QuickTime Pro on a Mac and MPEG movie formats. Finally, as time-lapse movies cannot be published in print, assembling still-image montages consisting of selected frames of the movies that represent the time-lapse events for the physical publication is recommended (see Figures 2.2a,b, 2.7b through f, 2.9f,g, 2.10a through g, 2.11b,c, and 2.13a through c). Publishers in gen-eral require that each submitted movie file is represented in the physical manuscript by at least one still image taken from the movie sequence (see the "Instructions to Authors" sections of individual journal/book publishers).

2.13   PRACTICAL CONSIDERATIONS FOR SETTING UP
A TIME-LAPSE RECORDING

For routine live-cell recordings of gap junctions, the cells should be passaged 1–2 days beforehand. If the cells stably express the fluorescent connexin (inducible or noninducible expression systems), one only needs to wait for the gap junction plaques to become visible under a conventional fluorescence microscope before beginning a time-lapse experiment. If the cells are transiently transfected, this should be done the day before the imaging for standard observations. Most connexins will form promi-nent gap junction plaques in cells within 16–24 hours (see Figure 2.8a). Transient transfections can be accomplished using a lipid-based transfection reagent kit such as Superfect (Qiagen) or Lipofectamine 2000 (Invitrogen). Prior to imaging, the cul-ture medium in the dish should be exchanged for fresh medium (not containing phe-nol red, see Section 2.2.1) to ensure that the cells have a sufficient supply of nutrients for the duration of the experiment. Dead and floating cells should be removed by washes with 1xPBS or culture medium, as these may obscure the imaging of healthy attached cells. Floating cells also often exhibit high autofluorescence, which can disguise the fluorescent signals in the attached cells.

If the cells are to be imaged in glass-bottom dishes, initial condensation tends to build up on the lid of the dish, which requires a few minutes for the conditions to sta-bilize. Automated image/incubation systems such as the Nikon BioStation will indi-cate to the user when the temperature in the chamber has equilibrated and stabilized (approximately 30 minutes). The stabilization period is required to ensure that the system will remain in focus throughout the recording period. Indeed, drifting out of focus is one of the main issues if the temperature in the imaging system is not well equilibrated. Thus, it is generally advisable to begin imaging during the day, periodically review the acquired image sequences, and refocus if necessary before leaving the system to image for extended periods of time, especially overnight. If the live-cell microscope system is equipped with an automatic X/Y stage, it allows the user to select, mark, and revisit multiple points of interest (gap junctions) for the entire duration of the time lapse. In addition, some advanced systems provide an infrared-based focusing system (autofocus) that keeps the distance between the objective lens and the cover glass constant. Of course, if the cells and the gap junctions move in Z, they still may move out of the focal plane.

To avoid the toxic effects resulting from short, high-energy wavelength exposure (e.g., the UV and the blue light used to excite CFP, GFP, etc.), it is advisable to observe the living cells with the lowest possible excitation light intensity and to keep the exposure times to the minimum that will still generate the desired image. The cells may tolerate the exposure to intense light for short times with negligible negative effects; however, in long recordings, the cells will be repeatedly exposed for many hours potentially generating nonphysiological cellular reactions (see Figure 2.2d). To image general gap junction dynamics, acquiring one image every 1–2 minutes while exposing for less than 2 seconds is a good start. In addition to fluorescence images, phase contrast or DIC frame acquisition is advisable, as overlaying the fluorescent and the contrast-enhanced white light images will generate a more comprehensive overview of the imaged cells (as in Figure 2.2a and b). It can also convincingly demonstrate that the suspected gap junction plaque is indeed residing in the lateral plasma membranes (as, e.g., shown in Figures 2.2a,b, 2.3g, 2.6e, 2.9a, and 2.11c).

2.14   CONCLUSIONS

Live-cell imaging of gap junctions is a powerful technique that is especially valuable if carried out together with other ultrastructural, biochemical, and molecular biology analyses. However, it is not a simple and quick technique. It requires dedication, planning, and a passion for microscopy. It is important to select the right imaging system and, as is true for all techniques, it is important to understand the system's abilities and limitations well enough to avoid potentially harmful pitfalls. However, the ability to observe the behavior of connexins and gap junctions in their natural environment, the wealth of acquired qualitative and quantitative spatiotemporal data, and the ability to detect connexins interacting with their various binding partners during their life cycle in real time, in our minds, is clearly worth the investment of time and resources. The continuous development of novel and improved fluorescent probes in combination with the development of enhanced and ever more affordable imaging systems (including superresolution techniques), promises to further increase our ability to investigate gap junctions in living cells in the near future. Finally, as fluorescence imaging is a highly visual technique, convincing data-reach, as well as esthetic images of cells are likely to excite life scientists at all stages of their careers. It is our hope that the numerous examples, tips, and procedures given in this chapter on how to perform live-cell imaging of connexins and gap junctions will contribute to increasing the excitement.

ACKNOWLEDGMENTS

MMF wishes to thank the many previous and current Falk lab members that over the past two decades have contributed in various successful ways to the live-cell imaging of gap junctions, their dedication, and their constructive criticism. In particular, we thank Dr. Ben N. G. Giepmans for performing the superb recording of an internalizing gap junction shown in Figure 2.11b when visiting my lab as a graduate student and Dr. Lynne Cassimeris, for critically reading and commenting on the paper. Work in the Falk lab is supported by the National Institutes of Health (NIH), National Institute of General Medical Sciences (NIGMS) Grant GM55725 and funds from Lehigh University.

REFERENCES

Baker, S. M., R. W. Buckheit III, and M. M. Falk. 2010. Green-to-red photoconvertible fluorescent proteins: Tracking cell and protein dynamics on standard wide-field mercury arc-based microscopes. *BMC Cell Biol.* no. 11:15.

Baker, S. M., N. Kim, A. M. Gumpert, D. Segretain, and M. M. Falk. 2008. Acute internalization of gap junctions in vascular endothelial cells in response to inflammatory mediator-induced G-protein coupled receptor activation. *FEBS Lett.* no. 582 (29):4039–46.

Bejarano, E., H. Girao, A. Yuste, B. Patel, C. Marques, D. C. Spray, P. Pereira, and A. M. Cuervo. 2012. Autophagy modulates dynamics of connexins at the plasma membrane in a ubiquitin-dependent manner. *Mol Biol Cell.* no. 23 (11):2156–69.

Berthoud, V. M., P. J. Minogue, J. G. Laing, and E. C. Beyer. 2004. Pathways for degradation of connexins and gap junctions. *Cardiovasc Res* no. 62 (2):256–67.

Boassa, D., J. L. Solan, A. Papas, P. Thornton, P. D. Lampe, and G. E. Sosinsky. 2010. Trafficking and recycling of the connexin43 gap junction protein during mitosis. *Traffic.* no. 11 (11):1471–86.

Bukauskas, F. F., K. Jordan, A. Bukauskiene, M. V. Bennett, P. D. Lampe, D. W. Laird, and V. K. Verselis. 2000. Clustering of connexin 43-enhanced green fluorescent protein gap junction channels and functional coupling in living cells. *Proc Natl Acad Sci U S A.* no. 97 (6):2556–61.

Chen, J., L. Pan, Z. Wei, Y. Zhao, and M. Zhang. 2008. Domain-swapped dimerization of ZO-1 PDZ2 generates specific and regulatory connexin43-binding sites. *EMBO J* no. 27 (15):2113–23.

Chudakov, D. M., S. Lukyanov, and K. A. Lukyanov. 2007. Tracking intracellular protein movements using photoswitchable fluorescent proteins PS-CFP2 and Dendra2. *Nat Protoc.* no. 2 (8):2024–32.

Falk, M. M. 2000. Connexin-specific distribution within gap junctions revealed in living cells. *J Cell Sci.* no. 113 (Pt 22):4109–20.

Falk, M. M. 2002. Genetic tags for labelling live cells: Gap junctions and beyond. *Trends Cell Biol.* no. 12 (9):399–404.

Falk, M. M., and N. B. Gilula. 1998. Connexin membrane protein biosynthesis is influenced by polypeptide positioning within the translocon and signal peptidase access. *J Biol Chem.* no. 273 (14):7856–64.

Falk, M. M., and U. Lauf. 2001. High resolution, fluorescence deconvolution microscopy and tagging with the autofluorescent tracers CFP, GFP, and YFP to study the structural composition of gap junctions in living cells. *Microsc Res Tech*. no. 52 (3):251–62.

Falk, M. M., S. M. Baker, A. M. Gumpert, D. Segretain, and R. W. Buckheit III. 2009. Gap junction turnover is achieved by the internalization of small endocytic double-membrane vesicles. *Mol Biol Cell*. no. 20 (14):3342–52.

Falk, M. M., L. K. Buehler, N. M. Kumar, and N. B. Gilula. 1997. Cell-free synthesis and assembly of connexins into functional gap junction membrane channels. *Embo J*. no. 16 (10):2703–16.

Falk, M. M., J. T. Fong, R. M. Kells, M. C. O'Laughlin, T. J. Kowal, and A. F. Thevenin. 2012. Degradation of endocytosed gap junctions by autophagosomal and endo-/lysosomal pathways: A perspective. *J Membr Biol*. no. 245 (8):465–76.

Falk, M. M., R. M. Kells, and V. M. Berthoud. 2014. Degradation of connexins and gap junctions. *FEBS Lett*. no. 588 (8):1221–9.

Falk, M. M., N. M. Kumar, and N. B. Gilula. 1994. Membrane insertion of gap junction connexins: Polytopic channel forming membrane proteins. *J Cell Biol*. no. 127 (2):343–55.

Fong, J. T., R. M. Kells, and M. M. Falk. 2013. Two tyrosine-based sorting signals in the Cx43 C-terminus cooperate to mediate gap junction endocytosis. *Mol Biol Cell*. no. 24 (18):2834–48.

Fong, J. T., R. M. Kells, A. M. Gumpert, J. Y. Marzillier, M. W. Davidson, and M. M. Falk. 2012. Internalized gap junctions are degraded by autophagy. *Autophagy*. no. 8 (5):794–811.

Fort, A. G., J. W. Murray, N. Dandachi, M. W. Davidson, R. Dermietzel, A. W. Wolkoff, and D. C. Spray. 2011. In vitro motility of liver connexin vesicles along microtubules utilizes kinesin motors. *J Biol Chem*. no. 286 (26):22875–85.

Fuller, S. D., R. Bravo, and K. Simons. 1985. An enzymatic assay reveals that proteins destined for the apical or basolateral domains of an epithelial cell line share the same late Golgi compartments. *EMBO J*. no. 4 (2):297–307.

Gaietta, G., T. J. Deerinck, S. R. Adams, J. Bouwer, O. Tour, D. W. Laird, G. E. Sosinsky, R. Y. Tsien, and M. H. Ellisman. 2002. Multicolor and electron microscopic imaging of connexin trafficking. *Science*. no. 296 (5567):503–7.

Giepmans, B. N. G., I. Verlaan, T. Hengeveld, H. Janssen, J. Calafat, M. M. Falk, and W. H. Moolenaar. 2001. Gap junction protein connexin-43 interacts directly with microtubules. *Curr Biol*. no. 11 (17):1364–8.

Griffin, B. A., S. R. Adams, and R. Y. Tsien. 1998. Specific covalent labeling of recombinant protein molecules inside live cells. *Science*. no. 281 (5374):269–72.

Gumpert, A. M., J. S. Varco, S. M. Baker, M. Piehl, and M. M. Falk. 2008. Double-membrane gap junction internalization requires the clathrin-mediated endocytic machinery. *FEBS Letters*. no. 582:2887–92.

Gurskaya, N. G., V. V. Verkhusha, A. S. Shcheglov, D. B. Staroverov, T. V. Chepurnykh, A. F. Fradkov, S. Lukyanov, and K. A. Lukyanov. 2006. Engineering of a monomeric green-to-red photoactivatable fluorescent protein induced by blue light. *Nat Biotechnol*. no. 24 (4):461–5.

Hesketh, G. G., M. H. Shah, V. L. Halperin, C. A. Cooke, F. G. Akar, T. E. Yen, D. A. Kass, C. E. Machamer, J. E. Van Eyk, and G. F. Tomaselli. 2010. Ultrastructure and regulation of lateralized connexin43 in the failing heart. *Circ Res*. no. 106 (6):1153–63.

Hunter, A. W., R. J. Barker, C. Zhu, and R. G. Gourdie. 2005. Zonula occludens-1 alters connexin43 gap junction size and organization by influencing channel accretion. *Mol Biol Cell*. no. 16 (12):5686–98.

Ivanov, A. I., I. C. McCall, C. A. Parkos, and A. Nusrat. 2004a. Role for actin filament turnover and a myosin II motor in cytoskeleton-driven disassembly of the epithelial apical junctional complex. *Mol Biol Cell*. no. 15 (6):2639–51.

Ivanov, A. I., A. Nusrat, and C. A. Parkos. 2004b. Endocytosis of epithelial apical junctional proteins by a clathrin-mediated pathway into a unique storage compartment. *Mol Biol Cell*. no. 15 (1):176–88.

Johnson, K. E., S. Mitra, P. Katoch, L. S. Kelsey, K. R. Johnson, and P. P. Mehta. 2013. Phosphorylation on serines 279 and 282 of connexin43 regulates endocytosis and gap junction assembly in pancreatic cancer cells. *Mol Biol Cell*. no. 24:715–33.

Jordan, K., J. L. Solan, M. Dominguez, M. Sia, A. Hand, P. D. Lampe, and D. W. Laird. 1999. Trafficking, assembly, and function of a connexin43-green fluorescent protein chimera in live mammalian cells. *Mol Biol Cell*. no. 10 (6):2033–50.

Kalmatsky, B. D., Y. Batir, T. A. Bargiello, and T. L. Dowd. 2012. Structural studies of N-terminal mutants of connexin 32 using (1)H NMR spectroscopy. *Arch Biochem Biophys*. no. 526:1–8.

Kalmatsky, B. D., S. Bhagan, Q. Tang, T. A. Bargiello, and T. L. Dowd. 2009. Structural studies of the N-terminus of connexin 32 using 1H NMR spectroscopy. *Arch Biochem Biophys*. no. 490 (1):9–16.

Kelly, S. M., J. K. Vanslyke, and L. S. Musil. 2007. Regulation of ubiquitin-proteasome system mediated degradation by cytosolic stress. *Mol Biol Cell*. no. 18 (11):4279–91.

Kimpe, T., and T. Tuytschaever. 2007. Increasing the number of gray shades in medical display systems—How much is enough? *J Digit Imaging*. no. 20 (4):422–32.

Koval, M. 2006. Pathways and control of connexin oligomerization. *Trends Cell Biol*. no. 16 (3):159–66.

Laing, J. G., and E. C. Beyer. 1995. The gap junction protein connexin43 is degraded via the ubiquitin proteasome pathway. *J Biol Chem*. no. 270 (44):26399–403.

Laing, J. G., P. N. Tadros, E. M. Westphale, and E. C. Beyer. 1997. Degradation of connexin43 gap junctions involves both the proteasome and the lysosome. *Exp Cell Res*. no. 236 (2):482–92.

Laird, D. W. 1996. The life cycle of a connexin: Gap junction formation, removal, and degradation. *J Bioenerg Biomembr*. no. 28 (4):311–8.

Laird, D. W., K. Jordan, T. Thomas, H. Qin, P. Fistouris, and Q. Shao. 2001. Comparative analysis and application of fluorescent protein-tagged connexins. *Microsc Res Tech*. no. 52 (3):263–72.

Lauf, U., P. Lopez, and M. M. Falk. 2001. Expression of fluorescently tagged connexins: A novel approach to rescue function of oligomeric DsRed-tagged proteins. *FEBS Lett*. no. 498 (1):11–5.

Lauf, U., B. N. Giepmans, P. Lopez, S. Braconnot, S. C. Chen, and M. M. Falk. 2002. Dynamic trafficking and delivery of connexons to the plasma membrane and accretion to gap junctions in living cells. *Proc Natl Acad Sci U S A*. no. 99 (16):10446–51.

Leithe, E., A. Kjenseth, S. Sirnes, H. Stenmark, A. Brech, and E. Rivedal. 2009. Ubiquitylation of the gap junction protein connexin-43 signals its trafficking from early endosomes to lysosomes in a process mediated by Hrs and Tsg101. *J Cell Sci*. no. 122 (Pt 21):3883–93.

Lichtenstein, A., P. J. Minogue, E. C. Beyer, and V. M. Berthoud. 2011. Autophagy: A pathway that contributes to connexin degradation. *J Cell Sci*. no. 124 (Pt 6):910–20.

Lippincott-Schwartz, J., and G. H. Patterson. 2009. Photoactivatable fluorescent proteins for diffraction-limited and super-resolution imaging. *Trends Cell Biol*. no. 19 (11):555–65.

Lopez, P., D. Balicki, L. K. Buehler, M. M. Falk, and S. C. Chen. 2001. Distribution and dynamics of gap junction channels revealed in living cells. *Cell Commun Adhes*. no. 8 (4–6):237–42.

Los, G. V., L. P. Encell, M. G. McDougall, D. D. Hartzell, N. Karassina, C. Zimprich, M. G. Wood et al. 2008. HaloTag: A novel protein labeling technology for cell imaging and protein analysis. *ACS Chem Biol*. no. 3 (6):373–82.

Maeda, S., S. Nakagawa, M. Suga, E. Yamashita, A. Oshima, Y. Fujiyoshi, and T. Tsukihara. 2009. Structure of the connexin 26 gap junction channel at 3.5 A resolution. *Nature*. no. 458 (7238):597–602.

Matlin, K. S., and K. Simons. 1983. Reduced temperature prevents transfer of a membrane glycoprotein to the cell surface but does not prevent terminal glycosylation. *Cell*. no. 34 (1):233–43.

Musil, L. S., and D. A. Goodenough. 1993. Multisubunit assembly of an integral plasma membrane channel protein, gap junction connexin43, occurs after exit from the ER. *Cell*. no. 74 (6):1065–77.

Nickel, B. M., B. H. DeFranco, V. L. Gay, and S. A. Murray. 2008. Clathrin and Cx43 gap junction plaque endoexocytosis. *Biochem Biophys Res Commun*. no. 374 (4):679–82.

Nimlamool, W., R. M. Kells Andrews, and M. M. Falk. 2015. Connexin43 phosphorylation by PKC and MAPK signals VEGF-mediated gap junction internalization. *Mol Biol Cell*. no. 26 (15):2755–68.

Piehl, M., C. Lehmann, A. Gumpert, J. P. Denizot, D. Segretain, and M. M. Falk. 2007. Internalization of large double-membrane intercellular vesicles by a clathrin-dependent endocytic process. *Mol Biol Cell*. no. 18 (2):337–47.

Purnick, P. E., D. C. Benjamin, V. K. Verselis, T. A. Bargiello, and T. L. Dowd. 2000. Structure of the amino terminus of a gap junction protein. *Arch Biochem Biophys*. no. 381 (2):181–90.

Qin, H., Q. Shao, S. A. Igdoura, M. A. Alaoui-Jamali, and D. W. Laird. 2003. Lysosomal and proteasomal degradation play distinct roles in the life cycle of Cx43 in gap junctional intercellular communication-deficient and competent breast tumor cells. *J Biol Chem*. no. 278 (32):30005–14.

Rhett, J. M., and R. G. Gourdie. 2012. The perinexus: A new feature of Cx43 gap junction organization. *Heart Rhythm*. no. 9 (4):19–23.

Rotundo, R. L., and D. M. Fambrough. 1980. Secretion of acetylcholinesterase: Relation to acetylcholine receptor metabolism. *Cell*. no. 22 (2 Pt 2):595–602.

Saraste, J., and E. Kuismanen. 1984. Pre- and post-Golgi vacuoles operate in the transport of Semliki Forest virus membrane glycoproteins to the cell surface. *Cell*. no. 38 (2):535–49.

Saraste, J., G. E. Palade, and M. G. Farquhar. 1986. Temperature-sensitive steps in the transport of secretory proteins through the Golgi complex in exocrine pancreatic cells. *Proc Natl Acad Sci U S A*. no. 83 (17):6425–9.

Schlingmann, B., P. Schadzek, F. Hemmerling, F. Schaarschmidt, A. Heisterkamp, and A. Ngezahayo. 2013. The role of the C-terminus in functional expression and internalization of rat connexin46 (rCx46). *J Bioenerg Biomembr*. no. 45 (1–2):59–70.

Schmoranzer, J., M. Goulian, D. Axelrod, and S. M. Simon. 2000. Imaging constitutive exocytosis with total internal reflection fluorescence microscopy. *J Cell Biol*. no. 149 (1):23–32.

Shaner, N. C., R. E. Campbell, P. A. Steinbach, B. N. Giepmans, A. E. Palmer, and R. Y. Tsien. 2004. Improved monomeric red, orange and yellow fluorescent proteins derived from Discosoma sp. red fluorescent protein. *Nat Biotechnol*. no. 22 (12):1567–72.

Shaner, N. C., G. H. Patterson, and M. W. Davidson. 2007. Advances in fluorescent protein technology. *J Cell Sci*. no. 120 (Pt 24):4247–60.

Shaw, R. M., A. J. Fay, M. A. Puthenveedu, M. von Zastrow, Y. N. Jan, and L. Y. Jan. 2007. Microtubule plus-end-tracking proteins target gap junctions directly from the cell interior to adherens junctions. *Cell*. no. 128 (3):547–60.

Shcherbakova, D. M., P. Sengupta, J. Lippincott-Schwartz, and V. V. Verkhusha. 2014. Photocontrollable fluorescent proteins for superresolution imaging. *Annu Rev Biophys*. no. 43:303–29.

Stout, R. F. Jr., E. L. Snapp, and D. C. Spray. 2015. Connexin type and fluorescent protein fusion tag determine structural stability of gap junction plaques. *J. Biol. Chem.* no. 290 (39):23497–514.

Thévenin, A. F., T. J. Kowal, J. T. Fong, R. M. Kells, C. G. Fisher, and M. M. Falk. 2013. Proteins and mechanisms regulating gap junction assembly, internalization and degradation. *Physiology.* no. 28 (4):93–116.

Thomas, T., K. Jordan, J. Simek, Q. Shao, C. Jedeszko, P. Walton, and D. W. Laird. 2005. Mechanisms of Cx43 and Cx26 transport to the plasma membrane and gap junction regeneration. *J Cell Sci.* no. 118 (Pt 19):4451–62.

Toomre, D., J. A. Steyer, P. Keller, W. Almers, and K. Simons. 2000. Fusion of constitutive membrane traffic with the cell surface observed by evanescent wave microscopy. *J Cell Biol.* no. 149 (1):33–40.

Wang, H. Y., Y. P. Lin, C. K. Mitchell, S. Ram, and J. O'Brien. 2015. Two-color fluorescent analysis of connexin 36 turnover: Relationship to functional plasticity. *J Cell Sci.* no. 128 (21):3888–97.

Xiao, D., S. Chen, Q. Shao, J. Chen, K. Bijian, D. W. Laird, and M. A. Alaoui-Jamali. 2014. Dynamin 2 interacts with connexin 26 to regulate its degradation and function in gap junction formation. *Int J Biochem Cell Biol.* no. 55:288–97.

3 FRAP for the Study of Gap Junction Nexus Macromolecular Organization

Randy F. Stout, Jr. and David C. Spray

CONTENTS

3.1   GENERAL INTRODUCTION

Gap junctions are structures that connect cells through formation of channels spanning adjacent cell's membranes. Gap junction channels are made up of dodecamers of connexin proteins—two hexamers dock end-to-end to form a full gap junction channel. The NH_2-terminus and the COOH terminus (CT) of each connexin are located in the cytoplasm with two extracellular loops extended outside of the cell leading to four transmembrane domains per connexin. In this chapter, we follow the most commonly used connexin nomenclature, where the molecular weight in kilodalton of the cDNA-predicted protein follows the connexin; in cases where mobility is measured of gap junction plaques made of multiple connexins, we refer

63

to the subfamilies (α, β, γ, δ, ε) to indicate the relationships. Gap junctions are, perhaps invariably, associated with the very tight clustering of individual channels into a structure termed the *gap junction plaque*. It is currently unknown if clustering (plaque formation) is required for the gap junction channel formation, or vice versa. Connexins within the gap junction plaque interact with many other cellular proteins, and this wider molecular complex is known as the gap junction nexus.[1]

In addition to acting as a pathway for the intercellular exchange of ions, metabolites, and signaling molecules, gap junctions have major roles in determining the arrangement and the activity of other cellular proteins and in regulating cellular processes such as cytoskeletal arrangement,[2] cell motility,[3,4] and autophagy.[5] The phosphorylation of specific residues in the CT of Cx43 controls channel function (see Marquez-Rosado et al.[6]), binding of cytoskeletal proteins,[7] and endocytosis of gap junction channels.[8,9] Some of these phosphorylation modifications have been found to be arranged into specific but modifiable patterns at the gap junction plaque structure.[10] Scaffolding proteins and cytoskeletal proteins associate with Cx43 gap junction plaques in specific patterns.[11] Even the addition of new and removal of old gap junctions is ordered with new channels added to the edge of the gap junction plaque and endocytic vesicles most often internalized from the interior of the gap junction plaque.[12]

It is clear that understanding the organization and dynamics of gap junction plaques will be important to understand how gap junctions modify and respond to cell and tissue physiologies. Several complementary technical approaches are useful in this area of gap junction research. We focus on FRAP in this chapter. Table 3.1

TABLE 3.1

Studies That Used FRAP to Study Connexin Macromolecular Organization or Plaque Formation/Stability in the Membrane

Study	Form of FRAP Technique Used	Gap Junction Nexus Components Studied	Selected Conclusions Based on FRAP Experiments
Lopez et al. 2001, *Cell Commun Adhes*	NA	Cx26-YFP, Cx43-GFP, Cx43-CFP	Mobility of the gap junction channels is slow and not dominated by diffusion.[17]
Thomas et al. 2001, *Cell Commun Adhes*	Gap junction reformation after whole gap junction bleach	Cx43-GFP, Cx26-YFP	Cx43 and Cx26 gap junction formation is dependent on actin; Cx43 requires microtubule (MTs) for traffic to the plasma membrane (PM).[18]
Lauf et al. 2002, *Proc Natl Acad Sci U S A*	FRAP on connexons; FRAP on gap junction plaques	Cx43-GFP	Undocked connexons are mobile in the plasma membrane; the local arrangement of the Cx43 gap junction channels is very stable within the gap junction plaques, while the gap junction plaque morphology is fluid.[19]

(Continued)

TABLE 3.1 (*Continued*)

Studies That Used FRAP to Study Connexin Macromolecular Organization or Plaque Formation/Stability in the Membrane

Study	Form of FRAP Technique Used	Gap Junction Nexus Components Studied	Selected Conclusions Based on FRAP Experiments
Thomas et al. 2005, *J Cell Sci*	Stability of the channel arrangement within the gap junction plaques, the connexon FRAP, the plaque stability	Cx43-GFP, Cx26-YFP	Both connexins have low mobility; Cx26 mobility is higher.[20]
Falk et al. 2009, *Mol Biol Cell*	Two-color FRAP within the gap junction plaques	Lipid dye (DiI) and Cx43-EGFP	The lipid dye can move between the gap junction channels.[12]
Simek et al. 2009, *J Cell Sci*	Intraplaque FRAP to test stability of channel arrangement within the gap junction plaques	Cx43-GFP, Cx26-GFP, truncated Cx43	Cx43 and Cx26 are mobile within the plaques; some plaques are stable, some are fluid, and more plaques are fluid when Cx43 is truncated.[21]
Bhalla-Gehi et al. 2010, *Am Soc Biochem Mol Biol*	FRAP on Panx1 and Panx2 channels in the plasma membrane (PM) and others	Panx1-EGFP, Panx2-EGFP	Panx1 and Panx2 are highly mobile within the membrane.[22]
Katoch et al. 2015, *J Biol Chem*	Whole plaque FRAP to test plaque mobility and turnover rate	Cx32-EGFP	The C-terminus of Cx32 regulates the gap junction size and turnover.[23]
Stout et al. 2015, *J Biol Chem*	Intraplaque FRAP, stability of the channel arrangement within the gap junction plaques	Cx43-msfGFP, truncated Cx43, Cx30-msfGFP, Cx26-msfGFP, and others	Cx43 channel arrangement is stable within the gap junction plaques and depends on the cytoplasmic CT of Cx43; Cx30 and Cx26 form very fluid plaques, and more.[24]
Kelly et al. 2015, *J Cell Sci*	FRAP with long time course 3D recovery acquisition to monitor plaque growth and turnover	Cx30-EGFP	Cx30 gap junctions have a greatly decreased turnover rate in comparison to Cx43 and some other connexins. The addition of new Cx30 gap junction channels occurs at the edge of gap junction plaques.[25]
Wang et al. 2015, *J Cell Sci*	Intraplaque FRAP with Halo-tagged Cx36	Cx36 with Halo tag with Oregon Green dye	Cx36 forms very fluidly arranged gap junction plaques.[26]

lists the published studies to date using FRAP to study the arrangement of the gap junction nexus components.

3.2 INTRODUCTION TO FRAP

The FRAP technique takes advantage of the capacity of many fluorophores to be inactivated through exposure to high-intensity light. High-intensity laser illumination is used to permanently disable the fluorophore of fluorescent proteins attached to the protein of interest within a spatially defined region of cells. Monitoring the movement of the unbleached proteins from outside into the bleach region allows assessment of the mobility characteristics of the protein of interest (Figure 3.1). Two references provide in-depth discussion of FRAP experimentation and analysis methods.[13,14] Researchers are encouraged to consult these references prior to planning FRAP experiments with

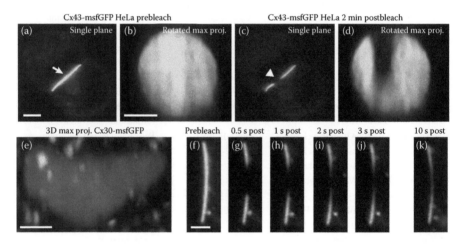

FIGURE 3.1 Illustration of the FRAP technique to study the arrangement of gap junction plaque components. Appearance of a Cx43-msfGFP gap junction plaque in HeLa cells that was vertically oriented with respect to the coverslip as imaged with a confocal LSM. (a) The plaque appears as a very bright line in a single-plane acquisition (arrow points to the plaque). The faint signal from intracellular Cx43-msfGFP can also be seen. (b) A 3D reconstruction, en face view of the plaque before photobleach (maximum intensity, viewed from above the cells in the Z axis). (c) A single confocal plane acquisition 2 minutes after a 2 μm section of the plaque was bleached (arrowhead). (d) A 3D reconstruction rotated to show the same gap junction plaque 2 minutes after bleaching. Notice that the bleach region that appears linear in the panel to the left is actually an hourglass shape due to the Gaussian shape of the bleaching laser. Also note that no recovery has occurred anywhere within the plaque at 2 minutes after the photobleach—indicating a stable structure. (e) A rotated 3D maximum intensity reconstruction of a human Cx30-msfGFP plaque between a pair of N2A cells. This gap junction plaque was oriented in a plane perpendicular to the cover glass growth substrate before a FRAP experiment. (f–k) Frames from a single Z-plane FRAP experiment on the plaque shown in the leftmost panel. The time points are indicated above each frame of the time lapse. Note that the recovery occurs inward from the edge of the bleach region and that the initially sharp border between the bleached and unbleached sections of the plaque becomes blurred over time—indicating a fluidly arranged structure.

(Continued)

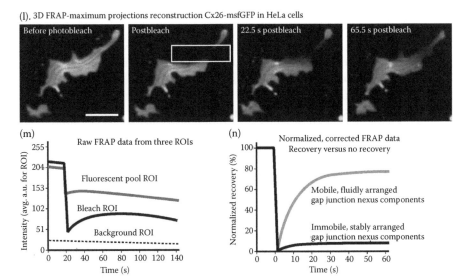

(l), 3D FRAP-maximum projections reconstruction Cx26-msfGFP in HeLa cells

FIGURE 3.1 (CONTINUED)   Illustration of the FRAP technique to study the arrangement of gap junction plaque components. (l) An example of 3D FRAP with a rectangular bleach region on a rat Cx26-msfGFP gap junction plaque between two HeLa cells. Maximum intensity reconstructions at ~4-second intervals were acquired. The plaque was bleached in the region indicated by the rectangle. Note that the unbleached Cx26-msfGFP has moved into the bleached area from the unbleached sections of the gap junction plaque within ~1 minute after the photobleach. (m) An illustration of the expected average intensity measurement for a FRAP experiment. The bleach event is at 20 seconds in this illustration. The large drop in the fluorescent pool available for recovery (fpF) at 20 seconds is due to the small overall size of the fluorescent pool, which is limited to the portion of the gap junction plaque that is visible within a single confocal plane. The bleaching was incomplete within the bleach region (bF) as shown by the black trace for the average intensity within the bleach region not reaching a value of zero. The slight downward slope of all the traces represents the acquisition bleach that is caused by the effect of the photobleach by illumination needed for time-lapse image acquisitions after the bleach event (acquisition bleach). (n) After scaling and normalization for the incomplete bleach and the variation in initial intensity and the correction for bleaching of the fluorescent pool and the acquisition bleach, an illustration of the expected normalized, corrected recovery curves is shown for a stably arranged (immobile) molecule (black trace) such as Cx43, and a fluidly arranged (mobile) molecule (gray trace) such as Cx30. Note that by correcting for the bleach of the fluorescent pool during the bleaching event, it increases the total percent recovery and the correction for acquisition bleach eliminates the downward slope once the plateau of recovery has been reached. The white scale bars represent 5 μm in (a) through (d) and 1 and 2 μm in (e) through (k).

connexins and pannexins. The focus of this chapter is to facilitate the adaptation of the FRAP technique for the study of the gap junction nexus.

Laser light near the maximal excitation wavelength of the fluorophore most efficiently induces bleaching, but shorter wavelengths also bleach longer wavelength-excitation fluorophores. The most efficient bleaching for a given fluorophore and sample needs to be empirically determined for each experimental design. The least extensive and most efficient bleaching possible should be used to minimize the undesired heat generation and oxidation of the proteins in live samples.[15,16] The following protocols detail the methods by which to obtain single and dual-label FRAP data sets on junctional and nonjunctional gap junction proteins and their associated binding partners and cooccupants of the gap junction plaque. We also provide step-by-step procedures for data analysis.

3.3 FRAP FOR STUDYING PLAQUE STRUCTURE

Use of the FRAP technique to study gap junction plaque structure and dynamics began soon after the initial generation of fluorescent protein-tagged connexins.[27-30] Along with the observation of gap junction plaque morphology in live cells, the early use of FRAP to study gap junctions indicated that connexins exhibit very different behavior when clustered into gap junctions compared with other transmembrane proteins.[19] Connexins localized to plaques did not exhibit mobility characteristic of diffusion-dominated processes but instead behaved in a manner suggestive of mobility controlled by the movements of the membrane in which they were embedded.[17] This is an important finding when considering the methods of analysis that can be used for FRAP data on gap junctions and which experimental groups can be compared, as discussed in later sections of this chapter. Later, Baker et al. used spatially restricted photoconversion of the photoconvertible protein Dendra2 to provide other important information regarding the dynamics of the gap junction plaque structure.[31] Several other studies by Thomas et al.,[20] Simek et al.,[21] Bhalla-Gehi et al.,[22] and Kelly et al.[25] have used the FRAP technique to study the structural dynamics of gap junction plaques and pannexin channels. These studies provided an indication that the connexin isoform that makes up the gap junction plaque, as well as the location of the fluorescent protein tag on the connexin, are major determinants of the mobility characteristics of the plaque. We recently explored this area further in a study that redefined the qualitative characteristics of several connexin types within the plaque and clarified the effects of fluorescent protein tags on connexin mobility.[24] A recent study by another group using a different fluorescent tagging strategy revealed that Cx36 gap junction plaques are fluidly arranged structures.[25] The studies using FRAP on gap junctions are in general agreement that the crowded protein environment of the plaque and the requisite linkage to the plasma membranes of two cells (along with as-of-yet unidentified other forces) produce interesting mobility characteristics for connexins within the gap junction plaque structure.

FRAP has also been used to reveal new aspects regarding the persistence of connexin proteins within a junctional area (often referred to as gap junction stability, not to be confused with the stability of the arrangement of channels within a gap junction plaque). The use of FRAP to measure large-scale movements and turnover of connexin proteins within the junctional membrane has revealed important information about gap junction biology,[23] but will not be discussed further in this chapter. We focus on the experimental approaches to use FRAP to examine the stability of the arrangement of the gap junction channels within the plaque structure and the mobility of the connexons (hemichannels) in the plasma membrane. The incorporation of fluorescent protein tagged connexins into gap junction plaques and the use of the technique described in this chapter is illustrated in the supplemental

FRAP in nonjunctional sfGFP-Cx43

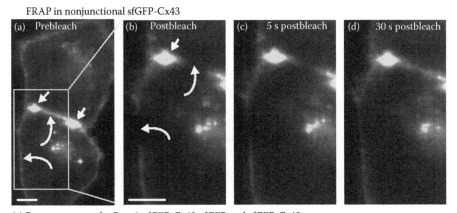

(e) Recovery curves for Panx1-sfGFP, Cx43-sfGFP, and sfGFP-Cx43

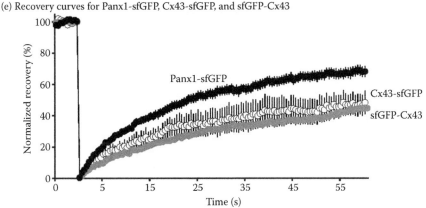

FIGURE 3.2 FRAP to study nonjunction connexin mobility. FRAP experiment with sfGFP-Cx43 with bleaching and recovery in the plasma membrane regions outside the gap junction plaque. (a) An image of HeLa cells expressing sfGFP-Cx43 with the pinhole open to the maximum prior to the photobleach. Two small gap junctions are present, connecting the two cells (straight, short arrows). The sfGFP-Cx43 in two sections outside of the gap junction plaque (curved arrows) will be bleached for this FRAP experiment. Note that the section of the membrane between the two plaques is roughly twice as bright as the lateral membrane, because there are two cells contributing sfGFP-Cx43 to the signal, but no gap junction plaque happens to be present in the location indicated by the upper curved arrows. (b) The 0.6-second post-photobleach at the locations indicated by the curved arrows. This image is from the same FRAP experiment time lapse as in a but zoomed to show the detail of the bleach. (c) A small amount of recovery is apparent at 5 seconds postbleach, as the border between light and dark at the bleach region edge becomes a less steep gradient, and the fluorescence signal appears to close in on the bleach region. (d) The majority of the recovery has occurred after 30 seconds postbleach. (e) Examples of the results displayed as normalized recovery curves for Pannexin1-sfGFP, Cx43-sfGFP, and sfGFP-Cx43, $n = 15$, 5, and 11, respectively; the error bars are standard errors of the mean (SEM). The ImageJ plug-in simFRAP[38] was used to calculate an effective diffusion coefficient for the two bleach regions shown in a through d. The calculated values for this example experiments are 0.116 μ^2/s for the upper appositional but nonjunctional bleach region and 0.134 μ^2/s for the lower left, nonappositional bleach in the sfGFP-Cx43 FRAP experiment shown in a through d. See Sections 3.9 through 3.11 in for information on this and other analysis methods.

animation available online (https://www.crcpress.com/Gap-Junction-Channels-and-Hemichannels/Bai-Sez/9781498738620). As described by others,[19] our tests show that nonjunctional connexin membrane mobility has characteristics of diffusion in a membrane, and the diffusion coefficients for nonplaque-localized Cx43 are in the range of other multimeric transmembrane proteins (Figure 3.2). The extreme differences in the protein density, the effective pool of unbleached fluorescent protein available to participate in the recovery, and the differences in membrane geometry make meaningful comparisons between the FRAP experiments on junctional and nonjunctional membranes difficult, if not impossible (see Sbalzarini et al.[32] and Mai et al.[33]).

3.4   EXPERIMENTAL SETUP FOR GAP JUNCTION FRAP

We found the following experimental setup to be useful for our studies on the intraplaque mobilities of gap junctions made up of Cx26, Cx30, Cx32, Cx43, and mutant connexins. Image interval, bleaching rate, and other parameters may need to be optimized for individual microscope system, lasers, and perhaps for connexin/cell type. This first procedure describes FRAP measurement of the gap junction arrangement stability in vertical plaques. This procedure will allow the quick and standard assessment of the stability of the arrangement of gap junction channels within the plaque structure. Vertical plaques (oriented perpendicularly to the substrate upon which the cells are cultured) are common in cells such as the HeLa and the neuro2a (N2A) cell lines, since the plasma membrane of these cells tend to extend in vertical planes when the cells are in contact (the conditions in which the plaques form). Vertical plaques more frequently form when nearly confluent (>80%) cells are transfected while they are attached to the growth surface upon which they will be imaged. Conversely, if horizontal (en face; aligned parallel to the growth surface) plaques are desired, the cells should be transfected prior to replating into the imaging chamber.

This total procedure takes 3 days, and FRAP experiments can be performed on days 3 through 7 after the procedure is initiated.

Step 1: Replate the cells using a standard procedure appropriate for the cell type (e.g., wash in Ca^{2+} and Mg^{2+} free PBS followed by trypsin for HeLa cells) to a culture dish of choice for imaging. For HeLa cells plated at high enough density to achieve 50% confluency upon cell attachment, it is best if the transfection is performed the following day. The plating density must be determined for each cell type and can be adjusted to the desired transfection and imaging schedule.

Step 2: Transfect the cells with a plasmid containing the sequence for a connexin-fluorescent protein fusion transgene with a promoter for mammalian cell expression such as the cytomegalovirus or the elongation factor 1 alpha promoters (or other promoter for the corresponding cell type). The plasmids for tagged connexin overexpression that were produced and used by the authors along with those from several other labs are available on the nonprofit plasmid repository Addgene.org. Multiple transfections can be useful as well as viral transduction when high transfection efficiency is needed. We most commonly use Optifect (Thermo Fisher, Lifetechnologies, Cat. No. 12579–017) for transfections for FRAP experiments. We describe the protocol used to transfect HeLa cells grown in an ibidi 8 well chambered slide for FRAP experiments:

a. The medium (DMEM [Dulbecco's modified eagle medium] + 10% fetal bovine serum [FBS] for HeLa cells) is removed from the cells growing in the 8 well dish and replaced with prewarmed Opti-MEM reduced serum free media (Fisher Cat. No. 31985062), 250 μL per well. Be sure to prevent the cells from drying during the media change. The cells are returned to the incubator until the transfection mixture (described later) is applied.

b. One microgram of DNA is placed into a separate tube with 50 μL of Opti-MEM if each well of the 8 well chambered slide will receive a different plasmid transfection. The DNA and Opti-MEM volumes can be scaled up if multiple individual chambers will receive the same plasmid.

c. Three microliters of optifect per 1 μg of DNA is placed into 50 μL of Opti-MEM media. For the transfection of an 8 well plate, 24 uL of Optifect is added to 400 uL of Opti-MEM media in a sterile microcentrifuge tube. The optifect/Opti-MEM mixture should be mixed by gently tapping the tube 2–3 times.

d. The tubes containing the optifect + Opti-MEM and the DNA + Opti-MEM are incubated at room temperature for 5 minutes, separately then combined 1:1. If each of the 8 wells on the slide will receive a different plasmid, then 50 μL of the optifect + Opti-MEM mixture is distributed to each of the tubes containing 50 μL of DNA + Opti-MEM. The resulting transfection mixtures (100 μL each) are then incubated at room temperature for 30 minutes.

e. After 30 minutes of incubation, 100 μL of the transfection mixture (from the previous step) is pipetted into each well of the 8 well chamber. The cells are then returned to the incubator for at least 3 hours.

f. After 3–16 hours, the transfection mixture should be replaced with the normal growth media for the cells (DMEM + 10% FBS for HeLa cells). The cells can be used for FRAP 24 hours after the transfection, but we wait for 48–96 hours before performing the FRAP, since the most numerous large gap junction plaques form during this posttransfection period.

Step 3: Prepare the cells for the FRAP experiments. At least 24 hours after the cells have been transfected and grown in normal media for at least 12 hours, they are ready to use for FRAP experiments. The media should be replaced with a pH-buffered solution at least 30 minutes but not more than 60 minutes before bringing the sample to the microscope. We normally use the phenol-free version of the normal growth media supplemented with FBS addition appropriate for the cell type (10% for HeLa and N2a cells) and 25 mM HEPES and 2 mM glutamine or GlutaMax™ (Fisher 35050061). We and other groups[20] have also used the Opti-MEM media without phenol red as the buffered external solution for imaging. These imaging media formulations are preferable over the simple buffered extracellular saline solutions because the FRAP experiment sessions can last for up to 4 hours, and the cells begin to endocytose the gap junctions due to serum starvation within this timescale.[5,34,35]

Step 4: Take the cells to the confocal laser-scanning microscope (LSM) for FRAP experiments. We use the following setup for imaging. The Zeiss 5Live microscope system is exquisitely suited for FRAP experiments. However, any advanced inverted, fast laser-scanning confocal microscope with a strong laser for bleaching in the 475–500 nm range and a means to maintain the cells at 37°C will work. A ×60 or ×100 oil immersion objective is required—objectives with ×40 and lower magnifications do not allow precise enough bleaching and observation of the gap junction plaque dynamics. Only a strong blue laser is needed for FRAP experiments with connexins tagged with GFP or YFP derivatives. Please see the List of Abbreviations at the end of this chapter for explanations of fluorescent protein and microscopy acronyms.

Step 5: Set up the imaging parameters. Gap junction plaques have a very distinct appearance when they are made up of connexins tagged with fluorescent proteins and are easy to identify for alpha and gamma class connexins that have so far been examined (Cx43 and Cx47). There are more intracellular clusters of fluorescent protein-tagged connexins for beta class connexins (Cx26, Cx30, and Cx32). These clusters likely represent the connexins localized to the ER and the Golgi apparatus. The FRAP experiments that assess the gap junction channel arrangement stability generally have three parts. Before the experiment, a serial section Z stack is acquired to obtain an image of the entire plaque. Then the FRAP experiment is performed (see the Section 3.5). Finally, a second serial section Z stack is acquired to visualize the amount of FRAP (or lack of recovery) in the entire plaque instead of the single plane acquired during the time-lapse acquisition.

Step 6: We use a scan speed of five scans per second to scan a 1024 × 512 XY serial Z-plane image stack with subairy disk Z-plane sampling (250 nm Z-step size). We set the first and last Z planes just above and below the vertical limits of the gap junction plaque structure. The laser attenuation is set to 1% transmission. The settings will need to be optimized for each confocal microscope. The goal will be to find the settings that minimize the phototoxicity (fast scan speed, lowest possible laser power, and highest detector gain) but allow fast enough image acquisition that the plaque does not move or change shape appreciably during the Z-stack acquisition. If the plaques that are expected to have fluidly arranged the gap junction channels (e.g., Cx26, Cx30) are to be compared to the plaques expected to have stably arranged the gap junction channels (e.g., Cx43). The parameters for the Z-stack acquisition should be set for the fluidly arranged plaques. In some cases, the entire width of the cells (from the top of the cells to the coverslip) can be imaged for presentation/publication quality images and to show the morphology of the cells, but in general, only the plaque which will be used in the FRAP experiment needs to be fully imaged.

Step 7: After a full-plaque serial Z-section image is acquired, the FRAP experiment is set up within the microscope control software as quickly as possible. The following setup procedure is specific to Zeiss confocal LSM systems using the older Zeiss software or Zen Black Edition 2009, 2012, and 2.1 version software with time lapse, region of interest (ROI), and bleaching module included.

- Time-lapse setup: Image interval, 500 milliseconds or 1 second; number of images total, 240 or 120.
- Bleaching setup: Scan speed, 5 fps; iterations, 3 (for GFP derivatives); laser power, 100%; bleach after 10 seconds (20 images for 500-millisecond

interval). The 80/20 beam splitter needs to be in place in our microscope for the bleaching event. A bleach region that will bleach a 1 or 2 µm section of the gap junction plaque works well. The same-sized bleach region needs to be used for groups to be compared.

- LSM acquisition setup (empirically adjusted for each microscope setup): Scan single XY plane every 500 milliseconds (or 1 second) 512 × 256 px; scan speed, 10 fps; detector gain, 5–10; offset, 0; laser attenuation, 0.8% transmission; pinhole set to minimum or 1 airy unit.

The microscope is initiated in continuous XY scan mode, a focal plane near to the center of the plaque is chosen (generally), and the microscope scanning is stopped once a focal plane is chosen. With that last scan displayed, a bleach ROI is generated using the rectangle or the polygon tool that will bleach 2 µm of the plaque, which appears as a two-dimensional (2D) curve or line in single XY-plane scans of vertically oriented plaques (see Figure 3.1). A 2 × 4 µm rectangle is usually used. Other bleach sizes can be used depending on the experimental design, but the same-sized bleach region should be used for all experimental groups to be compared. The gap junction plaque size (diameter and area) greatly varies within the same connexin type, cell type, and even within the same cells when a cell has multiple separate gap junctions with one or more contacting cells. This leads to systematic experimental measurement errors for some commonly used FRAP analysis techniques as discussed in later sections of this chapter.

Step 8: Run the FRAP experiment. During the initial stages of a study, it can be useful to empirically adjust the preceding parameters for a given connexin, cell type, and microscope. To do this, it is helpful to have a real-time readout of the FRAP in the bleach region and in a nonbleached region. If the microscope software allows, the average intensity of the bleach region of the gap junction plaque should be visualized in real time to get a rough idea of the percentage of bleaching that occurs during the bleach event. This real-time monitoring will also allow the experimenter to assess if the recovery has reached a plateau (steady state of maximal FRAP) by monitoring the recovery into the bleach region. If a plateau is not reached within the preset time-lapse period (120 second), this indicates that the recovery is occurring more slowly than it does for Cx26 and Cx30; in this scenario, the number of postbleach images should be increased. If there appears to be no recovery (or artifact recovery due to whole plaque movement) as with the full-length Cx43, then increasing the number of postbleach images will not help, and a qualitatively stable arrangement of channels within the plaque should be reported; in this scenario, the extension of the postbleach imaging duration will not be helpful, and a few pilot experiments with long-term imaging (full-plaque serial Z-section acquisitions every 5 minutes for >30 minutes) should be performed to confirm that the arrangement of the channels within the gap junction plaque is qualitatively stable.

Step 9: Once the single-plane time-lapse FRAP experiment is completed, the experimenter should reuse the serial Z-stack setup that was used prior to the FRAP experiment. The first and last acquisition frames (Z-plane) may need to be reset to account for plaque and cell movement during the time since the first Z stack was acquired.

Step 10: The three images for each FRAP experiment should then be saved as a separate trial. Eight to ten trials should be performed for each experimental group, if possible.

3.5 EXPERIMENTAL SETUP FOR NONJUNCTIONAL FRAP

FRAP can also be performed on fluorescent protein-tagged connexin and pannexin proteins in cellular locations outside of the gap junction plaque structure. We describe FRAP experiments to assess the mobility of nonjunctional connexons (a.k.a. hemichannels) and pannexin channels. The setup for these experiments is much more similar to the standard use of FRAP to study membrane proteins, and several in-depth reviews and methods articles have been published on this technique.[13,36] Researchers should be mindful that different connexins are localized to the nonplaque plasma membrane to greatly different degrees (e.g., monomerized superfolder green fluorescent protein [msfGFP]-Cx43 connexons localize to the membrane in much higher numbers than msfGFP-Cx32 connexons). Therefore, caution must be taken in comparing the connexon mobility between different connexin types. However, the use of analysis methods such as modeling (see Section 3.10) may avoid this issue by taking different starting densities of connexin fluorescent protein (the brightness of the fluorescent signal, and then fluorescent pool available for recovery) into account. Serial section Z-plane scans may or may not be needed for nonplaque FRAP experiments, depending on the experimental design/goals.

The microscope setup for nonplaque FRAP is similar to that for intraplaque FRAP (Section 3.4) with the following adjustments:

- Since the signal density is much lower for nonplaque localized connexins than for gap junctions, the signal intensity needs to be maintained by increasing the detector gain, the pinhole width, and/or the illumination power. The confocal pinhole should be opened to the maximum for the FRAP experiment (not for the pre-FRAP Z-stack acquisition, if it will be performed). This will lead to blurring of the resulting time-lapse images but allows faster scan speed and lower excitation laser power in order to avoid phototoxicity.
- It may be helpful to increase the length of the membrane to be bleached to at least 5 μm or more, but this value should be kept the same for all groups to be compared.
- An image interval of 500 milliseconds works well, although an interval of 1 second may provide usable FRAP data.

Representative results are shown in Figure 3.2. The diffusion coefficient modeling program developed by Siggia et al.,[37] or the newer modeling program that works as a plug-in for ImageJ developed by Blumenthal et al.,[38] are recommended for comparison between the groups for analysis of nonplaque FRAP, since these programs take initial brightness, bleach size, and recovery rate into account by directly drawing data from the images and matching them to a model curve, while most other FRAP analysis programs use the data extracted from the ROIs by the user as average intensity values with only numerical values entered into these more simple FRAP analysis programs. Other recently developed programs for FRAP experiments are also available.[39] The data analysis is discussed in later sections. It should be noted that in the case where FRAP is performed on areas where two cell membranes are in close apposition but no gap junction plaque is visible, there is no published method to exclude the possibility that single gap junction channels are present. Therefore, to be certain that FRAP experiments are being performed on undocked connexons (hemichannels), one would need to perform the bleaching on a section of a cell membrane not apposed to another cell.

3.6   EXPERIMENTAL SETUP FOR TWO-COLOR FRAP

Different connexin types can be coexpressed within the same cells by mixing the plasmid DNA prior to transfection. Information on the position and the mobility of two connexins within the same gap junction or other membrane compartment can be obtained when fluorescent protein tags with distinct excitation/emission spectra are fused to different connexins in the transgene coding sequence. We found that the photobleaching characteristics for enhanced blue fluorescent protein 2 (EBFP2) are well suited for two-color FRAP experiments in conjunction with monomerized superfolder green fluorescent protein (msfGFP) tags. EBFP2 and GFPs can be simultaneously bleached, or GFPs can be bleached separately from EBFP2. This opens up many possibilities to examine the mobilities of the components of multiprotein complexes such as the gap junction nexus. In this section and in Figure 3.3, we describe FRAP with EBFP2-Cx30 and nonmonomerized superfolder green fluorescent protein (sfGFP)-Cx43.

The cotransfection to set up two-color FRAP experiments is essentially the same as the transfection with a single plasmid with the following modifications:

- The amount of DNA (plasmid transgene vector) is reduced for each plasmid to keep the total amount of DNA applied to the cells at the same level as that used when individual plasmids are transfected. For the cotransfection of EBFP2-Cx30 and sfGFP-Cx43, each well of the ibidi chamber receives 0.5 μg of EBFP2-Cx30 and 0.5 μg of sfGFP-Cx43.
- The two plasmids to be transfected should be premixed before they are added to the Opti-MEM solution in Step 2b of Section 3.4. This has been found to increase the percentage of cells that take up both plasmids in the transfection.

The setup for the pre- and post-FRAP Z stack and the setup for the FRAP experiment are similar as described in Section 3.4 with the following modifications:

- A second channel is set up to image the EBFP2 with care to avoid the bleedthrough from the green channel used to image the sfGFP. We use 405 nm laser excitation for EBFP2 and for bleaching. We found that the 405 nm light also effectively bleaches the GFPs. We have found that 488 nm effectively bleaches the GFPs (including EYFP and monomeric venus fluorescent

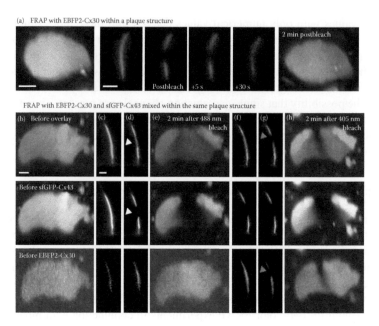

FIGURE 3.3 **(See color insert.)** Two-color FRAP. (a) A single-color FRAP experiment with EBFP2-Cx30 in HeLa cells shows that the Cx30 alone forms fluidly arranged gap junction plaques as evidenced by the rearrangement of unbleached EBFP2-Cx30 into the bleach region that is complete by 2 minutes (right panel). When EBFP2-Cx30 and sfGFP-Cx43 are cotransfected, they form gap junction plaques with the two gap junction channels types intermingled (not heteromeric or heterotypic—just intermingled dodecamers of each connexin). The columns in b through h from left to right are (b) A maximum projection 3D reconstruction of the plaque before any photobleach, (c) the single-plane acquisition from the FRAP experiment immediately before photobleach, (d) immediately after a 3 μm section near the top of the plaque was bleached with the 488 nm laser, (e) a 3D reconstruction 2 minutes after the 488 nm photobleach, (f) a single plane at the time point immediately before the second photobleach, (g) a single-plane acquisition immediately after a 1.5 μm section of the plaque was bleached with the 405 nm laser, and (h) a 3D reconstruction 2 minutes after the 405 nm bleach and ~4.5 minutes after the 488 nm bleach. When Cx43 and Cx30 are mixed (as in (b) through (h) within the same plaque, both connexins are immobile—Cx43 stabilizes the position of Cx30. This is evident in the rightmost panel showing a 3D reconstruction of the plaque 2 minutes after photobleach. The small 1.5 μm bleach region has not blurred, and the EBFP2-Cx30 has not recovered—in contrast to the near complete blurring and recovery shown for EBFP2-Cx30 alone (in (a)). This is an example of a sequential sfGFP FRAP followed by a EBFP2 FRAP experiment. Both fluorescent proteins (EBFP2 and sfGFP) can be bleached with 405 nm light, but simultaneous 405 nm and 488 nm bleaching is more efficient. Note that there is a dequenching effect in which the EBFP2 signal is enhanced when the sfGFP is bleached (most clearly visible in the bottom row, third panel from left). This is caused by the inner filter effect, whereby the sfGFP absorbs some of the blue EBFP2 signal until some of the sfGFP is bleached, and the full EBFP2 signal is revealed. Experimenters should be mindful of this effect anytime two fluorophores are being used. The intensity was scaled to enhance the visibility of the blue signal in the bottom row due to relatively less bright signal from EBFP2. The detector gain was slightly adjusted for the 3D reconstructions between the FRAP experiment for sfGFP and the one for EBFP2. The scale bars are 2 μm.

protein [mVenus] variants of GFP) but does not diminish the fluorescent signal from the EBFP2.

- When the simultaneous bleaching of both EBFP2-Cx30 and sfGFP-Cx43 is desired, we use 100% transmission of both 405 nm and 488 nm laser light.
- When selective bleaching of sfGFP-Cx43 is desired in the presence of EBFP2-Cx30, the 488 nm laser is applied to the bleach ROI with 100% transmission.
- Finally, if the GFP in a stably arranged structure has already been bleached (as in Figure 3.3d), we only apply the 405 nm laser at 100% transmission— leaving the 488 nm laser off in order to minimize phototoxicity.
- Cx43 and Cx30 do not form heteromeric or heterotypic channels,[40,41] but the individual homotypic channels do intermingle within single gap junction plaque structures (as long as monomeric fluorescent protein tags are used[24]). Therefore, the two-color FRAP approach allowed us to show that the presence of sfGFP-Cx43 gap junction channels stabilized the position of the EBFP2-Cx30 gap junction channels when the two connexin types were intermingled within the same plaque (Figure 3.3e and 3.3f). This previously unknown interaction may be important for regulating the in-plaque lifetime of Cx30 and the association of connexin-binding partners at the gap junction.

3.7   FRAP ON OTHER NEXUS COMPONENTS

Proteins such as ZO-1, DLG1/SAP97, CIP85, calmodulin, CaMKII, Ubiquitin, IP3Rs, motor proteins, autophagy-related proteins, beta catenin, claudins, occludin, clathin adaptor proteins, and tubulin have all been shown to interact with gap junctions through direct binding to the connexins (see Herve et al.[42]). These binding partners can also affect the posttranslational modifications on the connexins, their channel functions, and their localization within the cells. Gap junction channels are embedded in the plasma membranes of the cells they connect, and the membrane that comprises a part of the gap junction nexus has specialized properties.[43] FRAP has been used to compare the mobility of many of the proteins mentioned earlier in the context of other cellular junctions,[44] but the relative mobilities and dynamics of nexus components during the interaction with connexins are highly understudied. Moreover, although Falk et al.[12] did some tests on the mobility of membrane dyes within the gap jucntion Nexus, the dynamics of lipid movement within the plaque remains largely unexplored. Through the use of connexins with blue fluorescent protein tags on the NH_2-termini of Cx43, we have recently been able to examine the mobilities of fluorescent protein-tagged connexin-interacting proteins. We also adapted the two-color FRAP technique described in the preceding section to monitor the positions of a variety of connexin-binding partners as they interact with the gap junction plaque structure. As with two-color, dual-connexin FRAP, this can be done with simultaneous monitoring of the mobility of the connexins and the connexin-binding partners, or single nexus components can be individually monitored. In both single and two-color FRAP experiments, the fluorescent protein tag on the connexin serves to mark the location of the gap junction plaque.

Although we have not yet found red fluorescent protein variants with useful bleaching characteristics for FRAP on gap junction channels (we tested mCherry, mKusabiraOrange2, and mDsRed at time of writing), the lipophilic membrane indicator dye DiI is effectively photobleached with strong 560 nm (and shorter) wavelength light. Matthias Falk used this approach to bleach the DiI within a subregion of a Cx43- enhanced green fluorescent protein (EGFP) gap junction and showed that the DiI was mobile within the plaque. We confirmed these results and extended them to show that dye-labeled phospholipids, and other fluorescent molecules with lipophilic membrane tethering domains were also mobile within the gap junction plaque. An interesting area for future research is testing if the cell type, the physiological conditions, and the membrane makeup affect the mobility of the membrane and the membrane-interacting components of the nexus. The percent recovery and the time to half recovery will depend on the bleach size, the cell type, and the expression level of the connexin-interacting protein. The bleach size and the imaging setup will need to be empirically determined for each particular study.

3.8   SIMPLE ANALYSIS METHODS

Once a FRAP experiment has been conducted, the time-lapse image data can be analyzed in a number of ways. The following sections list the analysis methods that we have used, but it is not exhaustive. As with most experiments, quick visualization and comparison of the data can be helpful to assure that the experimental parameters are optimized and to help in choice of further analysis.

The most common form of analysis is to extract fluorescence intensity data from three regions of the image: (1) the bleach region(s) (bF), (2) the pool of fluorescence signal available for recovery (fpF), and (3) the background fluorescence (gF).

The correct identification of the fluorescent pool can affect the quantitative results. For FRAP on fluorescent protein-tagged connexins within gap junctions plaques, the fpF is limited to the single gap junction that will have a subregion photobleached (See note 2).

We discuss fpF in depth because single-plane FRAP on connexins and other gap junction nexus components have an unusually small fpF (in the imaged confocal volume of the gap junction plaque), and a large percentage of the fluorescent pool is not visible in single-plane confocal images that make up the FRAP experiment time lapse when plaques perpendicularly aligned to the image plane are examined. Serial Z-section three-dimensional (3D) time-lapse microscopy can eliminate this issue but requires very a fast image acquisition capability and a slow-moving protein of interest. Another approach is to perform our modification of previously used FRAP analysis methods that we call *bleach border blur analysis*—described in depth in Section 3.11. For standard FRAP data analysis, the data extracted from the bF, the fpF, and the gF ROIs can be normalized and transformed to partially correct for problems generated by the small overall fpF represented by the gap junction plaque and the large percentage of fpF outside of the image acquisition plane in confocal imaging.

The normalization and scaling of FRAP data is standard for the generation of recovery curves, which is a common method of display in scientific publications. Here the initial (prebleach, bFo) is scaled to 100% after background (gF) subtraction.

The first value postbleach (or the minimum bleach ROI value) is scaled to 0% (bFmin). This procedure normalizes and scales the data from 0% to 100% in order to display the percent recovery over time.

To help minimize the effect of the unusual aspects of the fpF in FRAP in gap junctions, a correction factor is calculated that accounts for the percentage or the fluorescent pool that is eliminated by the photobleach (approximate, since only small confocal slivers of both the fpF and the 3D bleach volume are visible in the acquisitions). The correction factor also partially accounts for any bleaching that occurs during the acquisition of the postbleach images. The acquisition bleach is very small for the FRAP on GFP-tagged gap junction channels but depends on the microscope and can be substantial for dimmer fluorophores. The correction factor should be calculated and used to transform the FRAP data after background sub-traction but before the basic scaling/normalization. The background at each time point (bFt) is subtracted from the value of the other 2 ROIs at each time point. The background- and nonbackground-subtracted ROI values are not used again, so the background-subtracted values for each time point are simplified to bF and fpF, from this point forward. The initial fluorescence bFo and fpFo are calculated by averag-ing the 10 data points prior to the photobleach time point. To create a correction factor, the fpFo is divided by the fpF at each time point; this represents the inverse of the fraction of the fluorescent pool eliminated by both the photobleach event and any acquisition bleach. This correction factor should be between 1 and 1.5 for postbleach time points for FRAP with gap jucntion channels. The correction factor is applied to the bF by multiplying it by the fraction of the bFo (fraction of FRAP) as calculated by dividing the bF/bFo. The data are then normalized to the baseline for incomplete bleach by subtracting the first corrected postbleach fraction of the Fo value (FoMin). This value is then divided by 1 − FoMin and multiplied by 100% to get a corrected scaled percentage FRAP at each time point.

The equation for the calculations:

$$FRAP\% = \left\{ \left[(fpFo/Fp) \times (bF/bFo) - bFomin \right] / (1 - bFoMin) \right\} \times 100\%$$

The templates for this analysis in Microsoft Excel™ are available from the corre-sponding author upon request. The data from the bleach ROI, the fluorescence pool ROI, and the background can be exported from image processing/analysis software such as FIJI/ImageJ. The following step-by-step procedure covers data extraction and recovery curve generation for a FRAP experiment on fluorescent protein-tagged gap junction channels within a plaque:

Step 1: The image stack is loaded into ImageJ,[45,46] Scroll through the frames to check if there is substantial movement of the cells in XY or Z direc-tions during the recorded FRAP experiment; check to make sure that the bleach ROI is close to the intended size and that it did not take more than 1.5 seconds to bleach the ROI. If these conditions are not met, the bleach-ing and/or the imaging setup need adjustment. The time to bleach can be determined by looking at the time stamp between the last prebleach and first postbleach image acquisitions.

Step 2: Next scroll to the first postbleach time point and select the three ROIs. These are made with the rectangle or the polygon ROI selection tool in ImageJ. The ROIs are added into the ROI manager plug-in (in the analysis → tools menu) with the Add ROI button in the ROI manager pop-up panel. The ROIs can be added in any order, but for easy export to the Excel template available from the authors, the ROIs should be made in the following order: (1) bleach ROI (bF), (2) fluorescent pool ROI (fpF), and (3) background ROI (gF). For the bleach ROI, a rectangle or a four-sided polygon should be drawn to encompass just the section of the plaque that was bleached and as little other area on either side of the gap junction as possible; add the bleach ROI to the ROI manager list. For the fluorescent pool ROI, a polygon tightly surrounding the fluorescent signal of the plaque is generated and added to the ROI manager (see note 3). For the background ROI, a polygon of any shape is made around a section of the image where there is no cell (no signal from the fluorescent protein-tagged connexin). Add the background ROI to the ROI manager list of ROIs (should now contain three ROIs in total).

Step 3: Make sure that the Mean Gray Value is the only measurement to be reported for the ROIs (Analyze → Set Measurements). Then select all three ROIs in the ROI manager list. Choose the More… button in the ROI manager panel, then choose Multimeasure, select Yes, and ImageJ will output the results in a new window. This will contain a column with a counting number list corresponding to the image frame number, then three columns, each with the mean gray value for an ROI at each time point in the FRAP experiment. Press Ctrl + A to select all the values and then press Ctrl + C to copy the values (press the corresponding keyboard commands if using a Mac). The values are now copied to the clipboard.

Step 4: Paste the values into the spreadsheet. If you are using the spreadsheet template provided by the authors, click to select cell B2 and then press Ctrl + V. The corrected, normalized, and scaled recovery curve will be calculated in column K, and the recovery curve will appear in the graph embedded in the sheet. If you are not using the template spreadsheet, perform the calculations to subtract the background, calculate the initial fluorescence for the bleach region and the fluorescent pool, and perform the correction for fluorescent pool bleach and scaling, if desired as described at the beginning of this section.

Step 5: Now copy and paste the normalized corrected data (column K in the authors' template) to a new sheet or analysis program as desired to generate an average curve and a standard deviation or a standard error mean (SEM). If you are copying to another sheet in Excel, choose to Paste Values instead of the normal Paste to avoid copying the equations to an improper context. Generate new spreadsheets in for each FRAP experiment. A shortcut to copy column K is included as a macro in the authors' template by pressing Ctrl + Shft + F.

This procedure is good for the quick visualization of the FRAP data. Several programs that use routines to run in the analysis platforms have been developed. Several programs have been developed that fit the intensity curve extracted by the user (usually the average fluorescent intensity for the three ROIs used in the simple analysis earlier) to the models of diffusion, but these programs do not perform simulations based on the real image data as is done in the analysis programs discussed in the next section of this chapter.

3.9   FRAP ANALYSIS WITH OTHER PROGRAMS

There are several programs for semiautomated analysis and FRAP curve fitting to model recovery curves.[37–39,47] We briefly discuss the analysis using one such program and refer the reader to the documentation and the previous step-by-step protocols for use of the programs. The program FrapCalc developed by Dr. Kota Miura, European Molecular Biology Laboratory (EMBL) Heidelberg, Germany, runs in IgorPro version 5 or later (Wavemetrics, Inc). FrapCalc works with data as exported in three columns as described in the previous section, fitting the data to the models of diffusion. The program also has the ability to perform other types of analysis and to analyze whole data sets in a semiautomated manner. This program is convenient for the quick calculation of the effective diffusion coefficient and the curve fitting. The analysis of intraplaque gap junction channel arrangement stability with the plaque FRAP is not recommended because the extraordinarily high-packing density of the gap junction channels within the plaque and the extreme anisotropy of the plaque structure lead to tenuous deviation from the assumed parameters for the model curves that are used to fit the data. In-depth documentation for the program is provided on the EMBL website (http://cmci.embl.de/downloads/frap_analysis) and other locations along with the source code for the program and the compilation instructions in IgorPro.

3.10   ANALYSIS BY FITTING TO SIMULATED FRAP MODELS BASED ON EXPERIMENTAL IMAGE DATA

A well-tested analysis program that has worked for both cytoplasmically localized proteins and integral membrane proteins is the stand-alone program for calculating the effective diffusion coefficient based on the experimental image data.[37] The use of this program is described in depth elsewhere[13] and is mentioned earlier with regard to nonjunctional FRAP analysis (Section 3.5). A strength of the diffuse program is that it performs simulations based on the user-supplied image data. A drawback of this program is that it currently only runs correctly in Macintosh operating system OS9, and processing the image files to the format needed for the program is labor intensive. Please refer to the publications from the Snapp and Lippincott-Schwartz laboratories for background, instructions for use, and instructions on how to acquire the code for the inhomogeneous diffusion simulation program.[13,37]

We have recently begun testing the utility of a newly published, freely available Java-based plug-in that runs in the ImageJ free image analysis program for use in the analysis of the FRAP data on gap junction nexus components.[38] A useful tutorial and documentation for this ImageJ plug-in is available bundled with the plug-in along with sample FRAP data available for download. This program uses image data loaded into ImageJ and fits the data to a 2D simulation of diffusion based on a user-defined bleach ROI, bleach cell ROI (fluorescence pool ROI), and reference cell ROI (this is not the same as the background ROI, see documentation). We used the supplied instructions with a slight modification to the ROI selection to match the fluorescent pool considerations for gap junction plaques and compared the effective diffusion coefficients calculated for the plaques aligned in a parallel plane as the coverslip and for the plaques aligned in a plane perpendicular to the coverslip. The results for the plaques formed of the mobile gap junction protein Cx30 are in general agreement and in the range calculated for the adhesive junction protein E-Cadherin.[48] We deviated from the selection of ROIs described in the plug-in documentation. We chose the bleached region of the plaque as the bleached area ROI but chose the bleached region and 2 μm of the plaque on either side of the bleach region as the FRAPed cell ROI, and a 2 μm section of the plaque distal from the bleach region (or when available on a completely separate gap junction plaque) as the reference cell ROI.

3.11   BLEACH BORDER BLUR ANALYSIS

To address the issues particular to gap junctions (anisotropy, variable size and orientation of the structure) and to minimize the assumptions needed to compare the gap junction nexus components, we developed a simple analysis method that quantifies well the recovery rates for plaques in any orientation. We focus on the earliest portion of the postbleach recovery (first 30 seconds after the photobleach) to minimize the effect of whole plaque movement. The bleach border blur analysis quantifies the sharpness of the border between bright and dark fluorescent proteins that is created by the bleach event. The steepest point is identified in the intensity profile across the area where the border was created. We identify the amplitude of the maximum differential of a line scan along the profile of the gap junction (see Figure 3.4 for illustration). This approach allows us to eliminate several sources of error resulting from whole plaque movement and cell shape change. The movement in the axis parallel to the membranes making up the gap jucntion will not affect the results. The movement in the axis perpendicular to the plaque can be accounted for by moving the ROI without biasing the results. These two sources of error can be eliminated since the value of the maximum differential (the sharpest point in the bleach border) is automatically identified through the algorithm of the analysis as opposed to the reliance on the investigator's manual identification. When FRAP is combined with time-lapse serial section Z stacks encompassing the entire gap junction plaque (see Section 3.12), the border analysis can be used to completely eliminate whole plaque movement as a source of error. Additionally, the recovery at different subplaque regions can be examined by combining 3D FRAP with bleach border blur analysis.

We describe the procedure we followed to perform the 2D border blur analysis. The software and the parameters we chose were based on the examination of the recovery curves for the connexins we were comparing. Other softwares such as MATLAB® (Mathworks, Inc.) may also be well suited for some experimental designs and may allow more automation of the analysis process. We extract the data

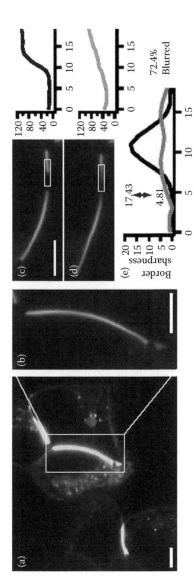

FIGURE 3.4 Illustration of bleach border blur analysis. (a) A zoomed-out single confocal plane image showing the morphology of HeLa cells expressing Cx43K258stop-msfGFP. The image intensity was rescaled to make the Cx43K258stop-msfGFP connexons in nonjunctional membrane visible and to outline the cells. (b) A zoomed-in image of the plaque that was photobleached during the FRAP experiment. (c) The plaque has been rotated, and an ROI (20 × 5 px) has been placed over the bleach border. The single-plane acquisition at the time point 0.6 seconds postbleach is shown. The 5X averaged line scan profile of intensity is shown on the right in arbitrary intensity units. (d) The plaque 30.6 seconds postbleach. Since this plaque was composed of Cx43 with the stabilizing C-terminus removed, the gap junction channels are fluidly arranged, and the border blurs by over 70% after 30 s. (e) The absolute value of the differential of the smoothed line scans is graphed with arbitrary units. The peak differential decreases from 17.43 at the first postbleach image to 4.81 at 30 seconds postbleach; these data compute to a normalized percent change of 72.4%. If, instead of the Cx43 truncated at amino acid 258, this plaque had been made up of full-length Cx43-msfGFP, the border would not significantly blur, and the peak differential would be very similar at the first postbleach time point and at the 30-second postbleach time point.[24] The scale bars are 5 μm.

in ImageJ, and then transfer it to IgorPro where the transformation and the analysis are performed. The resulting data are then further processed in Microsoft Excel and finally analyzed with a statistical software to compare the experimental groups. This workflow has worked well for us, but if an investigator wishes to examine large data sets, it may be beneficial to integrate the operations into a single software platform such as IgorPro or MATLAB.

Step 1: Load the FRAP time-lapse image set into ImageJ. Rotate the image until the 2D section of the plaque where the bleaching occurred is as close to horizontal as possible (parallel with the X axis in the image display; see Figure 3.4b and 3.4c. The image rotation can be achieved by selecting Image → Transform → Rotate, and then typing in a custom rotation value to bring the plaque to horizontal. Now create a rectangular ROI that spans the entire plaque in the Y direction and extending 1–2 px above and below the plaque. The ROI should be wide enough to span the bleach border even after it has blurred. For the connexins we have tested with our microscope setup (Zeiss 5Live with 63x objective, 200 nm pixel size), we found that a $5 \times 1 \mu m$ (XY) rectangle is appropriate. This ROI should be saved and reloaded in the ROI manager plug-in of ImageJ to keep the same dimensions for each experiment to be analyzed. Place the ROI across the bleach border as shown in Figure 3.4c.

Step 2: Extract the line scan data for the bleach border for the first postbleach image and the image acquired 30 seconds after the photobleach. Open the ImageJ analyze menu and then choose Plot Profile. Select Live in the plot profile pop-up window. Scroll through the image stack to the first postbleach image and make sure that the ROI is properly positioned to capture the bleach border. Care should be taken that if the plaque is slightly curved; it intersects with the lateral edges of the ROI and not the vertical edges. If the curvature of the plaque causes it to intersect with the vertical edges, it will create a false bleach border in the line scan. The ROI can be repositioned to avoid this issue, but if the gap junction plaque curvature is too high, the experiment would have to be excluded from the bleach border analysis. Once the ROI is positioned and the line scan appears in the plot profile pop-up, press the List button, and a list of intensity values (average of Y pixel columns) for each X position (pixel by pixel in the X axis) is displayed. Copy this list and paste it into a new sheet (new experiment) in IgorPro.

Step 3: Smooth the line scan and calculate the differential. The IgorPro procedure listed in Step 4 of this section (Section 3.11) can be used to perform the operations for this step by copying and pasting the text into the Procedure Window; however, this will only work with 20-column data generated from a line scan from a 20 px wide × 5 px high ROI. Individual data sets (columns of values in this case) are called *waves* in IgofPro. In IgorPro, perform a smoothing operation on wave1 and wave3; in the analysis menu, choose Smooth... and Smooth 3X (three-point binomial or moving average smoothing) with the generation of a new wave for each of the two smoothed waves. Three-point smoothing works well with our microscope's pixel size and may need empirical adjustment for different microscope systems. Differentiate the smoothed wave1 and smoothed wave3. Rename the resulting smoothed differentiated wave1 and wave3 to something simple such as TA and TB by choosing the Data → Rename menus. Generate two new waves with 19 data points by choosing the Data → Make waves, creating waves with the names FA and FB, and choosing

19 rows; then click the Do it button. Fill in the values for the new waves FA and FB with the absolute value of each point from TA and TB by typing "FA = abs(TA)," then pressing Enter (Return), and then "FB = abs(TB)," then pressing Enter (Return). Waves FA and FB may or may not be displayed depending on the choices selected by the user. FA and FB are the absolute value of the differentials of the smoothed wave at 19 points along the line scans.

Step 4: Identify the maximum of the absolute value of the differential of the line scan for the first image postbleach and 30 seconds after the postbleach. The simplest way to do this is to type the following commands into the procedure window of IgorPro: "WaveStats FA" then press Enter and "WaveStats FB" then press Enter—assuming the user named the waves FA and FB for the first postbleach and the 30-second postbleach line scans. The maximum value for the wave is listed as V_Max =, and the location in the X axis where this value occurred is listed as V_maxloc =. It is a good idea to visualize the final smoothed, differentiated, absolute value line scans (waves FA and FB) to make sure the that maximum value and the location of the maximum value make sense given the starting image data.

The following text can be copied and pasted into the procedure window of IgorPro to achieve the data operations described here, assuming a 20×5 ROI is used for the line scan:

```
Duplicate/O wave1,wave1_smth;DelayUpdate
Duplicate/O wave3,wave3_smth;DelayUpdate
Smooth 3, wave1_smth,wave3_smth
Differentiate wave1_smth/D = wave1_smth_DIF
Differentiate wave3_smth/D = wave3_smth_DIF
Rename wave1_smth_DIF,TA; Rename wave3_smth_DIF,TB;
Make/N = 19/D FA,FB
FA = abs(TA)
FB = abs(TB)
WaveStats FA
WaveStats FB
```

(press Enter to perform the procedure)

Step 5: Transfer the Vmax values to Microsoft Excel (or equivalent) and compute the normalized change in the peak border sharpness from time point A (postbleach) to time point B (30 seconds postbleach). To do this, copy the V_max value (max smoothed absolute differential value) into rows in Excel. Calculate the percent change from the first time point to the second by the following calculation: %border_blur = [(MaxValFA-MaxValFB)/MaxValFA] × 100%. This deltaF/Fo value is calculated for each FRAP experiment; then the groups can be compared with the statistics of choice.

3.12   THREE-DIMENSIONAL FRAP

Three-dimensional time-lapse imaging with photobleaching (3D FRAP) can be performed on plaques in any orientation (Figure 3.1c). The phototoxicity and the acquisition bleach are increased in the 3D FRAP, since 3–15 serial Z sections are acquired for each time point. This also makes longer intervals between the acquisitions necessary, depending on microscope characteristics such as detector sensitivity, scan speed. These issues are minimized for gap junction plaques that are aligned with the plane of the coverslip, since only 3–5 Z sections per time point are needed to encompass the Z range of the plaque and to account for slight plaque movement during the course of the FRAP experiment. Three-dimensional FRAP is useful for FRAP on such horizontally oriented plaques, since the plaque usually moves out of the confocal image plane over the course of 1–2 minutes, if only a single plane is acquired at each time point. FRAP on horizontal gap junction plaques can be used to test if the stability of the arrangement of the channels within the plaque varies in different regions such as the perimeter versus the center of the structure by bleaching a stripe across the plaque. Vertically aligned plaques are less useful for this purpose, since the bleach laser volume is hourglass-shaped with a gradient of bleaching in the Z-directions away from the focal plane.

3.13   FUTURE USES AND ALTERNATIVE APPROACHES

FRAP is just one tool in the optomanipulation repertoire available for studying the mobilities of gap junction nexus components. Other techniques such as fluorescence correlation spectroscopy,[49] single-particle tracking, either with antibody tagging of quantum dots or with photoactivation localization microscopy, and FLIP can provide complementary information. One technique that is similar in some ways to FRAP is the use of photoactivatible proteins such as PAmCherry and photoconvertible proteins such as Dendra2,[12,50] mEOS3.2,[51] and mEOS4b[52] to perform spatially restricted photoconversion followed by monitoring the movement of the converted protein to other cellular locations as discussed by Baker et al.[31] It will be important to use robustly monomeric fluorescent proteins for FRAP or photoconversion experiments since nonmonomerized (e.g., EGFP, sfGFP) or fluorescent proteins with a slight tendency to aggregate (e.g., mCherry) may affect the mobility of the connexins within the gap junction (see Stout et al.[24] and Costantini et al.[53]). Of the photoconvertible proteins, mEOS3.2 has been demonstrated to be robustly monomeric,[51] while mEOS4b was shown to be at least as monomeric as EGFP.[52]

We have not performed exhaustive testing on all available fluorescent protein tags to confirm efficient bleaching and lack of oligomerization within the highly ordered and tightly packed environment of the gap junction plaque. In our hands, mCherry does not bleach efficiently. The tags that we found to be suitable are EBFP2, msfGFP, mGFP, and mVenus. Other fluorescent protein tags may be useful,[53] and it will be of great interest to identify a red fluorescent protein suitable for FRAP on gap junction Nexus components—which would allow three-color channel FRAP experiments. Halo-tagged Cx36 and Cx43 were recently used with an Oregon Green dye label for the FRAP experiments.[26] It will be interesting to compare the mobility of Cx36 tagged by msfGFP to Cx36 tagged with the Halo-labeling approach.

New microscope technology may allow extended capabilities for FRAP and photoconversion experiments such as light sheet microscopy that reduces photoxicity and allows very fast 3D imaging in live samples see.[54,55]

In summary, there are many variations of the FRAP technique and forms of analysis to match a large number of experimental designs. The FRAP to study connexins and other components of the gap junction nexus molecular complex will be a powerful tool to allow better understanding of the role of gap junction in health and disease.

Protocol and material notes:

Note 1: Ibidi 8 well imaging dishes with ibidiTreat (Cat. No. 80826) work well for FRAP experiments using HeLa, N2a, rat insulinoma, primary and immortalized astrocytes, and brain endothelial cells. Matek 6 well imaging slides (Cat. No. P06G-1.5-10-F) and Cellvis imaging slides (Cellvis, Inc., Cat. No. C8-1.5H-N) provide excellent image quality but may require precoating with poly-L-lysine for cell attachment, depending on the cell type to be used. Precoated (collagen and poly-L-lysine) Matek glass bottom dishes worked well for astrocyte culture and imaging.

Note 2: The main source of FRAP for fluorescent protein-tagged conexins is the plaque and not the insertion of new gap junctions into the interior of the plaque on the time course of minutes (Figure 3.1). These data are in agreement with other groups who have used similar fluorescent protein tags and various tagging strategies.[12,25,56] Cx26 has been reported to transition from the cytoplasm to the interior parts of the gap junctions, but based on our FRAP results (see Supplemental movies S4 and S6 in Stout et al. 2015[24]) and included supplemental videos in that publication, the main source for the FRAP within gap junction plaques for this connexin is parts of the plaque outside the bleach region and not the cytoplasm.

Note 3: This is where the FRAP on gap junctions differs from the FRAP on diffuse intracellular proteins such as ER resident proteins and cytoskeleton, which generate what is often referred to as a *cell ROI* as the fluorescent pool ROI which encompasses the entire cell area. There are several reasons why the whole cell is not used for the fluorescent pool (fpF) in the FRAP analysis on gap junctions. The most important is that internalized gap junctions (AGJs[27,57,58]), connexins still localized to the ER/Golgi apparatus[59,60], and connexons in small vesicles attached to the microtubules through transport proteins[61–64] are not actually part of the fluorescence pool available for recovery within the plaque, within the timescale that can be measured with standard FRAP experimentation. Therefore, if the entire cell area was used as the fluorescent pool ROI (fpF), we would underestimate the percentage of the fluorescent pool that is bleached, and the correction factor would be smaller than it should be.

LIST OF ABBREVIATIONS

CT	COOH-terminus
Cx26	connexin 26
Cx30	connexin 30
Cx32	connexin 32
Cx43	connexin 43

Cx47	connexin 47
DMEM	Dulbecco's modified eagle medium
EBFP2	enhanced blue fluorescent protein 2
EGFP	enhanced green fluorescent protein
FBS	fetal bovine serum
FRAP	fluorescence recovery after photobleaching
GFP	green fluorescent protein
HEPES	4-(2-hydroxyethyl)-1-piperazineethanesulfonic acid
LSM	laser-scanning microscope
mVenus	monomeric venus fluorescent protein
msfGFP	monomerized superfolder green fluorescent protein
NA	numerical aperature
N2A	neuro2a
PBS	phosphate buffered saline
ROI	region of interest
sfGFP	nonmonomerized superfolder green fluorescent protein
YFP	yellow fluorescent protein.

ACKNOWLEDGMENTS

We thank the following individuals and organizations for their support: Julie Zhao, Marcia Maldonado, Julian Botta, Dr. Alfredo Fort, Dr. Erik L. Snapp, Dr. Peng Guo, and the Analytical Imaging Facility of Albert Einstein College of Medicine and the National Cancer Institute cancer center support grant (P30CA013330), and the Rose F. Kennedy Intellectual and Developmental Disabilities Research Center cell and molecular imaging core. mKusabiraOrange2-Cx26-7 was a gift from Michael Davidson (Addgene plasmid # 57865). This work was supported by the following grants: NIH grant R01 NS092466, NIH grant R01 AR057139.

REFERENCES

1. Duffy, H. S., Delmar, M., and Spray, D. C. Formation of the gap junction nexus: Binding partners for connexins. *J Physiol Paris* **96**, 243–249 (2002).
2. Naus, C. C., Zhu, D., Todd, S. D., and Kidder, G. M. Characteristics of C6 glioma cells overexpressing a gap junction protein. *Cell Mol Neurobiol* **12**, 163–175 (1992).
3. Cina, C. et al. Involvement of the cytoplasmic C-terminal domain of connexin43 in neuronal migration. *J Neurosci: Off J Soc Neurosci* **29**, 2009–2021 (2009).
4. Bates, D. C., Sin, W. C., Aftab, Q., and Naus, C. C. Connexin43 enhances glioma invasion by a mechanism involving the carboxy terminus. *Glia* **55**, 1554–1564 (2007).
5. Bejarano, E. et al. Connexins modulate autophagosome biogenesis. *Nature Cell Biol* **16**, 401–414 (2014).
6. Marquez-Rosado, L., Solan, J. L., Dunn, C. A., Norris, R. P., and Lampe, P. D. Connexin43 phosphorylation in brain, cardiac, endothelial and epithelial tissues. *Biochim Biophys Acta* **1818**, 1985–1992 (2012).
7. Saidi Brikci-Nigassa, A. et al. Phosphorylation controls the interaction of the connexin43 C-terminal domain with tubulin and microtubules. *Biochemistry* **51**, 4331–4342 (2012).
8. Johnson, K. E. et al. Phosphorylation on Ser-279 and Ser-282 of connexin43 regulates endocytosis and gap junction assembly in pancreatic cancer cells. *Mol Biol Cell* **24**, 715–733 (2013).

9. Nimlamool, W., Kells Andrews, R. M., and Falk, M. M. Connexin43 phosphorylation by PKC and MAPK signals VEGF-mediated gap junction internalization. *Mol Biol Cell* **26**,15 2755–2768 (2015).

10. Cone, A. C. et al. Protein kinase Cdelta-mediated phosphorylation of connexin43 gap junction channels causes movement within gap junctions followed by vesicle internalization and protein degradation. *J Biol Chem* **289**, 8781–8798 (2014).

11. Rhett, J. M., Jourdan, J., and Gourdie, R. G. Connexin 43 connexon to gap junction transition is regulated by zonula occludens-1. *Mol Biol Cell* **22**, 1516–1528 (2011).

12. Falk, M. M., Baker, S. M., Gumpert, A. M., Segretain, D., and Buckheit, R. W. III. Gap junction turnover is achieved by the internalization of small endocytic double-membrane vesicles. *Mol Biol Cell* **20**, 3342–3352 (2009).

13. Snapp, E. L., Altan, N., and Lippincott-Schwartz, J. Measuring protein mobility by photobleaching GFP chimeras in living cells. *Curr Protoc Cell Biol.* 2003; Chapter 21(Unit).

14. Day, C. A., Kraft, L. J., Kang, M., and Kenworthy, A. K. Analysis of protein and lipid dynamics using confocal fluorescence recovery after photobleaching (FRAP). *Curr Protoc Cytom* 2.19.11–12.19.29 (2012).

15. Schneckenburger, H. et al. Light exposure and cell viability in fluorescence microscopy. *J Microsc* **245**, 311–318 (2012).

16. Wagner, M. et al. Light dose is a limiting factor to maintain cell viability in fluorescence microscopy and single molecule detection. *Int J Mol. Sci* **11**, 956–966 (2010).

17. Lopez, P., Balicki, D., Buehler, L. K., Falk, M. M., and Chen, S. C. Distribution and dynamics of gap junction channels revealed in living cells. *Cell Commun Adhes* **8**, 237–242 (2001).

18. Thomas, T., Jordan, K., and Laird, D. W. Role of cytoskeletal elements in the recruitment of Cx43-GFP and Cx26-YFP into gap junctions. *Cell Commun Adhes* **8**, 231–236 (2001).

19. Lauf, U. et al. Dynamic trafficking and delivery of connexons to the plasma membrane and accretion to gap junctions in living cells. *Proc Natl Acad Sci U S A* **99**, 10446–10451 (2002).

20. Thomas, T. et al. Mechanisms of Cx43 and Cx26 transport to the plasma membrane and gap junction regeneration. *J Cell Sci* **118**, 4451–4462 (2005).

21. Simek, J., Churko, J., Shao, Q., and Laird, D. W. Cx43 has distinct mobility within plasma-membrane domains, indicative of progressive formation of gap-junction plaques. *J Cell Sci* **122**, 554–562 (2009).

22. Bhalla-Gehi, R., Penuela, S., Churko, J. M., Shao, Q., and Laird, D. W. Pannexin1 and pannexin3 delivery, cell surface dynamics, and cytoskeletal interactions. *J Biol Chem* **285**, 9147–9160 (2010).

23. Katoch, P. et al. The carboxyl tail of connexin32 regulates gap junction assembly in human prostate and pancreatic cancer cells. *J Biol Chem* **290**, 4647–4662 (2015).

24. Stout, R. F. Jr., Snapp, E. L., and Spray, D. C. Connexin type and fluorescent protein fusion tag determine structural stability of gap junction plaques. *J Biol Chem* **290**, 23497–23514 (2015).

25. Kelly, J. J., Shao, Q., Jagger, D. J., and Laird, D. W. Cx30 exhibits unique characteristics including a long half-life when assembled into gap junctions. *J Cell Sci* (2015).

26. Wang, H. Y., Lin, Y. P., Mitchell, C. K., Ram, S., and O'Brien, J. Two-color fluorescent analysis of connexin 36 turnover: Relationship to functional plasticity. *J Cell Sci* (2015).

27. Jordan, K., Chodock, R., Hand, A. R., and Laird, D. W. The origin of annular junctions: A mechanism of gap junction internalization. *J Cell Sci* **114**, 763–773 (2001).

28. Falk, M. M. Connexin-specific distribution within gap junctions revealed in living cells. *J Cell Sci* **113** (**Pt 22**), 4109–4120 (2000).

29. Martin, P. E. et al. Assembly of chimeric connexin-aequorin proteins into functional gap junction channels: Reporting intracellular and plasma membrane calcium environments. *J Biol Chem* **273**, 1719–1726 (1998).

30. Holm, I., Mikhailov, A., Jillson, T., and Rose, B. Dynamics of gap junctions observed in living cells with connexin43-GFP chimeric protein. *Eur J Cell Biol* **78**, 856–866 (1999).

31. Baker, S. M., Buckheit, R. W. III, and Falk, M. M. Green-to-red photoconvertible fluorescent proteins: Tracking cell and protein dynamics on standard wide-field mercury arc-based microscopes. *BMC Cell Biol* **11**, 15 (2010).

32. Sbalzarini, I. F., Mezzacasa, A., Helenius, A., and Koumoutsakos, P. Effects of organelle shape on fluorescence recovery after photobleaching. *Biophys J* **89**, 1482–1492 (2005).

33. Mai, J., Trump, S., Lehmann, I., and Attinger, S. Parameter importance in FRAP acquisition and analysis: A simulation approach. *Biophys J* **104**, 2089–2097 (2013).

34. Bejarano, E. et al. Autophagy modulates dynamics of connexins at the plasma membrane in a ubiquitin-dependent manner. *Mol Biol Cell* **23**, 2156–2169 (2012).

35. Fong, J. T. et al. Internalized gap junctions are degraded by autophagy. *Autophagy* **8**, 794–811 (2012).

36. Snapp, E. L., and Lajoie, P. Photobleaching regions of living cells to monitor membrane traffic. *Cold Spring Harbor Protoc* **2011**, 1366–1367 (2011).

37. Siggia, E. D., Lippincott-Schwartz, J., and Bekiranov, S. Diffusion in inhomogeneous media: Theory and simulations applied to whole cell photobleach recovery. *Biophys J* **79**, 1761–1770 (2000).

38. Blumenthal, D., Goldstien, L., Edidin, M., and Gheber, L. A. Universal approach to FRAP analysis of arbitrary bleaching patterns. *Sci Rep* **5**, 11655 (2015).

39. Ulrich, M. et al. Tropical—Parameter estimation and simulation of reaction-diffusion models based on spatio-temporal microscopy images. *Bioinformatics* **22**, 2709–2710 (2006).

40. Smith, T. D. et al. Cytoplasmic amino acids within the membrane interface region influence connexin oligomerization. *J Membr Biol* **245**, 221–230 (2012).

41. Orthmann-Murphy, J. L., Freidin, M., Fischer, E., Scherer, S. S., and Abrams, C. K. Two distinct heterotypic channels mediate gap junction coupling between astrocyte and oligodendrocyte connexins. *J Neurosci* **27**, 13949–13957 (2007).

42. Herve, J. C., Derangeon, M., Sarrouilhe, D., Giepmans, B. N., and Bourmeyster, N. Gap junctional channels are parts of multiprotein complexes. *Biochim Biophys Acta* **1818**, 1844–1865 (2012).

43. Locke, D., Liu, J., and Harris, A. L. Lipid rafts prepared by different methods contain different connexin channels, but gap junctions are not lipid rafts. *Biochemistry* **44**, 13027–13042 (2005).

44. Shen, L., Weber, C. R., and Turner, J. R. The tight junction protein complex undergoes rapid and continuous molecular remodeling at steady state. *J Cell Biol* **181**, 683–695 (2008).

45. Schneider, C. A., Rasband, W. S., and Eliceiri, K. W. NIH Image to ImageJ: 25 years of image analysis. *Nat Methods* **9**, 671–675 (2012).

46. Schindelin, J. et al. Fiji: An open-source platform for biological-image analysis. *Nat Methods* **9** (2012).

47. Jonsson, P., Jonsson, M. P., Tegenfeldt, J. O., and Hook, F. A method improving the accuracy of fluorescence recovery after photobleaching analysis. *Biophys J* **95**, 5334–5348 (2008).

48. Adams, C. L., Chen, Y. T., Smith, S. J., and Nelson, W. J. Mechanisms of epithelial cell-cell adhesion and cell compaction revealed by high-resolution tracking of E-cadherin-green fluorescent protein. *J Cell Biol* **142**, 1105–1119 (1998).

49. Gerken, M., Thews, E., Tietz, C., Wrachtrup, J., and Eckert, R. Diffusion behavior of gap junction hemichannels in living cells. *Curr Pharm Biotechnol* **6**, 151–158 (2005).

50. Chudakov, D. M., Lukyanov, S., and Lukyanov, K. A. Using photoactivatable fluorescent protein Dendra2 to track protein movement. *Biotechniques* **42**, 553, 555, 557 passim (2007).

51. Zhang, M. et al. Rational design of true monomeric and bright photoactivatable fluorescent proteins. *Nat Methods* **9**, 727–729 (2012).

52. Paez-Segala, M. G. et al. Fixation-resistant photoactivatable fluorescent proteins for CLEM. *Nat Methods* **12**, 215–218 (2015).

53. Costantini, L. M. et al. A palette of fluorescent proteins optimized for diverse cellular environments. *Nat Commun* **6**, 7670 (2015).

54. Liu, Z., Lavis, L. D., and Betzig, E. Imaging live-cell dynamics and structure at the single-molecule level. *Mol Cell* **58**, 644–659 (2015).

55. Majoul, I. V. et al. Fast structural responses of gap junction membrane domains to AB5 toxins. *Proc Natl Acad Sci U S A* **110**, E4125–4133 (2013).

56. Gaietta, G. et al. Multicolor and electron microscopic imaging of connexin trafficking. *Science* **296**, 503–507 (2002).

57. Gumpert, A. M., Varco, J. S., Baker, S. M., Piehl, M., and Falk, M. M. Double-membrane gap junction internalization requires the clathrin-mediated endocytic machinery. *FEBS Letters* **582**, 2887–2892 (2008).

58. Piehl, M. et al. Internalization of large double-membrane intercellular vesicles by a clathrin-dependent endocytic process. *Mol Biol Cell* **18**, 337–347 (2007).

59. Yeager, M., Unger, V. M., and Falk, M. M. Synthesis, assembly and structure of gap junction intercellular channels. *Curr Opin Struct Biol* **8**, 517–524 (1998).

60. Musil, L. S., and Goodenough, D. A. Multisubunit assembly of an integral plasma membrane channel protein, gap junction connexin43, occurs after exit from the ER. *Cell* **74**, 1065–1077 (1993).

61. Hesketh, G. G., Van Eyk, J. E., and Tomaselli, G. F. Mechanisms of gap junction traffic in health and disease. *J Cardiovasc Pharmacol* **54**, 263–272 (2009).

62. Martin, P. E., and Evans, W. H. Incorporation of connexins into plasma membranes and gap junctions. *Cardiovasc Res* **62**, 378–387 (2004).

63. Laird, D. W. et al. Comparative analysis and application of fluorescent protein-tagged connexins. *Microsc Res Tech* **52**, 263–272 (2001).

64. Fort, A. G. et al. In vitro motility of liver connexin vesicles along microtubules utilizes kinesin motors. *J Biol Chem* **286**, 22875–22885 (2011).

4 Patch Clamp Analysis of Gap Junction Channel Properties

Donglin Bai and John A. Cameron

CONTENTS

4.1 INTRODUCTION

4.1.1 GAP JUNCTIONS

Gap junction channels span through two plasma membranes electrically and meta-bolically synchronizing the neighboring cells via a rapid exchange of ions and small metabolic/signaling molecules (Saez et al. 2003; Goodenough and Paul 2009). It is well documented that gap junctional intercellular communication is crucial in many important physiological processes. The gap junction channels are composed of connexin molecules (Sohl and Willecke 2004). Twenty-one different connexin genes are identified in the human genome, and they are named by their respective calculated molecular weight in kilodaltons. Six same or different connexins oligo-merize to form a homomeric or a heteromeric gap junction hemichannel (also known as connexon), respectively. Two hemichannels dock end to end to form a whole gap junction channel. If the docked hemichannels are identical, then the formed chan-nel is a homotypic gap junction channel. If different hemichannels dock together, heterotypic gap junction channels are formed (Bai and Wang 2014). Because most tissue cells express more than one connexin, it is theoretically possible to form homomeric and heteromeric hemichannels and homotypic and heterotypic gap junc-tion channels in many tissues. Considerable experimental evidence indicates that all these different types of gap junction channels and hemichannels do exist in native tissues (Sosinsky 1995).

4.1.2 MEASURING GAP JUNCTION FUNCTION

Several methods are frequently used to study gap junction function including measuring the transfer of small dye molecules (often fluorescent) via scrape load-ing (el-Fouly, Trosko, and Chang 1987; Yum et al. 2007), microinjection (dye transfer from one cell to many) (Woodward et al. 1998), or fluorescence recov-ery after photobleach (FRAP, dye transfer from many cells to one) (Yum et al. 2007; Yum, Zhang, and Scherer 2010). In this chapter, we will describe two patch clamp methods to estimate the gap junction coupling conductance (G_j). There are several fundamental differences between the dye coupling (transfer) and patch clamp (electrical coupling) methods. First, dye coupling measures the ability of the gap junctions to pass dye molecules (mostly fluorescent molecules, but can also be nonfluorescent dyes). It is heavily dependent on the availability of the dyes (brightness, rate of photobleach, size, charge, shape, etc.). It is also important to know that many gap junctions are selective and only allow certain dyes to pass through. Thus, for a given gap junction channel, the absence of dye coupling does not necessarily indicate a lack of gap junction coupling. On the other hand, patch clamp methods measure the ability of gap junctions to conduct ions. Under normal physiological conditions, the ions passing through gap junctions are K^+, Cl^-, Na^+, etc. It is generally safe to claim that no electrical coupling means no gap junc-tion function, because these ions are much smaller than the dye molecules, and all known gap junctions are permeable to ions. Second, the driving force of dye coupling is the chemical gradient, which is hard to control, while the driving force

TABLE 4.1

Comparison of Dye Coupling versus Single and Dual Patch Clamp Methods

	Single Patch Clamp	Dual Patch Clamp	Dye Coupling
Measuring	Ability to conduct ions	Ability to conduct ions	Ability to pass dyes
Common ions/dyes	K^+, Na^+, Cl^-	K^+, Na^+, Cs^+, Li^+ Cl^-, Br^-, I^-, relatively easy to manipulate via manipulations in ICS	Fluorescent dyes: Lucifer yellow, Alexa Fluor®, ethidium bromide, propidium iodide, DAPI Other dyes: neurobiotin
Driving force	Electrical gradient	Electrical gradient	Chemical gradient
Regulation of driving force	Limited	Can be easily controlled	Difficult to control
Applications	Indirectly estimating gap junction coupling in cell clusters and confluent culture	Used in isolated cell pairs, capable of directly measuring gap junction conductance and characterizing V_j gating and unitary channel properties	Scrape loading is the simplest method; microinjection and FRAP are also commonly used in studying gap junction dye transfer

of electrical coupling using the dual patch clamp method (electrical gradient) can be easily controlled. Finally, an important advantage of dual patch clamp is the quantitative characterizations of detailed gap junction channel properties (e.g., V_j gating) at macroscopic and unitary channel current levels. Table 4.1 summarizes the key differences among these methods and their applications. In this chapter, we will provide detailed protocols for single and dual patch clamp methods, as well as highlighting their primary applications and limitations.

4.1.3 Measuring Native Tissue Cell Gap Junction Coupling Conductance and Dye Transferring Are Difficult and Rare

Studying gap junction coupling in native tissue cells in situ or in vivo is always in demand, but they are very rare and are often at qualitative levels. Many reasons are behind this. First, most tissue cells are 3D with multiple layers of gap junction-interconnected cells. These cells are often tightly packed with enriched extracellular matrices and are largely inaccessible to quantitative techniques, such as patch clamp. Freshly isolated tissue cells (cell pairs or clusters) can partially overcome this difficulty, but the isolation procedure is likely to present strong mechanical and chemical stresses to affect the gap junction function of these isolated cells. Second, in some tissues, the gap junctions are too low in expression or too distally located to be measured for coupling conductance (e.g., the electrical synapses between neurons often show low coupling conductance and are frequently located at distal dendritic sites), while others are too high in coupling conductance to be precisely measured (e.g., coupling between astrocytes or cardiomyocytes; some limitations on measuring highly coupled cells are discussed in the following Section 4.3.7). Third, the native tissue cells often express multiple connexins, which could form multiple homo- and heteromeric gap junction channels. It is very difficult to assess the relative contributions of these different types of gap junctions and the unique properties of each gap junction channel type. Fourth, many other ion channels including voltage- and ligand-gated ion channels and intercellular communications (e.g., via synaptic transmission and paracrine signaling) can potentially contribute to the propagation of calcium waves, which is often believed to be due to gap junction coupling. The temporal changes of the calcium waves in tissues/cultured cells could be slow enough to fully exclude the contribution of these other forms of intercellular communication. To avoid these issues and to study gap junction coupling in a properly controlled manner, many gap junction researchers employ recombinant expression systems (in *Xenopus* oocytes or in mammalian cell lines deficient of gap junctions, i.e., N2A, HeLa, and rat insulinoma [RIN]) to study the gap junction channel properties of individual connexins. Here we will describe two patch clamp methods to study the gap junction channel properties.

4.2 SINGLE PATCH CLAMP FOR ESTIMATING GAP JUNCTION COUPLING CONDUCTANCE

4.2.1 OVERVIEW OF THE APPLICATION

Patch clamp whole-cell recording under voltage clamp configuration to record the membrane capacitance can be used for a quantitative estimation of gap junction coupling conductance in cell pairs, clusters, or confluent cells. This method was developed by Harks et al. (2001) and de Roos, van Zoolen, and Theuvenet (1996) at the University of Nigmegen and can provide a good estimate of the gap junction coupling using a relatively simple patch clamp setup.

4.2.2 THEORETICAL BACKGROUND

Figure 4.1a shows an equivalent circuitry of a single patch clamp recording under voltage clamp whole-cell recording mode. A 10 mV depolarization voltage pulse is used to generate a capacitive current with a peak (I_{peak}) and a steady-state current (I_{ss}; Figure 4.1a). The patch pipette access resistance and the cell membrane resistance can be derived from the equations provided in Figure 4.1a. When the patched cell is in a cluster of (or confluent) cells interconnected by gap junctions (Figure 4.1b and c), the equivalent circuitry can be simplified as shown in Figure 4.1d, where G is the estimated gap junctional conductance between the patched cell and its surrounding cells and can be calculated by the equation provided (Figure 4.1d). The details on the equation are described earlier (Harks et al. 2001; de Roos, van Zoelen, and Theuvenet 1996).

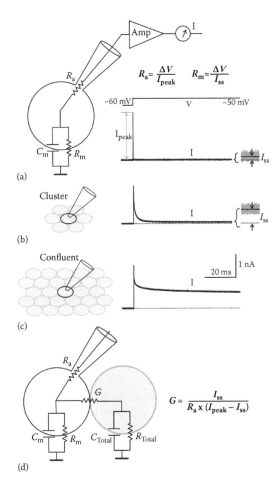

FIGURE 4.1 Single patch clamp capacitance measurement to estimate the gap junction coupling in cluster/confluent cells. (a) An equivalent circuitry of a single patch clamp on a single isolated cell is shown on the left panel. Patch pipette access resistance (R_a), cell plasma membrane resistance (R_m) and capacitance (C_m), as well as patch clamp amplifier (Amp) are shown. Under the voltage clamp whole-cell recording mode, a capacitive current (with a peak at the beginning, [I_{peak}] and a steady-state current [I_{ss}] at a later time) can be recorded in response to a 10 mV voltage step (from −60 to −50 mV). (b, c) The I_{peak}, the I_{ss}, and the rate of decay of the capacitive currents are different in gap junction-coupled cluster/confluent cells. To better illustrate the differences between I_{ss} for a single cell and a cluster, current traces from a and b were vertically expanded and offset to the right of the original traces. (d) The coupling conductance (G) between the patched cell and its surrounding cells can be simplified into a circuitry as shown (Harks et al. 2001) and can be calculated according to the equation shown on the right.

4.2.2.1 Practical Tips for Capacitive Current Recording

- A high-quality patch seal without any leak at the patch pipette tip is very important. This seal needs to withstand the process of breaking into the whole-cell recording mode. In real experiments, it is hard to tell the difference between the leakage between the patch pipette tip and the plasma membrane or perfect the whole-cell recording in high gap junction-coupled large clusters/confluent cells. One way to double-check is to add a gap junction blocker (carbenoxolone, FFA, etc.) in the bath medium at the end of an experiment (Tong et al. 2006).

- Obtaining accurate capacitive peak and steady-state current amplitudes is crucial for this measurement. To have an accurate estimation of the peak capacitive current, the patch pipette capacitance need to be cancelled (compensated) on the patch clamp amplifier, when the gigaohm seal is formed (the cell-attached mode).
- Cells that endogenously express gap junctions (normal rat kidney, rat epidermal keratinocytes [REK], endothelial cells, etc.) are good candidates for using this technique.

4.2.3 REQUIREMENTS

Here is a list of required equipment: a patch clamp amplifier (Axopatch 200B) (**Note:** In this and future brackets, we provide the equipment name/model on our setup; other equivalent equipment may also serve the same purpose), a vibration-isolation air table (Technical Manufacturing Corporation [TMC]), a micromanipulator (Narishige MWO-3), a patch pipette puller (Narishige, PC-10), an analog-to-digital/digital-to-analog ADDA-converter (Digidata 1200A or higher models), a PC and the relevant software (pClamp 6 or higher versions of pClamp software package) to design the voltage clamp protocols, an inverted microscope, and a holder for the dish (or a recording chamber to hold the coverslips).

4.2.4 DETAILED PROTOCOL

1. Culture the cells of interests on a culture dish (or coverslips); rinse the cells at least three times with saline prior to patching to remove the culture medium. The saline contains (in mM) 140 NaCl, 5.4 KCl, 1 $MgCl_2$, 1.8 $CaCl_2$, and 20 HEPES, and pH 7.4.
2. Transfer the dish (or a coverslip in a recording chamber) on an inverted microscope, connecting the ground electrode in the bath.
3. Fill the patch pipette with intracellular solution (ICS) (Harks et al. 2001; Tong et al. 2006) and position near a cell in the middle of a cluster of (or confluent) cells. Under voltage clamp mode, a 5 mV seal test-induced current pulse will be observe. Slowly lower down the patch pipette aiming the patch pipette tip at the middle of a cell until a 5–10% reduction of the current pulse amplitude (an increase in pipette resistance) is observed. Apply suction via a 1 mL syringe to form a patch seal. After a high gigaohm seal is formed, fully compensate the patch pipette capacitance with the patch clamp amplifier. It is very important to form a tight gigaohm seal for at least 2 minutes prior to trying to break into the whole-cell recording mode.

4. The voltage clamp holding potential (V_H) should be close to the resting membrane potential (e.g., $V_H = -60$ or $-70\,mV$).
5. Use a small depolarization voltage pulse of $10\,mV$ and 120-millisecond duration to generate a capacitive current. Tip: a smaller depolarizing pulse or a hyperpolarizing pulse can be used to avoid the activation of voltage-gated ion channels on the plasma membrane.
6. Measure the peak (I_{peak}) and the steady state (I_{ss}) of the capacitive current. The currents are high-cut filtered at $10\,kHz$ and digitized at $100\,kHz$ via an ADDA converter to a PC via a pClamp software, Clampex. **Note:** The filter setting is crucial for this experiment—you need to follow the suggested conditions to obtain the proper reading on the current peak amplitude. Do not perform whole-cell capacitance compensation!
7. Calculate the estimated conductance (G) between the patched cell and its surrounding cells according to the equation in Figure 4.1d.

4.2.5 LIMITATIONS AND ISSUES

- The measured capacitance could only be determined qualitatively due to space clamp problems (de Roos, van Zoelen, and Theuvenet 1996).
- There is limited control over the transjunctional voltage. Larger voltage steps are not recommended due to potential activation/deactivation of the voltage-gated ion channels.
- A single isolated cell (without any gap junction coupling) will show as a low level of coupling conductance (G_j) (a basal level, determined by the single cell capacitance).
- The estimated conductance (G_j) is an underestimation of the actual coupling conductance (Harks et al. 2001).

4.3 DUAL PATCH CLAMP TECHNIQUE

4.3.1 OVERVIEW OF THE APPLICATION

The dual patch clamp voltage clamp method was first used to characterize gap junction properties in isolated cell pairs by Harris, Spray, and Bennett (1981a,b) at the Einstein College of Medicine in the early 80s. The advantages of this method include accurate measurement of the coupling conductance and easy/flexible control of the transjunctional voltage (V_j). These features enable the detailed characterization of V_j-dependent gating and unitary channel properties of gap junction channels. It is the most widely used and accepted method to study the normal gap junction channel properties as well as to characterize the possible defects of disease-linked gap junction mutants. It is generally used on pairs of isolated tissue cells (with native gap junctions) or model cells expressing known wild-type/mutant connexins. Gap junction-deficient model cells, such as N2A cells, RIN cells, or HeLa cells, are often used as a convenient vehicle to express connexins (Cottrell and Burt 2001; Valiunas 2002; Bai et al. 2006; Palacios-Prado et al. 2009; Xin, Gong, and Bai 2010). Expressing connexins in Xenopus oocytes and performing two-electrode voltage clamp in each of the paired oocytes can also be used to study gap junction macroscopic properties (Verselis, Ginter, and Bargiello 1994; Foote et al. 1998). The details of how to study gap junctions in paired oocytes are not included in this chapter.

4.3.2 Circuitry and G_j Measurement

The equivalent circuitry of the dual patch clamp on an isolated cell pair can be simplified as shown in Figure 4.2a (assuming the seal between the patch pipette and the cell plasma membrane is in the multigigaohms range). Each cell in the pair is under a voltage clamp with holding potentials at 0 mV (no transjunctional voltage or $V_j = (V_0 - V_1) = 0$). To reduce the voltage clamp errors, low patch pipette access resistance (R_{a0} and R_{a1}) and high membrane resistance (R_{m0} and R_{m1}) for both cells are needed (tips to achieve this are discussed in the following Section 4.3.2.1). If the R_{m0} (and R_{m1}) $\gg R_{a0}$ (and R_{a1}), respectively, then the equivalent circuitry of the dual patch clamp can be simplified into two pipette access resistance in a series connected by the gap junction resistance (R_j or conductance G_j; Figure 4.2b). A voltage pulse is generated in one of the cells (Figure 4.2c, cell 0), while the other cell remains at the holding potential (0 mV) (Figure 4.2c; cell 1). This voltage pulse establishes a V_j (20 mV) between cell 0 and cell 1. If functional gap junction channels exist at the cell–cell interfaces between this pair, then the transjunctional current ($I_j = -I_1$) will be recorded in cell 1 (Figure 4.2c). According to Ohm's law, the gap junction coupling conductance can be calculated by I_j/V_j. If the $R_j \gg R_{a1}$ and R_{a0} (or $G_j \ll G_{a0}$ and G_{a1}), then this calculation is a good estimation of the gap junction conductance, but when R_j is low and comparable to R_{a1} or R_{a0}, then the corrections to eliminate the R_{a1} and the R_{a0} will be needed (for details see below).

4.3.2.1 Practical Tips for Measuring G_j

- Improve the patch quality, including higher seal resistance (freshly prepared patch pipette, filtered ICS, longer waiting time prior to the break into whole-cell recording configuration, etc.), lower access resistance (clean break into the whole-cell recording mode), and try to obtain a stable recording (make minimum changes on the seal and the access resistance and avoid touching the recording setup).
- Access resistance depends on the tip size of the patch pipette and the quality of the break into the whole-cell recording mode. Both larger patch pipettes (pipette resistance between 2–4 MΩ) and better break into the whole cell (in our setup, 0.3–0.5 mL suction via 1 mL syringe) are strong enough to obtain a sudden break into the whole-cell recording mode. This way is usually better than the lower suction combined with ZAP (large electrical oscillations), which tend to end up with higher R_a and reseal problem).
- When G_j is high, the access resistance can cause substantial errors in the measured coupling conductance. Two ways can be used to correct this; the first one is to use the patch clamp amplifier series resistance compensation for each patch pipette. Pros: This is online compensation, and the obtained I_j

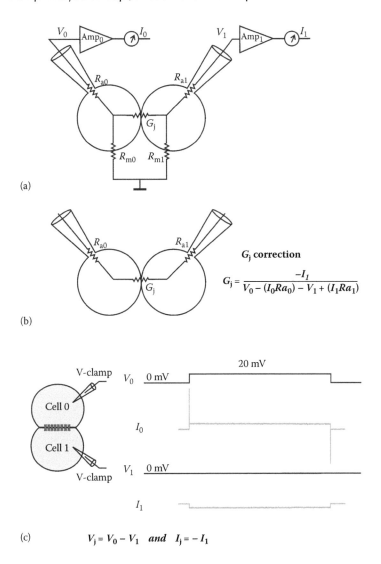

FIGURE 4.2 Dual patch clamp method to measure the transjunctional current in isolated cell pairs. (a) The equivalent circuitry of dual patch clamp recording under the voltage clamp whole-cell recording mode on an isolated pair of gap junction-linked cells. R_m and R_a represent the membrane and the access resistance, respectively, and G_j represents the gap junctional conductance. When R_m is sufficiently higher than R_a, a simplified equivalent circuitry may be used, as shown in b, where the current passing through the cell plasma membrane is relatively small and therefore omitted. In these cases, only R_{a0} and R_{a1} need to be compensated using the access resistance compensation equation shown on the right. The equation provided is used when channel 1 is the recording channel ($I_j = -I_1$). The corrected conductance (G_j) takes into account the current injected into both the pulsing and the recording channel (I_0, I_1), the voltage applied to each cell (V_0, V_1), as well as the patch pipette access resistance (R_{a0}, R_{a1}). (c) A 20 mV voltage pulse was administered to cell 0, while cell 1 was held at 0 mV. The transjunctional current trace is shown. I_0 is the current recorded from cell 0.

is already corrected. Cons: In the process of series resistance compensation, large oscillations can occur, which can cause the loss of the patch seal. To avoid this, researchers tend to use lower levels of compensation (i.e., ~85%). The second way is to correct the G_j off-line after recording. Pros: Because this is off-line, it does not take valuable recording time (data collection), and there is no risk of losing the recording. Cons: The method needs to obtain access resistance (see Section 4.2 for more details) and current (I_0 and I_1) from both channels, which is time consuming. The equation used for correction is given in Figure 4.2b (Van Rijen et al. 1998; Musa et al. 2004).

- High cell membrane resistance (R_{m0} and R_{m1}) is needed to ensure better voltage clamp. Including K^+ channel blockers, such as Cs^+, TEA, Ba^{2+}, in ICS (for details see Tables 4.2 and 4.3) and extracellular solution (ECS) will help block most of the resting leak K^+ channels, which substantially increases the R_m. It is recommended to have $R_m \geq 500\,M\Omega$ to achieve a good voltage clamp (Van Rijen et al. 1998).

TABLE 4.2
ICS for Dual Patch Method

Name	MW	Final (mM)	To Make 25 mL from Stock (mL)	Stock (mM)
CsCl	168.4	130	3.25	1000
EGTA	380.4	10	2.5	100
CaCl$_2$	147.0	0.5	2.5	5
MgATP	507.2	3	2.5	30
Na$_2$ATP	551.1	2	2.5	20
Hepes	238.3	10	2.5	100
ddH$_2$O			8	

Note: Mix and adjust the pH to ~7.20–7.25 with 1 M of CsOH. Measure and adjust the osmolarity with ddH$_2$O to ~290–300 mOsm. Filter with a 0.45 μm syringe filter and store in a freezer (−20°C).

TABLE 4.3
Stock Solutions

Name	MW	Stock (mM)	Volume (mL)	Weight (g)	Storage
CsCl	168.4	1000	50	8.420	Fridge
EGTA	380.4	100	50	1.902	Fridge
CaCl$_2$	147.0	5	50	0.037	Fridge
MgATP	507.2	30	10	0.150	Freezer
Na$_2$ATP	551.1	20	10	0.110	Freezer
HEPES	238.3	100	50	1.192	Fridge

• When using gap junction-deficient model cells, fluorescent proteins are often used to help identify the cells expressing connexins. The reporter fluorescent proteins (e.g., GFP, RFP, YFP, DsRed) are either tagged in frame onto connexins (mostly on the carboxyl terminus) or untagged via an internal ribosome entry site (IRES) vector (e.g., IRES-2-EGFP, IRES2-DsRed2). The cell pairs expressing tagged connexins can be directly visualized for their localization in the cytosol and at the cell–cell interfaces in live cells (Sun et al. 2013, 2014), while the cell pairs expressing untagged connexins (Figure 4.3d) are better suited for a functional study without any possible interference of the fluorescent proteins on the carboxyl terminus (Tong et al. 2014).

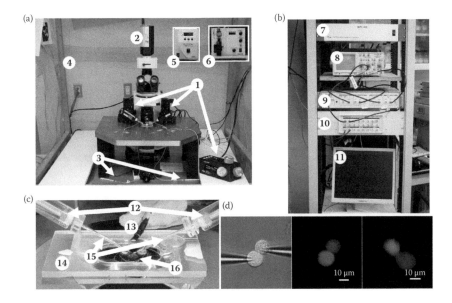

FIGURE 4.3 (See color insert.) Equipment required for dual patch clamp technique. (a–c) A dual patch clamp setup in our laboratory on an upright microscope with water immersion lens. The key components are labeled by numbers and explained in the following. (d) Left panel showed two patch pipette patched on a pair of GFP-positive N2A cells. This pair was expressing Cx50-IRES-GFP. Note the level of uneven levels of GFP expression, which could be a good sign that the cell pair does not have a cytoplasmic bridge. Another heterotypic cell pair with one expressing Cx40-IRES-GFP and the other expressing Cx43-IRES-DsRed, which can be achieved by the transfection of each vectors individually in different dishes and after overnight culture, mix two cultured cells together for a few hours for a patch clamp study of heterotypic cell pairs. 1, Sutter manipulators and controller (ROE-200). 2, CCD camera. 3, Syringe and tubings for two patch pipettes. 4, Faraday cage. 5, UV light source (X-cite series 120PC). 6, Two-stage patch pipette puller (Narishige PC-10). 7, Sutter MPC-200 micromanipulator power source. 8, Oscilloscope. 9, Multiclamp 700A patch clamp amplifier. 10, ADDA converter (Digidata 1322A). 11, PC running PClamp software. 12, Patch pipette holders. 13, Ground electrode (AgCl pellet). 14, Recording chamber to hold the coverslips with cells. 15, Patch pipettes. 16, A glass coverslip with cells.

4.3.3 Transjunctional Voltage-Dependent Gating (V_j Gating)

For most gap junctions, the macroscopic junctional current (I_j) amplitude usually does not change with time at low V_j (e.g., ±10–20 mV, see Figure 4.4a), but when the absolute V_j are increased to higher levels (±40–120 mV), the amplitude of I_j decline with time due to the transjunctional voltage-dependent gating (V_j gating) of the gap junction channels (Figure 4.4a). Optimal voltage clamp conditions are required to study the kinetics and the steady-state V_j gating of the gap junction channels. In addition to improving the patch quality, low levels of G_j (<5–7 nS) are needed to reduce the voltage clamp error (Wilders and Jongsma 1992; Moreno

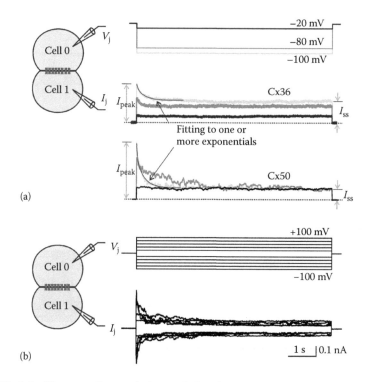

FIGURE 4.4 The recording and the analysis of macroscopic transjunctional current obtained by the dual patch clamp method. (a) The macroscopic transjunctional current ($I_j = -I_1$) recordings of isolated cell pairs expressing either Cx36 or Cx50 in response to a series of V_j pulses applied to cell 0 (−20, −80, −100 mV). By comparing I_{peak} to I_{ss}, the V_j gating was seen to a greater extent at larger V_j for both types of connexins. Cx50 had both a higher rate of gating and a larger extent of gating than Cx36. The V_j gating kinetics can be quantitatively described by fitting the current traces with one or more exponential functions as shown. (b) An example of a voltage protocol used on an isolated cell pair-expressing mouse Cx50. One cell of the pair (cell 0) was administered a series of voltage pulses (7 seconds in duration) from ±20 to ±100 mV in 20 mV increments, while cell 1 was held at 0 mV. Because the cells were linked by gap junctions, a transjunctional current ($I_j = -I_1$) was recorded, and V_j gating was observed, as shown by the current traces.

(Continued)

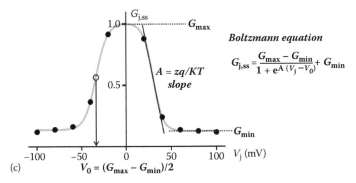

(c)

FIGURE 4.4 (CONTINUED) (c) The Boltzmann equations can be used to fit the normalized average $G_{j,ss}$ (I_{ss}/I_{peak}, filled circles) versus V_j plot in Cx50 gap junctions ($n = 5$). The Boltzmann equation is given on the right, in which V_0 is the voltage at which the conductance is reduced by half [$(G_{max} - G_{min})/2$]; G_{max} is the maximum normalized conductance, G_{min} is the normalized voltage-insensitive residual conductance, and parameter A, describing the slope of the fitted curve, which reflects the V_j sensitivity ($A = zq/KT$) in terms of the number of equivalent gating charges, z, moving through the entire applied field, where q is the electron charge, and K and T are the Boltzmann constant and absolute temperature, respectively.

et al. 1994). Depending on the connexins expressed, V_j gating properties are different in different gap junction channels (as shown in Figure 4.4a for Cx36 and Cx50 gap junction channels). The signature V_j gating properties include the kinetics, which quantitatively describes the rate of the macroscopic I_j declining (or V_j gating, one or more time constants are often needed (Figure 4.4a), and the extent of V_j gating, which is often reflected as the ratio of I_{ss}/I_{peak} at different V_j. The normalized steady-state to the peak conductance is $G_{j,ss}$, which can be plotted with V_j in the entire tested range. This $G_{j,ss}/V_j$ relationship can normally be described by a two-state Boltzmann equation as given in Figure 4.4c. Cx50 gap junction channels show prominent V_j gating and are well described by the Boltzmann equations (Figure 4.4c). The $G_{j,ss}/V_j$ relationship of most gap junctions can be fitted to the Boltzmann equations with their distinct set of parameters (Bukauskas and Verselis 2004; Gonzalez, Gomez-Hernandez, and Barrio 2007). The detailed definition of each Boltzmann parameter is given in Figure 4.4c.

4.3.3.1 Practical Tips for Studying V_j Gating

- Improve the patch quality, including higher seal resistance (freshly prepared patch pipette, filtered ICS, longer waiting time prior to the break into the whole cell, etc.), lower access resistance (clean break into the whole-cell recording mode by only using a suction), and try to obtain stable recording (make minimum changes on the seal and the access resistance and avoid touching the setups).
- Work on cell pairs with G_j in the range of <5–7 nS to avoid errors of voltage clamp. Note that too low of a G_j can also be an issue for gap junctions with large unitary channel conductance, where opening/closing of one channel can profoundly change the ratio of I_{ss}/I_{peak}.

- As noted earlier, the ideal G_j in a pair of cells is <5–7 nS for characterizing their V_j gating properties. To increase the probability of obtaining cell pairs in this range, one can increase/decrease the expression levels (for recombinant expression studies) by using different expression vectors, amount of vectors, amount of transfection reagent and duration, replating time, and combinations of these factors. If studying endogenously expressed gap junctions, the reduction of G_j may be required by reducing the replating time, the physiological ways to reduce coupling (manipulating intracellular pH, Ca^{2+}, etc.), or the use the pharmacological method (adding moderate concentration of gap junction blockers). It is not fully clear if these physiological/pharmacological manipulations may modify the V_j gating properties, so avoid using these if possible.
- Customize the V_j protocols to fit your studies on V_j gating. We use V_j pulses (7-second duration and every 15 or 30 seconds to repeat) to study the V_j gating properties of Cx50 and its mutants. We also alternate the V_j polarity and start from low V_j, e.g., the V_j pulse sequence is −20 mV, +20 mV, −40 mV, +40 mV, …, −100, +100 mV (Xin, Gong, and Bai 2010; Tong et al. 2014). Depending on the gating kinetics and the recovery time of different connexins in the gap junction, other types of V_j protocols are also used by different researchers, such as slow ramp V_j protocols (Musa et al. 2004; Paulauskas et al. 2009).

4.3.4 RECORDING UNITARY GAP JUNCTION CHANNEL CURRENTS

Gap junctional currents drastically vary in amplitude in cell pairs expressing connexins. In the poorly coupled cell pairs, unitary channel currents (i_j) can be identified especially in moderate to high levels of V_j (±40–100 mV). An example is shown in Figure 4.5a, which displays i_j of at least three Cx50 gap junction channels (at the beginning of the V_j pulse). Each of these different open states (from close to open1 and amplitude between two adjacent open states) shows roughly equal amplitude. A long-lived residue state is identifiable in between open1 and closed state in this record (labeled as residue state in Figure 4.5a). Gap junctions of different connexins show drastically different unitary channel conductance, from 300 pS (Cx37), 200 pS (Cx50), to 5–10 pS (Cx36) homotypic gap junction channels (Veenstra et al. 1994; Srinivas et al. 1999; Teubner et al. 2000; Moreno et al. 2005; Gonzalez, Gomez-Hernandez, and Barrio 2007; Xin, Gong, and Bai 2010; Xin, Sun, and Bai 2012). Most gap junction channels also display one or more residue-conducting state (also known as subconducting state or substate; Figure 4.5a,b residue state). It usually requires sophisticated software to analyze i_j with multiple channels. We focus our unitary channel analysis on i_j with apparently only one channel in operation (Figure 4.5b). Single gap junction channel currents offer detailed information on the channel properties, including unitary channel conductance (γ_j, also known as main conductance or $\gamma_{j,main}$), residue conductance (or subconductance, $\gamma_{j,sub}$ or $\gamma_{j,res}$) (Bai et al. 2006; Xin, Gong, and Bai 2010; Xin et al. 2012), probability of open/residue/closed state (P_{open}, P_{res} or P_{sub} and P_{closed}, respectively) (Tong et al. 2014), and open dwell time analysis (Xin, Sun, and Bai 2012; Tong et al. 2014).

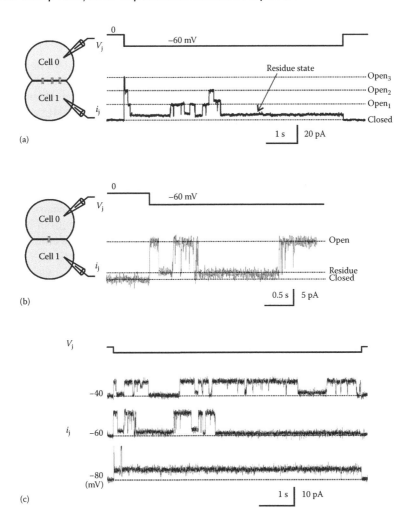

FIGURE 4.5 Recording and analysis of unitary gap junction channel currents obtained by the dual patch clamp method. (a) An isolated cell pair expressing at least three Cx50 gap junction channels. In response to a voltage pulse applied to cell 0 (−60 mV), three near identical unitary current (i_j) levels (labeled open$_{1-3}$) can be identified with spontaneous opening/closing transitions during the voltage pulse; at a later time, the junctional current settled at an apparent residue state until the end of the V_j pulse. In this case, the elucidation of single-channel conductance would be challenging and require advanced mathematical models. (b) An isolated cell pair apparently coupled by only one Cx50 gap junction channel. When a V_j pulse of −60 mV was applied to cell 0, the current trace of cell 1 showed distinct main open conducting states (open) as well as a substate (residue state). Single-channel conductance (γ_j) could be easily determined using i_j/V_j because the main channel openings and the residue conductance state(s) were easily distinguishable. (c) Single gap junction channel events of an isolated cell pair at increasing transjunctional voltages ($V_j = −40, −60,$ or −80 mV). By comparing the current traces at increasing voltages, it was clear that the amplitude of i_j increase with the increase of V_j. In addition, the P_{open} (for the main open state) was inversely proportional to V_j.

4.3.4.1 Practical Tips for Single Channel Recording

- Improve the patch quality as described earlier.
- Single-channel currents (i_j) are usually in the pA range with a lot of background noise associated (Figure 4.5b). Increasing the signal/noise ratio will help to visualize/resolve unitary channel events, and several methods can help increase this ratio. (1) Low-pass filters between 200–500 Hz can be used for gap junction channels with long openings (with tens to hundreds of millisecond duration). Be careful reducing low-pass filters because they can change the kinetic properties of the onset/offset of the channel transitions. (2) An increase in V_j will increase i_j, making it easier to observe channel events (Figure 4.5c). However, for most gap junction channels, that P_{open} decreases with the increase of V_j (Figure 4.5c), which means much less events to be analyzed. (3) Sometimes 60 (or 50) Hz noise (from building power supplies) can be in the record, which will need to be eliminated via a notch filter. It is better to remove this noise online prior to the i_j data collection. This noise can also be minimized by the use of a Faraday cage (Figure 4.3a, 4).
- Single-channel data contain a lot of information. Some of this information is directly measurable, and others are not very easy to obtain for many different reasons. Compared to $\gamma_{j,sub}$, $\gamma_{j,main}$ is relatively easy to obtain. For open dwell time analysis, you need to accumulate hundreds to thousands of events to have a good estimation of this parameter during the steady state (not in the beginning of the V_j pulse). For Cx50 gap junction channels, dwell time analysis is possible for V_j of ±40–60 mV.
- Manipulate your expression system to increase the probability of observing single-channel currents, including reducing the amount of plasmid/vector, using low expression level of vectors, and reducing the transfection time and/or the replating time.

4.3.5 REQUIREMENTS

Figure 4.3 shows one of our setups for a dual patch clamp, including patch clamp amplifier (one Multiclamp 700A or two Axopatch 200B), a vibration-isolation air table (TMC), two micromanipulators (Sutter MPC-200 or two Narishige MWO-3), a patch pipette puller (Narishige PC-10), an ADDA-converter (Digidata 1322B or more recent models), a PC and the relevant software (pClamp 9.2) to design voltage clamp protocols, a fluorescent microscope (Olympus BWI51 or Leica DMIRB), and a holder for the culture dish (or a recording chamber to hold the coverslips). The setup on an upright microscope (Olympus BWI51) and the associated equipment are shown in Figure 4.3.

4.3.6 DETAILED PROTOCOL FOR DUAL PATCH CLAMP

1. Prepare the ECS and the ICS prior to the experimental day. Double check if the osmolarity of ECS (310–320 mOsm; Table 4.4) and ICS (290–300 mOsm; Table 4.2) are in an acceptable range after warming these solutions to room temperature.

TABLE 4.4
ECS for Dual Patch Clamp Method

Name	MW	Final (mM)	Make 1 L (g)
NaCl	58.44	135	7.889
KCl	74.55	5	0.373
HEPES	238.3	10	2.383
MgCl$_2$	203.31	1	0.203
CaCl$_2$	147.0	2	0.294
BaCl$_2$	244.28	1	0.244
CsCl	168.4	2	0.337
Na pyruvate	110.0	2	0.220
D-Glucose	180.16	5	0.901

Note: Mix and adjust the pH to ~7.4 with 1 M of NaOH. Dissolve these chemicals in 0.9 L of double distilled water (ddH$_2$O), and use ddH$_2$O to adjust the osmolarity to ~310–320 mOsm. Filter with a 0.45 μm bottle-top vacuum filter and store in ECS in a fridge (4°C). MW: molecular weight.

2. Pull a sufficient number of patch pipettes before recording. Adjust the puller settings to ensure that the patch pipette resistance is in the range of 2–4 MΩ.
3. Turn on the required instruments for dual patch clamp setup (Figure 4.3a and b).
4. In the culture hood, take a coverslip out of the culture dish, rinse the coverslip with ECS twice to thoroughly remove the culture medium, and then place it onto the recording chamber prefilled with ECS. Push the coverslip to the middle of the chamber and then firmly press it down (to avoid floating!).
5. Position the chamber in the center of the microscope stage and put the grounding electrode into the chamber directly in contact with the ECS (Figure 4.3c).
6. Identify a cell pair that is fully isolated from other cells on the coverslip and position the pair in the center of the viewing field of the microscope. For fluorescent protein-tagged (or untagged) connexins, ensure that you have observed the appropriate fluorescence in the pair with correct excitation-emission filters.
7. Fill a patch pipette with ICS (make sure that no air bubbles are near the tip of the pipette). Then, slide the patch pipette onto the electrode holder (making sure that one of your bare hands is grounded all the time during this operation). Push the pipette until the AgCl-coated silver wire is in good contact with the ICS and then tighten the pipette holder to ensure that the patch pipette is firmly held in the holder, and there is no leak of any air in any part of the holder. A fresh coat of AgCl on the day of the recording will help to establish a stable baseline current.

8. Add a small amount of positive pressure to the pipette holder by attaching a syringe to the patch system. Positive pressure is important to keep the patch tip clean during the process of lowering the pipette into the ECS and near the cell pair.

9. Use a micromanipulator to lower the pipette into the recording chamber and continue to lower the pipette tip aiming at the middle, but slightly above (~40 μm) a cell in the pair. Keep the seal test on throughout this process to monitor the pipette resistance. Set the current baseline to zero on the patch amplifier control panel.

10. Repeat steps 7–9 for the other patch pipette and position the tip above the other cell in the pair. Double-check the resistance given by the seal test for each of the patch pipettes and ensure that it is stable.

11. Using fine control of the micromanipulator, continue lowering one patch pipette to the middle of one cell of the pair (~5–10 μm above). Readjust the pipette position with the fine controls of the manipulator to aim at the center of the cell. Slowly lower the patch pipette while monitoring its seal test-induced current pulse on the PC screen. When the tip touches the cell plasma membrane, the seal test current pulse becomes slightly smaller, which indicates an increase in the pipette resistance by ~0.1–1 MΩ. Stop lowering the patch!

12. Slowly pull the syringe plunger to create a gentle negative pressure (suction)—this forms a tight seal between the plasma membrane and the tip of patch pipette. The seal test current pulse will quickly decrease during the sealing process. Remove the negative pressure by disconnecting the syringe and reapplying suction can help to seal. Applying a holding potential of −60 mV is usually helpful to reach GΩ seal quicker and better.

13. Wait for approximately 1 minute to allow the seal to become very strong, reaching the several gigaohm ranges (this short waiting time is very important for a long-lasting stable recording!).

14. To break the patch into whole-cell recording configuration, reattach the syringe. Slowly increase the negative pressure by pulling the syringe plunger until the seal test current suddenly jumps up into whole-cell recording mode (with much larger capacitive current to the test seal voltage pulse). If the capacitive current is stable, remove the negative pressure by disconnecting the syringe. If a holding potential was used, turn it off at this point.

15. Repeat steps 11–14 for the other patch pipette to form the whole-cell recording on the other cell of the pair.

16. Once both patches formed good whole-cell recording, the voltage pulse protocols can be applied to one cell of the pair while the other cell is held at 0 mV. Reversing the pulsing channel from one to the other is recommended.

17. Depending on the initial G_j level, determine if the pair is suitable for characterizing the properties of macroscopic V_j gating, unitary channels, or just quantifying G_j. Use different sets of voltage pulse/ramp protocols for specific questions.

18. At the end of an experiment (or when one of the patches lost seal/whole cell), sequentially lift each patch pipette up to the highest position. Remove the

patch pipette from its holder and put it into waste sharps container. Get ready for patching a new cell pair.

19. Repeat steps 6–17 to obtain the recordings from more cell pairs on the same coverslip. Change to a new coverslip every 1–2 hours or whenever a gap junction blocker is used. Rinse the chamber at least three times to ensure no gap junction blocker is present for new measurements.

4.3.7   LIMITATIONS AND ISSUES

- For better voltage clamp, isolated cell pairs in a culture or freshly isolated tissue cells are needed for a dual patch clamp study and are sometimes unlikely to be in physiological relevant conditions. Confluent, monolayer or large clusters of cells in a culture are better studied with the single patch clamp for the estimation of gap junction coupling conductance.
- The overexpression of connexins in gap junction-deficient model cells as well as in isolated tissue cell pairs often yield high gap junction coupling (G_j is ≥ 40–$100 \, nS$). At this high level of G_j, voltage clamp errors are large, which diminish/reduce V_j gating (Van Rijen et al. 1998). In addition, high-coupled pairs are often very difficult to differentiate from mitotic cytoplasmic bridge (not yet fully proliferating into two cells, and the cytoplasm is continuous between the apparent cell pair). To confirm that cell pairs with high G_j are due to gap junctions, additional tests are needed, including large gap junction-impermeable fluorescent dyes (Dextran Texas Red, MW 10,000 Da) (Sun et al. 2013) and/or use of gap junction blockers (carbenoxolone, flufenamic acid, mefluoqin) (Tong et al. 2006; Sun et al. 2013) or conditions (such as lowering the pH of ECS) (Peracchia, Wang, and Peracchia 2000; Peracchia et al. 2003).

4.4   CONCLUSION

Gap junctions are a unique group of channels located between cells. Functional studies with dye transfer provide a simple, in most cases, qualitative measurement for gap junction function. To characterize gap junction coupling in detail, single and dual patch clamp methods should be used to give quantitative measurements/estimations. For detailed characterization of gap junction gating properties at macroscopic and microscopic (single-channel) currents, we recommend using the dual patch clamp technique. With many new advances in microscopy, quantitative fluorescent measurement, and other techniques, we hope to see their direct applications on gap junction research or their combinations with single or dual patch clamp in functional studies of gap junction channels.

ACKNOWLEDGMENTS

We thank all Bai lab members for their contributions in improving the quality of the patch clamp technique. We also thank David Spray and Miduturu Srinivas for teaching us the dual patch clamp technique.

REFERENCES

Bai, D., and A. H. Wang. 2014. Extracellular domains play different roles in gap junction formation and docking compatibility. *Biochemical Journal* 458 (1):1–10.

Bai, D., C. del Corsso, M. Srinivas, and D. C. Spray. 2006. Block of specific gap junction channel subtypes by 2-aminoethoxydiphenyl borate (2-APB). *J Pharmacol Exp Ther* 319 (3):1452–8.

Bukauskas, F. F., and V. K. Verselis. 2004. Gap junction channel gating. *Biochim Biophys Acta* 1662 (1–2):42–60.

Cottrell, G. T., and J. M. Burt. 2001. Heterotypic gap junction channel formation between heteromeric and homomeric Cx40 and Cx43 connexons. *Am J Physiol Cell Physiol* 281 (5):C1559–67.

de Roos, A. D., E. J. van Zoelen, and A. P. Theuvenet. 1996. Determination of gap junctional intercellular communication by capacitance measurements. *Pflugers Arch* 431 (4):556–63.

el-Fouly, M. H., J. E. Trosko, and C. C. Chang. 1987. Scrape-loading and dye transfer: A rapid and simple technique to study gap junctional intercellular communication. *Exp Cell Res* 168 (2):422–30.

Foote, C. I., L. Zhou, X. Zhu, and B. J. Nicholson. 1998. The pattern of disulfide linkages in the extracellular loop regions of connexin 32 suggests a model for the docking interface of gap junctions. *J Cell Biol* 140 (5):1187–97.

Gonzalez, D., J. M. Gomez-Hernandez, and L. C. Barrio. 2007. Molecular basis of voltage dependence of connexin channels: An integrative appraisal. *Prog Biophys Mol Biol* 94 (1–2):66–106.

Goodenough, D. A., and D. L. Paul. 2009. Gap junctions. *Cold Spring Harb Perspect Biol* 1 (1):a002576.

Harks, E. G., A. D. de Roos, P. H. Peters, L. H. de Haan, A. Brouwer, D. L. Ypey, E. J. van Zoelen, and A. P. Theuvenet. 2001. Fenamates: A novel class of reversible gap junction blockers. *J Pharmacol Exp Ther* 298 (3):1033–41.

Harris, A. L., D. C. Spray, and M. V. Bennett. 1981. Kinetic properties of a voltage-dependent junctional conductance. *J Gen Physiol* 77 (1):95–117.

Moreno, A. P., V. M. Berthoud, G. Perez-Palacios, and E. M. Perez-Armendariz. 2005. Biophysical evidence that connexin-36 forms functional gap junction channels between pancreatic mouse beta-cells. *Am J Physiol Endocrinol Metab* 288 (5):E948–56.

Moreno, A. P., M. B. Rook, G. I. Fishman, and D. C. Spray. 1994. Gap junction channels: Distinct voltage-sensitive and -insensitive conductance states. *Biophys J* 67 (1):113–9.

Musa, H., E. Fenn, M. Crye, J. Gemel, E. C. Beyer, and R. D. Veenstra. 2004. Amino terminal glutamate residues confer spermine sensitivity and affect voltage gating and channel conductance of rat connexin40 gap junctions. *J Physiol* 557 (Pt 3):863–78.

Palacios-Prado, N., S. Sonntag, V. A. Skeberdis, K. Willecke, and F. F. Bukauskas. 2009. Gating, permselectivity and pH-dependent modulation of channels formed by connexin57, a major connexin of horizontal cells in the mouse retina. *J Physiol* 587 (Pt 13):3251–69.

Paulauskas, N., M. Pranevicius, H. Pranevicius, and F. F. Bukauskas. 2009. A stochastic four-state model of contingent gating of gap junction channels containing two fast gates sensitive to transjunctional voltage. *Biophys J* 96 (10):3936–48.

Peracchia, C., X. G. Wang, and L. L. Peracchia. 2000. Chemical gating of gap junction channels. *Methods* 20 (2):188–95.

Peracchia, C., K. C. Young, X. G. Wang, J. T. Chen, and L. L. Peracchia. 2003. The voltage gates of connexin channels are sensitive to CO(2). *Cell Commun Adhes* 10 (4–6):233–7.

Saez, J. C., V. M. Berthoud, M. C. Branes, A. D. Martinez, and E. C. Beyer. 2003. Plasma membrane channels formed by connexins: Their regulation and functions. *Physiol Rev* 83 (4):1359–400.

Sohl, G., and K. Willecke. 2004. Gap junctions and the connexin protein family. *Cardiovasc Res* 62 (2):228–32.

Sosinsky, G. 1995. Mixing of connexins in gap junction membrane channels. *Proc Natl Acad Sci U S A* 92 (20):9210–4.

Spray, D. C., A. L. Harris, and M. V. Bennett. 1981a. Equilibrium properties of a voltage-dependent junctional conductance. *J Gen Physiol* 77 (1):77–93.

Spray, D. C., A. L. Harris, and M. V. Bennett. 1981b. Gap junctional conductance is a simple and sensitive function of intracellular pH. *Science* 211 (4483):712–5.

Srinivas, M., R. Rozental, T. Kojima, R. Dermietzel, M. Mehler, D. F. Condorelli, J. A. Kessler, and D. C. Spray. 1999. Functional properties of channels formed by the neuronal gap junction protein connexin36. *J Neurosci* 19 (22):9848–55.

Sun, Y., X. Tong, H. Chen, T. Huang, Q. Shao, W. Huang, D. W. Laird, and D. Bai. 2014. An atrial-fibrillation-linked connexin40 mutant is retained in the endoplasmic reticulum and impairs the function of atrial gap-junction channels. *Dis Model Mech* 7 (5):561–9.

Sun, Y., Y. Q. Yang, X. Q. Gong, X. H. Wang, R. G. Li, H. W. Tan, X. Liu, W. Y. Fang, and D. Bai. 2013. Novel germline GJA5/connexin40 mutations associated with lone atrial fibrillation impair gap junctional intercellular communication. *Human Mutat* 34 (4):603–9.

Teubner, B., J. Degen, G. Sohl, M. Guldenagel, F. F. Bukauskas, E. B. Trexler, V. K. Verselis et al. 2000. Functional expression of the murine connexin 36 gene coding for a neuron-specific gap junctional protein. *J Membr Biol* 176 (3):249–62.

Tong, X., H. Aoyama, T. Tsukihara, and D. Bai. 2014. Charge at the 46th residue of connexin50 is crucial for the gap-junctional unitary conductance and transjunctional voltage-dependent gating. *J Physiol* 592 (Pt 23):5187–202.

Tong, D., J. E. Gittens, G. M. Kidder, and D. Bai. 2006. Patch-clamp study reveals that the importance of connexin43-mediated gap junctional communication for ovarian folliculogenesis is strain specific in the mouse. *Am J Physiol Cell Physiol* 290 (1):C290–7.

Valiunas, V. 2002. Biophysical properties of connexin-45 gap junction hemichannels studied in vertebrate cells. *J Gen Physiol* 119 (2):147–64.

Van Rijen, H. V., R. Wilders, A. C. Van Ginneken, and H. J. Jongsma. 1998. Quantitative analysis of dual whole-cell voltage-clamp determination of gap junctional conductance. *Pflugers Archiv Eur J Physiol* 436 (1):141–51.

Veenstra, R. D., H. Z. Wang, E. C. Beyer, S. V. Ramanan, and P. R. Brink. 1994. Connexin37 forms high conductance gap junction channels with subconductance state activity and selective dye and ionic permeabilities. *Biophys J* 66 (6):1915–28.

Verselis, V. K., C. S. Ginter, and T. A. Bargiello. 1994. Opposite voltage gating polarities of two closely related connexins. *Nature* 368 (6469):348–51.

Wilders, R., and H. J. Jongsma. 1992. Limitations of the dual voltage clamp method in assaying conductance and kinetics of gap junction channels. *Biophys J* 63 (4):942–53.

Woodward, T. L., M. A. Sia, O. W. Blaschuk, J. D. Turner, and D. W. Laird. 1998. Deficient epithelial-fibroblast heterocellular gap junction communication can be overcome by co-culture with an intermediate cell type but not by E-cadherin transgene expression. *J Cell Sci* 111 (Pt 23):3529–39.

Xin, L., X. Q. Gong, and D. Bai. 2010. The role of amino terminus of mouse Cx50 in determining transjunctional voltage-dependent gating and unitary conductance. *Biophys J* 99 (7):2077–86.

Xin, L., S. Nakagawa, T. Tsukihara, and D. Bai. 2012. Aspartic acid residue d3 critically determines cx50 gap junction channel transjunctional voltage-dependent gating and unitary conductance. *Biophys J* 102 (5):1022–31.

Xin, L., Y. Sun, and D. Bai. 2012. Heterotypic connexin50/connexin50 mutant gap junction channels reveal interactions between two hemichannels during transjunctional voltage-dependent gating. *J Physiol* 590:5037–52.

Yum, S. W., J. Zhang, and S. S. Scherer. 2010. Dominant connexin26 mutants associated
    with human hearing loss have trans-dominant effects on connexin30. *Neurobiol Dis*
    38 (2):226–36.
Yum, S. W., J. X. Zhang, V. Valiunas, G. Kanaporis, P. R. Brink, T. W. White, and S. S. Scherer.
    2007. Human connexin26 and connexin30 form functional heteromeric and heterotypic
    channels. *Am j Physiol Cell Physiol* 293 (3):C1032–C1048.

5 What Do You Need to Measure Gap Junctional Permselectivity?

José F. Ek Vitorín

CONTENTS

5.1   INTRODUCTION

This chapter complements previously published methods (Ek-Vitorin and Burt 2005; Ek-Vitorin et al. 2006, 2016; Heyman and Burt 2008; Heyman et al. 2009) by providing more detailed and practical instructions to assess gap junction permselectivity. Included are hints originated from problems or questions encountered during student training, and from previously unprinted personal observations. Although a comprehensive account of technical aspects was attempted, a basic awareness of the physical concepts (e.g., Ohm's law, permeability, diffusion), the methods touched on here (e.g., voltage clamp, sterile techniques), the equipment required (e.g., how to use Bayonet Neill–Concelman [BNC] cables, alligator clips, microscopes, micromanipulators, software), as well as the essential literature on connexins and gap junction channels, is expected from the reader. As suitable for a practical approach, a minimal theoretical background is integrated here, and the works cited are but a few chosen examples to illustrate specific points. More information on gap junctions and patch clamp can be found in accompanying chapters (e.g., Chapter 4) and in the extensive field literature.

5.2   THEORETICAL FRAMEWORK

5.2.1   Are Gap Junction Channels Pipes or Sieves?

It is widely recognized that gap junctions support the functional synchronization of tissues by allowing the diffusion of atomic ions, small metabolites and signaling molecules between neighboring cells. Intuitively, it seems obvious that direct cell-to-cell communication would underlie the harmonious function(s) of tissues and organs. In actuality, reconciling this simple notion with the complexity of gap junction function and regulation (Bruzzone, White, and Paul 1996) is not as straightforward. Indeed, a thoughtful consideration of the very existence and nature of junctional coupling (as well as published data) leads to the realizations that along with its clear benefits, such an intimate form of cell communication holds potential inconveniences, as well as significant limitations, and that, with a few exceptions, the exact mechanism(s) by which gap junctions provide for functional synchrony remain unknown. In other words, our understanding of the intercellular coupling provided by gap junctions is incomplete. To illustrate this contention, let us ponder the following:

- The identity of molecules whose transit through gap junctions produces meaningful functional changes is generally not known.
- Tissues work more coordinately when their cells are coupled by gap junctions (Vozzi et al. 1995), and pathological conditions have been linked to lack/sparsity of gap junctions (Trosko et al. 1990) and to dysfunctional gap junctions (Kelsell et al. 1997). Thus, tissues rich in gap junctions may be thought of as homogeneous cell populations, somehow akin to syncytia; however, despite high junctional coupling, cells within tissues or in culture can be found at diverse stages of the cell cycle, i.e., not synchronized (Burt et al. 2008).
- Gap junction channels do not allow the indiscriminate diffusion of all cellular components: molecules above the size limit of their pores will not diffuse from cell to cell. This would preserve some cell individuality, perhaps in detriment of synchronicity.
- Junctional permeability is determined by the connexin isotype(s) composing the channels (Elfgang et al. 1995; Kanaporis et al. 2008). Since not all connexins form highly permeable gap junction channels, the molecular size limit for junctional diffusion must be smaller (more restrictive) in some tissues than in others.
- Because junctional exchange occurs by simple diffusion, a given signal might be diluted and rendered ineffectual as it goes from cell to cell, more so if the gradient of the messenger molecule was initially small.
- The electrical conductivity of junctional channels formed by different connexins is not predictive of their permeability to bigger molecules (Veenstra et al. 1995; Weber et al. 2004). To further expound, Cx43 and Cx37 display full open channel conductances of ~100–120 and 350–400 pS, respectively (Burt et al. 2008; Ek-Vitorin et al. 2006); however, using the techniques described here, the permeability of Cx37 was found at least one order of magnitude lower than that of Cx43 (Ek-Vitorin and Burt 2005). Yet, even a low expression of Cx37 (but not Cx43) strongly decreases growth of a tumor cell line, suggesting that Cx37 (but not Cx43) channels sustain the spread of a growth-regulating (i.e., synchronizing) signal.
- For a given connexin, the junctional permeability to endogenous substances may change with the regulation of the protein (Ek-Vitorin et al 2006;

Goldberg, Lampe, and Nicholson 1999), as the channel configurations resulting from such normal regulation may display lower permeability to a given permeant. For instance, it was shown that a positive and a negative dye that readily permeate the full open channels of Cx43 do not permeate the voltage-dependent residual state of these channels (Bukauskas, Bukauskiene, and Verselis 2002).

- Junctional coupling may result in the rescue of failing cells by metabolites from healthy neighbors, but the junctional exchange is bidirectional, and healthy neighbors may be damaged by lethal wastes from dying cells (Mesnil, Piccoli, and Yamasaki 1997; Kanno et al. 2003).
- The synchronous heartbeat is the best and most cited example of gap junctions' electrical function. The membrane voltage (V_m) changes that are collectively known as an action potential (AP) passively (electrotonically) propagate from a group of cells to the neighboring cells. This instantaneous (faster than the AP) propagation exclusively depends on the presence of gap junction channels. However, despite the overall high gap junction coupling in cardiac tissues, the distance that an AP could electrotonically travel within the heart is small (in the order of 0.5 mm) (Weidmann 1970; Kleber and Riegger 1987) (cf. Weidmann 1952; Keung, Keung, and Aronson 1982) and may decrease in pathological conditions (Rohr, Kucera, and Kleber 1998; Gutstein et al. 2001; Vaidya et al. 2001). Then, how is the whole heart stimulated? The passively propagated depolarization of an AP brings sodium channels in the neighbor cells to their threshold, initiating new APs. Even in a normal myocardium, V_m gradients between adjacent heart regions can be documented during AP propagation (Morley, Vaidya, and Jalife 2000; Tamaddon et al. 2000). Therefore, despite the very low resistance of gap junctions, APs require a finite amount of time to spread to adjacent regions, and effective electrical propagation in the whole heart also depends on the renewability of APs.

Many questions arise upon these considerations. How much junctional coupling is required to support the propagation of APs? What is the significance of gap junctions in tissues with coupling levels much lower than those of cardiac cells or in tissues coupled by connexin isotypes with lower conductive properties? How far can the depolarization from damaged cells travel within a wounded nonexcitable tissue, i.e., in the absence of a mechanism to regenerate the electrical changes? Do electrical changes regulate the metabolic coupling capabilities of gap junctions? Do electrical changes modulate the gap junction's metabolic adaptations needed for tissue repair or synchronized secretion? Conversely, do metabolic changes (e.g., kinase activity) affect the electrical properties of gap junctions? How far and how abundantly must metabolites travel for cellular synchronization? How much gap junction metabolic coupling is required for normal secretion, wound healing, and other coordinated tissue responses to physiological/pathological stimuli? Does the presence of various cell cycle stages within a tissue indicate a lack of synchronization (e.g., are cells not coupled by gap junction channels, or do they reside too far from each other, or does the particular connexin(s) expressed severely limit the diffusion of synchronizing signaling substances?). Are the properties of gap junction channels (e.g., permeability) regulated to meet the emergent needs of the living tissue? Or, on the contrary, is the functional state (e.g., differentiation, cycle phase) of cells regulated by the diffusion of signaling molecules within the tissue, and thus by the unique properties of the specific connexin(s) there expressed?

At this stage, it seems safe to assert that gap junctions allow the diffusion of some molecules and limit the diffusion of others (thus acting as molecular sieves and not simply as inert pipes joining adjacent cytoplasmic spaces). Such attribute may eventually permit to identify the junctional permeants that are also key molecular messengers, and with this knowledge, to address some of the questions earlier. The methods described here aim toward these long-term goals. For this purpose, electrical coupling is explored by providing current-carrying ions that readily permeate all gap junctions, and metabolic coupling is assessed by fluorescent molecular probes representing the size and the charge of possible junctional permeants. The superposition of these two measurements (together with other techniques described in accompanying chapters) can provide operational profiles of each existing connexin (e.g., Ek-Vitorin, Pontifex and Burt 2016).

Other laboratories have used similar approaches and published results consistent (Eckert 2006) or in contrast (Valiunas, Beyer, and Brink 2002) with ours. However, because we are not apprised of the details of their protocols, the methods described here are based on our own experience. Before diving into the matter, a few words justifying these methods and clarifying the nomenclature used are in order.

5.2.2 WHAT IS PERMSELECTIVITY?

Permeability is the property of substances (here specifically a membrane) to let another substance pass through it. A membrane permeable to only some of the present substances (e.g., anions vs. cations, or vice versa) is said to be *semipermeable* or to display selective permeability or permselectivity (permeability + selectivity) (cf. Oxford and Webster dictionary). When a semipermeable membrane sits between two different solutions, the separation of charges proceeds until a state of equilibrium, an electrochemical potential, is reached; the ultimate strength and polarity of this voltage relates to the specific ionic permeability (permselectivity) of the membrane and the initial concentration of the permeant ion(s) on either side (cf. Nernst potential) (Hille 2001). Junctional membranes (i.e., containing gap junction channels) represent a special case of semipermeable membrane (actually two apposed lipid bilayers traversed by intercellular channels) through which the prospective separation of charges would occur between two virtually identical compartments. The permeability coefficient P (or simply the permeability) is the flux of material through a membrane per unit driving force per unit membrane thickness. For junctional membranes, the thickness is assumed constant, and the driving forces, in principle absent, are provided by creating a dye concentration gradient (causing dye diffusion) or a transjunctional voltage difference, V_j (causing ionic current), between contacting cells. With these intercellular electrochemical gradients, the junctional permselectivity can be probed by measuring the diffusion and/or the current for two (A and B) or more species, and obtaining the ratio P_A/P_B. The following procedures further assume that (1) the dyes are used at concentrations linearly correlated with their fluorescence intensity values; (2) the dyes diffuse passively (no strong intercellular electrical gradient applied) between two intracellular compartments of similar size, and equalize in concentration over time; (3) the functional state of gap junction channels is, on average, unchanged during the recording period; and importantly, (4) the junctional electrical conductance, g_j, is a direct index of the gap junction channel permeability to current-carrying ions (Ek-Vitorin and Burt 2005).

Of note, gap junction channel charge selectivity for current-carrying ions was previously explored (e.g., Beblo and Veenstra 1997; Wang and Veenstra 1997; Heyman and Burt 2008) using ionic substitutions. From these and several similar studies, it seems that the charge selectivity of gap junction channels for atomic ions is weak compared to that of Na^+ or K^+ channels, but the selectivity of gap junction channels for molecules approaching the size limit of the channel pore may be significantly stronger (see Ek-Vitorin and Burt 2013 for a review and Ek-Vitorin, Pontifex and Burt 2016 for recent data). It is this possibility that can be addressed with the equipment, material, and methods described in the following sections.

5.3 EQUIPMENT

A good electrophysiology setup (or rig) must be designed to maximize the experimental success while minimizing the physical effort required from the researcher in charge of the often long trials. The essential components (the equipment or the permanent elements) of an optional arrangement are listed, described, and illustrated in the following sections.

5.3.1 ELECTRICAL RACK CABINET OR TOWER

An electrical rack cabinet or tower is a sturdy metallic structure to hold all electronics in a tiered order and to facilitate connections and relocation. Some racks have fitted metal shelving; others are simpler and have adaptable frames with perforations in the front pillars where standard size electronics or handmade shelves can be fixed in place with screws (Figures 5.1 and 5.2). Wheels in the base for easy displacement are desirable.

5.3.2 FLOATING TABLE (AIR TABLE) FOR ELECTROPHYSIOLOGY

A floating table (air table) refers to an isolation system to minimize the vibrations transmitted from the floor to the electrodes (Figures 5.1 and 5.3). It is virtually impossible to obtain gigaohm (GΩ) seals or to keep a stable attachment of the cell membrane to the electrode glass, or microelectrodes in place, without a vibration isolation system: cell membranes are easily torn by a shaky electrode. Various isolation systems are sold by TMC™ (Technical Manufacturing Corporation, Peabody, Massachusetts).

5.3.2.1 Hints
- The floating table surface should be large enough to house the microscope and the micromanipulators with their numerous mountings.

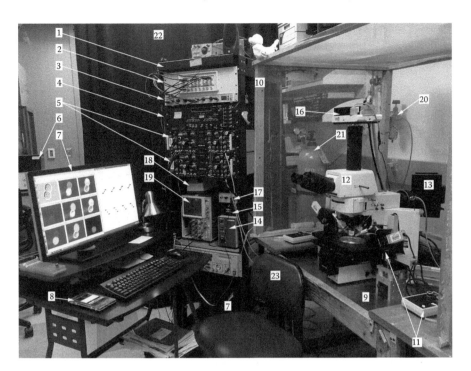

FIGURE 5.1 Basic equipment to measure the permselectivity of gap junctions. 1, rack. 2, Digidata. 3, low-pass filter. 4, timer. 5, patch-clamp amplifiers. 6, pipette puller. 7, computer and screen. Displayed are representative images of NBD transfer on a cell pair (false color) and current traces. 8, electrode storage. 9, antivibration (floating) table connected to the building compressed air supply. 10, handmade Faraday cage. 11, micromanipulator and control pad (right side). 12, inverted microscope. 13, fluorescence lamp house. 14, fluorescence lamp power source. 15, shutter driver. 16, CCD camera. 17, battery-operated pulse generator (controlled by 4). 18, microscope light power source. 19, oscilloscope. 20, compressed air valve (to floating table). 21, emergency nitrogen tank (for floating table). 22, fire-resistant black curtain. 23, antistatic (grounded) chair. The equipment array around the seating should maximize easiness of experiments. Other supplies (e.g., syringes and containers with solutions, distilled water, alcohol, halothane solution, etc.) are in a separate small table facing the rig.

- A surrounding metal or wooden frame built over the table (without contact with the table surface) should accommodate, at a workable height, lateral shelves (armrests) to place manipulator controls, superfusion syringes, and a few other elements. Although armrests can partially hold cables to and from the rack (thus helping to isolate mechanical disturbances from components outside the table), most cables should go straight from the rack to the table, then to the component inside the cage (e.g., microscope, headstages).
- An efficient antivibration device can be built with a large, heavy marble or granite slab standing on tennis balls over a sturdy table.

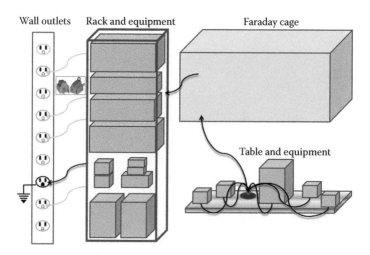

FIGURE 5.2 Proper grounding. A single loop (signaled in darker arrows) must close the electrical circuit of the whole rig (draining the electrical noise to a single outlet or other real ground). Standard three-prong to two-prong adapters (inset) should be used for other outlets.

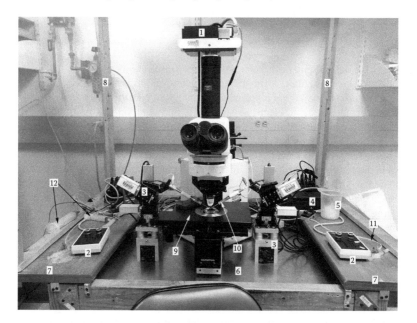

FIGURE 5.3 Manipulation and superfusion setup. 1, CCD camera. 2, micromanipulators control pads. 3, manipulators with four-axis movement (left side) and magnetic bases (right side). 4, manipulator power source/controls. 5, glass disposal. 6, floating table metallic surface. 7, handmade armrests on a wooden table that also sustain the Faraday cage. 8, wood and mesh wire Faraday cage. 9, microscope stage with recording chamber in place. 10, right-side electrode holder. 11, syringe and tubing of pressure system (right side). 12, syringes for superfusion. Notice that the manipulator pads and the tubing suction syringes are at hand reach, but no unnecessary equipment/material is present on the armrests.

5.3.3 FARADAY CAGE

A Faraday cage (Figures 5.1 and 5.3), a cubical enclosure of conductive material covering all parts on the floating table, functions to block static and non-static electrical fields, including electromagnetic radiation (e.g., radio) from the environment; electric charges are collected and redistributed on the whole cage, decreasing the electromagnetic gradients and thus minimizing the interference (electrical noise) in the signals recorded within the enclosure. Grounding the Faraday cage (see Section 5.5.4) will send excess charges to the earth and keep the cage neutral. Metallic Faraday cages can also be bought from TMC (see Section 5.3.2).

5.3.3.1 Hints

- A very good cage can be made by affixing an electrically continuous fine metal wire mesh over a cubical wooden (Figure 5.1) or metallic frame, at least on four sides including the top. The front side of the cage can remain open for easy access to the working space, but having closable front doors will improve the cage efficiency.

5.3.4 MICROSCOPE WITH FLUORESCENCE IMAGING CAPABILITIES

The reader must peruse the manufacturer manuals for microscope features and operation. The following advice is relevant to the proper assembly of the imaging and electrophysiological parts of the experimental setup. Inverted or upright microscopes (e.g., Olympus IX71 or BX50WI, with UV light source) can be used for electrophysiology recordings. Inverted microscopes offer more working space for electrode placing and are easier to use for other applications, but sometimes the electrodes can be difficult to find with their inverted optics and short focus depth. Upright microscopes (Figures 5.1 and 5.3) have a small working space between the cells and the water immersion objectives (Figure 5.4), but the electrodes seem easier to find, and the images obtained without interfering diffractive phases (glass or plastic) are crisper. The right filter (excitation and emission wavelengths) must be selected for the dye(s) to be used. A well-chosen filter can be used for multiple dyes with similar excitation/emission profiles. However, these profiles must be well separated when using two dyes simultaneously. Some microscope models have space for up to six filter cubes in a rotating wheel. The UV light from an arc lamp provides strong illumination; to limit the exposure of cells and decrease phototoxic effects, a shutter driver (e.g., Uniblitz model D122, Vincent Associates, Rochester, New York) controlled by the camera software can open the light pathway only for the time needed for snapshots. A software to handle multiple image file types (e.g., V++, Digital Optics Ltd, Auckland, New Zealand) is required. While newer and better cameras are in the market every season, collecting absolute fluorescence intensity values requires a CCD camera (not a video camera). To decrease stray light collection, the room lights should be turned off during the image taking, particularly if using an upright microscope.

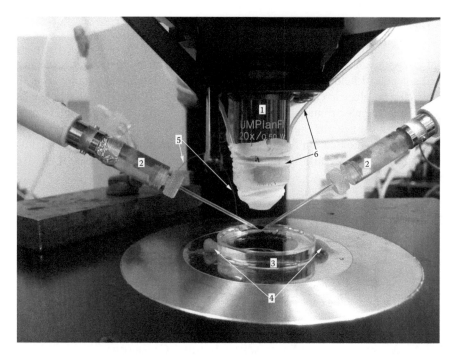

FIGURE 5.4 Close up: working space on an upright microscope stage. 1, water immersion objective. 2, electrode holders with glass patch pipettes in place. 3, cap of 35 mm cell culture dish illustrating recording chamber. 4, beeswax to stabilize the chamber. 5, ground wire. 6, right-side arm of the superfusion system held to the objective with a latex glove finger. The electrodes and the objective will be submerged in the chamber solution, reducing the working space to less than one-half the chamber height. A water immersion objective has longer focal distance when submerged; therefore, the chamber should be initially filled completely (without overflowing) to get the most working space. The holders' inclination angles illustrated allow the electrode tips to reach the field center without contacting the objective or the chamber border.

5.3.4.1   Hints

- During dual dye diffusion studies, manual rotation of the filter wheel may be necessary. To decrease the mechanical disturbance caused by swapping filter sets, a microscope company technician can adjust (loosen) the wheel settings, so that the filters fall in place softly rather than with a click. Strong clicks from the shutter or the camera can also be transmitted to the microscope stage.
- A dark fire-resistant curtain (Figure 5.1) can reduce the exterior light from glass doors or windows. The Faraday cage can also be wrapped with a dark cloth or other flexible materials, preferentially fire resistant (according to university regulations).

5.3.5   Voltage (Patch) Clamp Amplifiers

To measure g_j, continuous (e.g., Axopatch-2D, Molecular Devices, Sunnyvale, California) or discontinuous (switching clamp; e.g., SEC05LX, npi electronic GmbH,

Tamm, Germany) (Figure 5.1) voltage clamp amplifiers can be used. Compatibility between different systems can be achieved by using the same type of analog–digital converters and computer software (i.e., Digidata series and pClamp, Molecular Devices) (Figure 5.1) to collect and analyze data from different setups. Independent of the system(s) available, accurately measuring g_j with a single-electrode voltage clamp is essential, and all effort must be made to obtain clean cell access (see Section 5.5.7.3). Pipette holders can receive either 1.0 or 1.2 mm glass; likewise, headstages may have either a 1.0 or a 1.2 mm pin size hole to receive the pipette holder's connecting pin. In our laboratory, PPH-IP-BNC (ALA Instruments) and QSW-A12P (Warner Instrument) holders have been in use.

5.3.6 MICROMANIPULATORS

There are several types of manipulators. Whether a classical hydraulic (e.g., three-axis WR-6, Narishige International, Long Island, New York), a mechanically based (e.g., four-axis motorized MX7600, Siskiyou, Berkeley, California) (Figure 5.3), or a newer (Takanome™ MTK-1 four-axis hanging joystick oil hydraulic, Narishige) system is preferred, the absolute requirements for patch-clamp work are that manipulators provide for both coarse and fine-motion control and electrode stability. Protect all manipulator parts (particularly any control pad) from liquid spills, as salts from solutions may affect their mechanism.

5.3.6.1 Hints

- The manipulators should be initially set at the middle of each motion range and the electrode holders placed at an angle and at a distance that will allow the electrode tips to arrive at the cells without traveling to the limit of a motion range.

5.3.7 COMPATIBLE COMPUTER(S)

Although smaller and faster computers become continually available, the selected system must accommodate the hardware and the software needed (Figure 5.1). Notice, for instance, that some digitizers (Digidata 1322 and older) and graphics programs use standard size peripheral component interconnect (PCI) cards requiring full-height PCI slots, which may not be available on the newer, slim computers; also, pClamp8 will not run on Windows 7 and later operating systems. In contrast, Digidata 1550 and pClamp10 can use built-in USB ports and run on Windows 7 (but not the Home edition).

5.3.8 PIPETTE PULLER

Consistency of the pipette's tip shape and electrical resistance (R_{EL}) is a strong determinant of success in patch-clamp work. Electrode pullers (e.g., P87 or P1000, Sutter Instrument, Novato, California) require little maintenance, and pull protocols can work for periods of months. However, electrode variability can arise from temperature and humidity changes, and more often from wear and deformation of the heating filament or eccentricity of the microelectrode glass within it, all resulting in inhomogeneous heat application. Manufacturer manuals (e.g., *Sutter's Pipette Cookbook 2015 P-97 and P-1000 Micropipette Pullers*) must be referred to for corrective actions and for detailed pipette-making instructions. In our experience, the best electrodes for dual dye injection are those with resistances between 20 and 40 MΩ ("sharp" electrodes, but not microelectrodes) with a gradual but not too long taper, more akin to a bee stinger (see *Sutter's Pipette Cookbook*). For patch-clamp work, "blunt" electrodes or micropipettes are best between 3 and 11 MΩ (with higher R_{EL} favoring a gigaohm seal formation; very best ~7 MΩ), with relatively short taper and slightly barrel-like tip, features that seem to decrease the sealing over of the ruptured membrane borders. Instructions for short-tapered electrodes from particular types of glass can be found in the mentioned puller manual, but in general several (3–4) pulls of 90–100% of the glass-melting temperature (heat setting), rather than a single pull with higher temperature, work well.

5.3.8.1 Hints

- A barrel-like tip can be obtained by polishing low resistance (3–4 MΩ) electrodes, which is done by placing the electrode tip back into the heating element and applying ~60–80% of the glass-melting temperature for 1–2 seconds: the glass smoothing and slight reduction of the tip opening thus obtained promote membrane adhesion and faster gigaohm seal formation, respectively.
- Of note, polishing is not an absolute requirement, and excellent seals can be obtained with R_{EL} as low as 1–2 MΩ if the cells are large enough and other conditions (e.g., cleanliness, see Section 5.5.7.2.1) are met; on the other hand, excessively long or high heating polishing will completely occlude the tip, rendering the electrode useless.

5.4 MATERIALS

This section lists the consumables (not permanent elements) required for the experiments described here, and some indications pertinent to their proper use. It is important to remember that substitutions may be done according to the research preferences or the requirements of the planned study.

5.4.1 MICROELECTRODE GLASS AND SILVER WIRE

1. The choice of capillary glass length, thickness, and hardness depends on experimental preparation and researcher preferences. Borosilicate (soft) glass with inner filament that facilitates pipette filling by capillary action is usually favored for isolated cell work.
2. Microelectrodes suitable for dye injection can be made with thin-wall glass 4 inches in length and 1.0 mm outer diameter (OD) × 0.75 mm inner diameter (ID) (Cat # 615000, A-M Systems, Sequim, Washington) or similar dimensions from other sources.
3. Patch-clamp pipettes (Figure 5.4) can be made from 4 inches, 1.2 mm OD × 0.68 mm ID (Cat # 602000, A-M Systems), 10 cm, 1.2 mm OD × 0.60 mm ID (Item # BF120–60–10, Sutter Instrument), or similar from other sources.

4. Silver (Ag) wire is used to manufacture electrode and ground wires; a diameter fitting well within the capillary glass, not too thin or too thick (e.g., from 203.20 µm [0.008 inches], Cat # 782000, up to 381.000 µm [0.015 inches]), Cat # 783000, A-M system) should be chosen.

5. All silver surfaces that will come in contact with the solutions must have a uniform coat of silver chloride (AgCl) to avoid polarization of electrodes and baseline drifts.

5.4.1.1 Hints

- Occasionally, visually inspect a capillary glass sample under a stereoscope to confirm cleanliness. Although infrequently, batches may contain contaminant particles that will impair gigaohm seal formation. Keep the capillary glasses in their original containers protected from dust, and wash hands before handling.

5.4.2 CHEMICALS: HOW TO MAKE EXTERNAL AND INTERNAL SOLUTIONS

1. To prepare physiological solutions resembling extracellular and intracellular ionic compositions, the amount of each chemical in grams per liter is calculated as $g/L = (mM \times MW)/1000$, where mM is the desired millimolar concentration, MW is the chemical molecular weight, and 1000 is the conversion factor from molar to millimolar.

2. To ease the solution making, make a list of chemicals as in Tables 5.1 and 5.2, which illustrate our most used formulas for 2, 1, or 0.5 L.

TABLE 5.1
NaCl External Solution (142.5 mM)

Chemical	mM	MW	g/1 L	g/2 L	Actual
NaCl	142.5	58.4430	8.3281	16.6563	
KCl	4.0	74.5513	0.2982	0.5964	
MgCl$_2$[a]	1.0	203.2914	0.2033	0.4066	
Glucose	5.0	180.1559	0.9008	1.8016	
Na pyruvate	2.0	110.0000	0.2200	0.4400	
HEPES	10.0	238.3012	2.3830	4.7660	
CsCl	15.0	168.3600	2.5254	5.0508	
TEACl	10.0	165.7000	1.6570	3.3140	
BaCl$_2$[b]	1.0	244.2636	0.2443	0.4885	
CaCl$_2$[b]	1.0	147.0100	0.1470	0.2940	

[a] Hexahydrate.
[b] Dihydrate.

TABLE 5.2

KCl Internal Solution (124 mM)

Chemical	mM	MW	g/1 L	g/0.5 L	Actual
KCl	124.0	74.5513	9.2444	4.6222	
MgCl$_2$[a]	3.0	203.2914	0.6099	0.3049	
Glucose	5.0	180.1559	0.9008	0.4504	
HEPES	9.0	238.3012	2.1447	1.0724	
EGTA	9.0	380.4000	3.4236	1.7118	
CsCl	14.0	168.3600	2.3570	1.1785	
TEACl	9.0	165.7000	1.1913	0.7457	
Na$_2$ ATP	5.0	551.1000	2.7555	1.3778	
CaCl$_2$[b]	0.50	147.0100	0.0735	0.0368	

[a] Hexahydrate.
[b] Dihydrate.

3. Track the accuracy step by step by writing the actual amount of the chemical weighed and added to the nanopure water (~80% of final volume) continuously stirred in a suitably sized Erlenmeyer flask.
4. Notice that both our external and internal solutions contain K$^+$ channel blockers (tetraethylammonium [TEA] Cl and CsCl) that will reduce membrane permeability and polarization, thus helping to decrease the electrical noise. However, notice that Cs$^+$ permeates well gap junction channels.
5. If membrane depolarization is not desirable, all K$^+$ channel blockers can be dispensed with and their contribution to the solution molarity kept by exactly increasing the main salt (NaCl or KCl); conversely, the molarity accrued by additional elements should be subtracted from the main ions.
6. Our solutions generally set on the acid range after all salts have dissolved, and must be adjusted to pH = 7.2 (or other designed value) using concentrated (e.g., 3 M, 10 N, 1 N) stock solutions of NaOH or KOH, or other stock solution as appropriate for the main ion of choice.
7. Add the concentrated alkaline salt solution by measured, repeated dripping while continuously monitoring the pH. If unintentionally set above 7.2, the pH can be readjusted with 1 N HCl as necessary.
8. While still monitoring (and adjusting as needed) the pH, add nanopure water until it is near to 90–95% of the desired volume and determine the osmolarity (refer to the manufacturer manual for the proper use of osmometer [e.g., Vapor Pressure Osmometer Vapro 5520, Wesco Inc., Logan, Utah]).
9. Carefully continue the dilution in a stepwise fashion until the osmolarity is within ~300–330 mOsm. Our solutions are preferably set as close as possible to 315 mOsm.
10. With aseptic technique (e.g., cell culture hood, gloves), vacuum filter each solution through a bottle-top filter (e.g., Nalgene Rapid-Flow, 0.2 uM pore

size filter, Cat# 291–4520, Thermo Fisher Scientific Inc.) and aliquot them into several (250 or 1000 mL as preferred) sterile bottles.

11. Firmly close each bottle and wrap the gap between the cap and the bottle-neck with Parafilm® (Bemis Company, Inc., Oshkosh, Wisconsin) to help prevent evaporation.

12. Clearly label each bottle with the solution's name, pH, osmolarity, date, your initials, and other pertinent observations (e.g., cells to be used with). Keep the bottles at 4°C for frequent use and at −20°C for long-term storage.

5.4.2.1 Hints

- Due to the difficulty of disposal, $BaCl_2$ was recently eliminated from our formulation with no apparent change in recording quality.
- Accurately adjusting the pH of physiological solutions using concentrated (alkaline/acid) salt stock solutions becomes more challenging when approaching the desired pH value, as it is easy to overshoot, that is, to set the pH beyond the designed value. To avoid this, a 5- to 10-fold dilution of the concentrated salt solution made in a chemical weighing boat can be used to continue the adjustment when the pH is within 0.5 units of the goal value. Be patient.
- The pH is not changed by the addition of nanopure water, but might be lowered by rinsing the salts that had remained on the flask neck.
- The osmolarity may also be arduous to adjust. Ideally, external and internal solutions must be accurately matched, but it is acceptable if the internal solution has a slightly lower osmolarity than the external solution, as this will prevent cell swelling.

5.4.3 FLUORESCENT DYES: WHAT IS AVAILABLE?

1. Few positively charged but many negatively charged fluorescent dyes are available to study the gap junction permeability in living cells. Accordingly, one cationic (N,N,N-trimethyl-2-[methyl(7-nitrobenzo[c][1,2,5]oxadiazol-4-yl)amino]ethanaminium iodide [NBD-m-TMA]) (Aavula et al. 2006) and several anionic fluorophores, including calcein, Lucifer yellow (LY), and some of the Alexa Fluor series (Molecular Probes®-Life Technologies, Grand Island, New York) are familiar to us.

2. The experiments described here require dyes that readily permeate the (particular connexin) gap junction under study. For instance, Cx43 channels do not display charge selectivity and may allow the diffusion of molecules approaching minor diameters of 1.0 nm (Valiunas et al. 2005). In contrast, Cx40 and Cx37 display cation selectivity and low permeability to probes with minor diameters <0.45 nm (Ek-Vitorin and Burt 2005; Heyman et al. 2009).

3. Adding an impermeant dye (e.g., Rhodamine-labeled dextran 3000 Da, Molecular Probes) that will stay in the injected cell helps to identify nonjunctional dye diffusion occurring between two emerging daughter cells of an incomplete mitosis, i.e., a cytoplasmic bridge. Kariokinesis usually produces

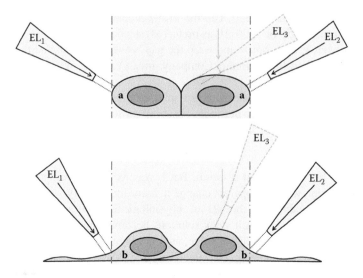

FIGURE 5.5 Approaching cells. (a) Round (chubby) or (b) flattened (sunny side-up eggs) cells can be approached at such angles (black dotted lines) that the circumferential plane of the electrode tip is near parallel (or the axis of the electrodes, black arrows in EL_1 and EL_2, is near perpendicular) to an amenable patch of membrane; this allows the membrane to completely cover the tip upon contact and facilitate gigaohm seal formation; in addition, the peripheral position of the electrodes will interfere less with the cell fluorescence (light gray area between the vertical dash and the dotted lines denotes the projection of usable fluorescent image under UV light). In contrast, approaching the cell center (EL_3 and gray dotted lines) is not conducive to good membrane–tip contact, unless the membrane is pushed down vertically (gray arrows), which will often damage the flat cells by pressing them to the chamber bottom and/or will distort the measurements both by the presence of electrode fluorescence and by the compression of cell volume.

cells of equal volume; thus, when selecting pairs of equally sized cells, dye transfer through cytoplasmic bridges is observed with some frequency.

4. Make stocks of selected dyes (e.g., 5 mM NBD-m-TMA, 5 mM Alexa350, 10 mg/mL Rhodamine-dextran 3000 [Rhodex3000]) by diluting them in internal solution, aliquot and store at −20°C in dark containers and boxes, where their fluorescence efficiency remains constant for several months.

5. For dual dye injection with high resistance (sharp) electrodes, combine two permeant dyes at 1:1 ratio (or dilution factor at 1:2; e.g., 10 uL of 5 mM each of NBD-m-TMA and Alexa350 or Alexa350 and Alexa480) in a dark plastic vial to achieve a final concentration of 2.5 mM each; add Rhodex3000 (0.5–1.0 uL of 10 mg/mL stock) to a final concentration of around 0.25 or 0.5 mg/mL.

6. For patch-clamp-style dye injection with low resistance (blunt) electrodes, dilute the permeant dye in internal solution to a final concentration of 250 µM (or higher if preferred) and add Rhodex3000 to approach anywhere between 0.01 and 0.1 mg/mL. For instance, 10 µL stock NBD-m-TMA + 190 uL of internal solution + 1.0 µL stock Rhodex3000.

5.4.3.1 Hints

- The intracellular brightness of the slowly moving Rhodex3000 seems to increase with time, possibly by cytoplasmic binding, thus clearly labeling the injected cell(s).
- Rhodex3000 signal may appear low if little of it was loaded, the membrane patch was incompletely broken, the membrane clogs the electrode (increased series resistance), or the dye fluorescence was bleached in storage.
- The 3000 MW dextran preparations as provided by the manufacturer contain polymers ranging from 1500 to 3000 Da (see material description provided by Molecular Probes). In addition, some age-dependent breakdown of Rhodex3000 may occur, increasing the ratio of smaller size dextran molecules and/or free rhodamine. Thus, a small percentage of rhodamine fluorescence might be seen diffusing through gap junction channels; particularly with very highly coupled cells, this might go beyond the expected 1–5% of the fluorescent intensity of the injected cell. However, Rhodex3000 diffusion through cytoplasmic bridge is more dramatic and very clear (see (Ek-Vitorin and Burt 2005).

5.4.4 CELLS: WHERE DO YOU FIND GAP JUNCTIONS?

1. Gap junctions can be studied in many vertebrate cell types, either primary cultures or immortalized lines.
2. The cells should be maintained under sterile conditions at 37°C in a 5% CO_2-humidified incubator and in their recommended medium (e.g., RPMI [Roswell Park Memorial Institute] or DMEM [Sigma-Aldrich, St. Louis, Missouri] supplemented with 10% FBS, 300 µg/mL Penicillin G, and 500 µg/mL Streptomycin, and the adequate inducing/selecting antibiotics) refreshed at least every third day to prevent starving/death.

5.4.4.1 Hints

- RIN cells (Gazdar et al. 1980) are preferentially used by the authors, because these cells lack endogenous connexin (functional) expression (Vozzi et al. 1995) and will therefore express only the exogenous connexin(s) we provide. RIN cells tend to be round (favoring seal formation) and display only modest size variability (a condition for our analysis).
- We have explored the amenability of a number of cell lines (American Type Culture Collection®, Manassas, Virginia) for permselectivity studies. Reliable permselectivity data were collected in normal rat kidney [NRK] cells, although they tend to flatten after 8 hours of plating.
- Endogenous connexins may decrease the junctional permselectivity of Cx43 in the neuroblastoma cell line N2A and may explain an observed anionic selectivity in TPA-treated (mostly uncoupled) epithelial Madin–Darby canine kidney (MDCK) cells; also, obtaining gigaohm seals on MDCK cells has proved difficult, possibly due to their thick glycocalyx (Kovbasnjuk and Spring 2000).

- We have also tested 293T and HeLa cells, with satisfactory results; notice, however, that both lines require handling with extreme caution due to their human origin and the presence of a papiloma virus (HPV18) in the latter. We have also documented frequent endogenous coupling in HeLa cells.

5.5   METHODS

This section, containing the actual procedures to measure the gap junction selectivity, is divided into three parts. First is a list of common preparatory techniques; second is the description of dual dye injection with sharp electrode, and the third is the simultaneous recording of dye and electrical coupling.

5.5.1   STERILE TECHNIQUE

Cell culture work (e.g., passing, transfection, plating for electrophysiology) must be performed under sterile conditions. The reader should consult the abundant literature available on the theme. At a minimum, a microbiological safety cabinet and personal protective equipment (e.g., sterile gloves, laboratory coat) should be used when handling cells, exercising special care to avoid their contamination from bacteria, fungi, and mycoplasma and cross-contamination with other cell lines or clones.

5.5.2   CELL TRANSFECTION

1. Transfections are routinely performed with Lipofectamine® (Invitrogen-Life Sciences) following the manufacturer instructions. For transient transfections, a marker like red or green fluorescent protein (RFP) or GFP(Life Technologies) can be cotransfected with the connexin of interest; RFP/GFP positive cells would have more likely also received the connexin gene.
2. Stable connexin-expressing cells are obtained after transfection, using cell dilution and specific antibiotic selection of connexin-positive cell clones. Although the clone selection requires several weeks, the reliability of stable expression considerably improves the speed of acquisition of electrophysiological data.
3. We use a tetracycline-controlled transcriptional activation (Tet-on plasmid, Clontech Laboratories, Mountain View, California) for connexin mutants that inhibit cell growth (Burt et al. 2008), or otherwise decrease cell survival; the expression of such mutants is induced 24–72 hours before the recordings. Viral technologies show higher transfection efficiency both for transient and stable expressions and are being incorporated into our methodology.

5.5.2.1   Hints

- The emission spectra of GFP and NBD-m-TMA overlap; thus, for permselectivity measurements with NBD-m-TMA (or fluorescein or LY), RFP is preferred. RFP and Rhodex3000 also emit in a similar range. To avoid dye emission overlaps, cells stably transfected with connexins but no fluorescent markers are a better choice.

5.5.3 PLATING CELLS ON COVERSLIPS OR ON CULTURE DISHES

The objective of this procedure is to obtain glass coverslips sparsely populated with cell pairs amenable for dye injection/electrophysiological recordings.

1. Use cells from a nonconfluent (60–80% of the bottom surface covered) 100 mm culture dish and work under sterile conditions.
2. To start lifting the cells from the culture dish, first discard the culture medium (using a Pasteur pipette and a suction system with a vacuum and a bottle trap).
3. Add ~5–6 mL of a Ca^{2+}- and Mg^{2+}-free salt solution (PBS-Deficient or PD [in mM]: 137.56 NaCl, 1.47 KH_2PO_4, 9.55 Na_2HPO_4, 2.7 KCl, and 0.0282 phenol red [~2 mL of 5% stock for 1 L] as pH indicator) and allow it to cover the cells by gently rocking the dish for ~20 seconds, then discard.
4. Add 1–2 mL of trypsin 0.05% solution (prepared by diluting in PD a 0.5% stock, Gibco Cat # 15400–054) to the dish and gently rock it to entirely cover the cells for ~10–20 seconds.
5. Extract the excess trypsin solution (do not dry out cells completely) and store the dish in an incubator (37°C) for 1–5 minutes.
6. Take the cells out of the incubator. If needed, rap the dish against a flat surface to complete the lifting and add 6–8 mL of the culture medium (this will practically stop the trypsin digestion).
7. Thoroughly dislodge the cells still attached to the bottom and stir the medium by repeatedly filling and emptying a 10 mL pipette upon the dish bottom. Pipetting up and down a number of times further separates the cells and improves the likelihood of obtaining cell pairs.
8. Use the pipette to deposit one or two drops of this cell suspension in the center of several previously autoclaved 25 mm coverslips separately placed at the bottom of a 100 mm or several individual 35 mm culture dishes. Add a sufficient medium to cover the cells (100 mL or 2 mL according to the dish).

5.5.3.1 Hints

- The incubation time for trypsin digestion depends on the cell type and their confluence: 293T and RIN cells are easily lifted, but MDCK cells take longer, particularly when highly confluent.
- Rinsing twice and leaving a small amount of PD solution before adding trypsin seem to facilitate trypsin digestion.
- Since transient transfections are usually performed in small (35 mm) culture dishes, the volumes of trypsin and medium used must be scaled down to the smaller dish capacity.
- When very few cells are available for the study, as from the result of a small batch or a low-efficiency transfection, the coverslips can be stored with only a small drop of cells at the center in a well-humidified incubator for a couple of hours (or until the cells adhere to the bottom), and then add the medium to total 10 mL (or 2 mL for 35 mm dishes).

5.5.4   Grounding: To Get the Best Recordings

This is arguably the most important condition to obtain clean and reliable electrical recordings. *To ground* means connecting all electrical equipment (the system) to the earth surface, which closes the electrical circuit. Shunting the current to the earth, preventing it from looping within the system and adversely influencing the signal measurements. All conductive elements (microscope, manipulator bases, etc.) are antennae apt to collect electrical noise, while any equipment using alternating current will bring electromagnetic fields inside the cage. Thus, any recognized source of noise must be grounded, i.e., connected to the real ground through the electrical installation.

1. The grounding circuit must be a single loop (Figure 5.2): all antennas inside the cage should share one grounding point on the table (for instance, a metal piece standing on the table or the armrest with perforations apt to receive alligator clips or banana connectors from multiple grounding cables), and from there to the Faraday cage to the rack to a single outlet in the wall, thus providing a single continuous pathway for stray currents.
2. All other outlets must be used ungrounded with three-prong to two-prong adapters (e.g., AXIS 45086 or Coleman Cable 9901).
3. Most electronic devices screwed to the rack are well grounded by that contact; others, like the UV light source, may require special grounding to the rack.

In the face of persistent noise despite apparent good grounding, suspect hidden electrical loops and reset the system as follows:

1. Dismantle all electrical connections.
2. With a voltmeter, check all your BNC cables for continuity between the two pins and between both the extremes of the grounding shield, as well as for the presence of shunts between the pins and the grounding shield; discard any cable displaying improper continuity (high resistance) or shunt (zero resistance).
3. Select a single grounding outlet and isolate the rest with three-prong to two-prong adapters, as indicated earlier.
4. Connect and turn on your amplifiers and note the root mean square (RMS) noise level in the Axopatch (it should be between 0.3–0.5). Start reassembling the connections one by one while monitoring the RMS noise, until an increase of RMS values identify the offending connection (i.e., the source of the noise) or all the connections have been restored without significant noise increase.
5. A similar procedure is doable for the switching-clamp (SEC05LX) amplifiers, but the noise must be monitored on an oscilloscope or a computer with a continuous trace at the highest available sampling rate from each channel involved, in bridge mode, at an output of 2–5 kHz and a hardware low-pass filter (if available) set at 1–2 kHz.

5.5.4.1   Hints

- Note that (unlike with the Axon amplifiers) the normal function of the SEC (switching clamp) amplifiers may add a high frequency noise to the recordings. When using the current clamp (CC) or the voltage clamp (VC) mode, the switching frequency (SwFx) must be finely adjusted on the front panel of the amplifier to achieve the lowest noise level, and it may be necessary to repeat this adjustment, even more delicately, during the course of an experiment. Because the SwFx values are often above 15 kHz, this adjustment and the output filtering (between 10 and 2 kHz) on the same amplifier can be enough to decrease the inherent amplifier noise. However, a 4- or 8-Bessel low-pass filter is an advisable extra precaution, as changes in access resistance, particularly during junctional uncoupling (from very high to single channel level conductances), may also increase the electrical noise in the condition of unchanged settings. Thus, visualizing single channels may require extra filtering.
- A piece of buried copper or old plumbing running underground has been previously used very effectively to "drain" electrical noise.

5.5.5   Chloriding Electrode and Ground Wires

Various types of Ag/AgCl reference (ground) electrodes are available (see A-M Systems), but a well-chlorided silver wire is a good and renewable choice. Chloriding prevents the electrode polarization due to current-passing and reduces the baseline drift, but keeps R_{EL} sufficiently low to allow the passing of current, and supports a clean recording. To obtain a good AgCl coat (or plate), the reader must do as follows:

1. Remove the old coat with a quick flaming or a very fine sandpaper (or use new wire).
2. Grounding is better if an ample surface contact exists between the ground wire and the external solution. This can be achieved if the length of the wire that will be immersed in the external solution is shaped into a tight coil that will rest on the chamber bottom. To get a regular coil shape, firmly twist the wire around a paper clip (or other similar small cylinder) before coating. If necessary, this tight coil can be flattened, to be completely covered by the chamber solution.
3. Clean the wires with ethanol (to remove grease) and rinse with nanopure (deionized) filtered water to remove the ethanol.
4. The easiest (and quite effective) chloriding method is to immerse the Ag wire in full-strength household bleach (sodium hypochlorite, NaClO) for at least 15 minutes (1–2 hours if possible) or until the wire is purple gray. Do not chloride the wire end that will be in continuity with the ground cables.
5. A more laborious method, electroplating, requires a voltage source to apply a positive current (<0.5 mA) through the wire immersed in NaCl (0.9%) or KCl (3 M) for several minutes, until the purple gray color is seen. Initially

applying a negative current helps in discarding the old coat, and alternating the polarity during the electroplating seems to produce a better coat. In some laboratories, a bigger piece of Ag is used both as ground and AgCl donor during coating.

6. The electrode wires must be kept as straight as possible, and the areas that will be in contact with the internal solution must also be chlorided. Do not chloride the end of the wires that will be in contact with the electrode holder gold pellet.

5.5.5.1 Hints

- Imperfectly chlorided or scratched wires can cause electrical noise and baseline drift. Changes on the chamber solution level that decrease the immersion of the ground wire may also cause baseline drifts.

5.5.6 RECORDING AND ANALYSIS OF JUNCTIONAL DYE PERMEABILITY (CHARGE AND SIZE SELECTIVITY)

5.5.6.1 Overview

Dye injection with a sharp electrode is used to simultaneously explore the permeability of junctions to two dyes introduced in a single bolus (dual dye injection) and define the charge selectivity (with dyes of opposite charge) or size selectivity (with dyes varying in molecular weight/mass/radii).

5.5.6.2 Recording: Step by Step

1. Turn on the UV lamp before placing a coverslip on the recording chamber to allow for (~10–15 minutes) warm up.
2. Cover the upper side of the chamber perforation rim with 100% white petroleum jelly (Vaseline®, Unilever) or silicone grease liberally applied with a ≤10 mL syringe and a blunted needle of sufficiently wide caliber (≤26).
3. Using microsurgery forceps, handle a coverslip by the edge and carefully dry the bottom (noncell side) with low-lint paper tissue (e.g., Kimwipe, Kimtech Science).
4. With the forceps, press the coverslip borders against the Vaseline frame, gently but sufficiently to eliminate water channels through the Vaseline.
5. Rinse the cells with the external solution to eliminate debris and excess grease.
6. Fill out the chamber with the external solution using a syringe or a small beaker and place the chamber in the light path of the microscope stage.
7. Find a suitable cell pair and center it in the view field of a ×40 objective.
8. Dip the back of the electrode in the dye mix in a small vial for 2–5 minutes or until the dyes can be clearly seen filling the electrode tip/shank.
9. Using a 2 mL syringe and a fine needle (metallic: metal hub 26–33 gauge, Hamilton, Reno, Nevada; plastic: MicroFil 26–31 gauge, World Precision Instruments, Sarasota, Florida), fill the electrode to half or two thirds of its

length with the internal solution or high molar KCl (as preferred) and place in the electrode holder; notice that the displacement of the internal solution by the electrode wire will complete the electrode filling.

10. Move the electrode toward the cells.

11. While applying short (e.g., 60–100 mS) small pulses (e.g., 100–200 pA in CC mode or 200 ms, ±10 mV in VC mode) several times per second (e.g., 2–5 Hz) and monitoring the resistance changes (both electrically and by the amplifier audio monitor), further lower the electrode toward one cell (the donor) of the selected pair until any or all the following changes occur: higher sound pitch, increased resistance or noise in the pulse traces, dimple in the cell surface.

12. Notice that for the dye injection with a sharp electrode, the membrane must be pierced. This is accomplished by slightly pushing the electrode into the membrane and applying one or several electrical discharges (zaps/buzzes) that will put the tip inside the cell (the zap strength is set high for Axopatch; buzz, to modest or low levels for SEC0.5LX amplifier) and simultaneously initiate the dye injection.

13. If needed, capacitance overcompensation (ringing) can be used to inject more dye.

14. When the bright signal indicates abundant dye entering the donor cell, swiftly withdraw the electrode with a fast oblique movement and start taking timely fluorescent images of both the donor and the recipient cell.

15. A suitable scheme is taking fluorescence images of one of the dyes (e.g., NBD) at 0, 10, 20, 30, 40, 50, and 60 seconds during the first minute, then at 2, 3, 5, 7, 13, 17, and 23 minutes, or more if necessary for very low coupling levels. The images from the second dye (e.g., Alexa350) can be taken with a 5-second delay (i.e., at 5, 15, 25, 35, 45, 55, 65 seconds, then 2 minutes and 5 seconds, 3 minutes and 5 seconds, and so on) with respect to every snapshot of the first dye. This requires that (1) the filter wheel be automatically or manually shifted to the appropriate set before every snapshot and (2) an automated program will take the snapshots at the exact time point and for the adequate sampling time for each dye (e.g., Alexa350 is usually brighter than NBD at similar concentrations: 100–50-millisecond sampling of Alexa350 will produce values comparable to 250 ms sampling of NBD-m-TMA).

16. Make sure that the UV light path is closed between the snapshots (to minimize photobleaching) and opens during the snapshots for long enough to collect fluorescence.

5.5.6.2.1 Hints

- On an inverted microscope stage (Figure 5.4), the recording chamber should be completely filled; this allows the use of the longer focal distance that the water immersion objectives have within the liquid and prevents accidentally pushing the electrode against the coverslip. Once the electrode is near the cells, the solution level can be decreased if desired. However, make sure that the grounding wire always remains under the liquid to prevent baseline drifts or loss of electrical continuity.

- Because, in an inverted microscope, cell fluorescence is better observed through glass than through plastic, the cells are preferentially plated in glass coverslips. Commercially available recording chambers accommodate coverslips of various sizes. The caps of 35 mm plastic cell culture dishes, which have a lower height than the dishes themselves and offer more working space, make for excellent recording chambers, but they must be adapted to receive the coverslips. First, a perforation of a smaller diameter than the coverslip should be made in the center of the cap. To do this perforation, step drill bits work best, but dremel conical sanding bits (consult your hardware store) can be used too. Second, the upper rim of this perforation should be covered by a narrow but sufficient amount of Vaseline as described earlier; the coverslip is set upon this border.

- Melting the Vaseline into the syringe will cause it to strongly stick to the syringe wall, becoming practically immovable; use a spatula or other means to fill the syringe.

- Because observing and measuring fluorescence is of the essence, know that paper lint and Vaseline have strong autofluorescence, and they must be absent from the field of view.

- The contact of the electrode wire and the internal solution should be ample. However, overfilling the electrode may easily cause the internal solution to go into the pressure tubing system; the salts remaining in the tubing system may be a source of increased osmolarity of the internal solution in future experiments (cells will swell) or worse, they may feed microorganisms.

- To facilitate finding the electrodes, first use a low-amplification objective and place the electrode tip in the center of the view field; then, use higher amplification and lower your electrodes straight down toward the chamber solution until contact is indicated by the potential readout in the amplifier (value goes nearer zero).

5.5.6.3 Analysis

1. The average fluorescence values from the noncellular area are subtracted at every time point from the average of each cell (total) area. The values thus obtained from the donor and the recipient are plotted and fit with the equation $k = -\ln [(F_{eq} - F_{(t)}/F_{total(t/t0)}/(F_{eq} - F_{(0)})]/T$ (cf. Heyman and Burt 2008), where k is the rate constant, F_{eq} is the total cell fluorescence at equilibrium, $F_{(t)}$ is total cell fluorescence at time t, $F_{(0)}$ is total cell fluorescence at time 0 and T is time (Figure 5.6).

2. A correct analysis requires that identical cell area/volume be measured for both dyes at every time point. This is readily accomplished for early time points when the cells remain quite stable (Figure 5.6).

3. If dye equilibrium is reached in seconds (as it happens with some frequency), enough points for a proper fit may not be acquired, rendering the experiment futile.

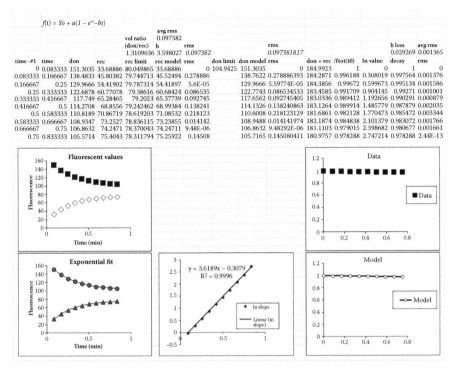

FIGURE 5.6 Illustration of the fitting procedures to obtain a rate constant of transjunctional dye diffusion after a bolus injection with sharp electrode. The several columns include among other values time (real and after adjusting to zero start), donor and recipient fluorescent values (after background subtraction), cell dye content (real and modeled after volume adjustment), logarithm conversion of recipient values and calculated dye loss. For this pair, the rate constant (b) is 3.598 for a donor/recipient volume ratio ~1.31, and the overall dye loss during the time displayed is ~3%. To find the best-calculated values that closely reproduce the experimental measurements, the solver feature of Excel is used. The plots show fluorescent values and exponential fit for the donor and the recipient (left), linearization of exponential function for the recipient values (center), and overall dye loss (right) during the displayed period. Notice that the slope value of the linear fit (~3.61) obtained with the Excel linear fit closely resemble that obtained from the exponential fit with the solve feature.

4. In contrast, very low diffusion rates can be difficult to analyze because the more prolonged the recording, the more likely the cell shape/position may change (due, for instance, to stage displacements when changing the filter, to cell shape changes, or to small displacements of electrode tip (in patch mode).

5. In addition, when working with flat cells, their volume is not always comparable.

6. To address some of these concerns, the fluorescence from whole cell profiles and a factor to account for the volume differences are used in the calculations of the rate constants for dual dye transfer (Heyman and Burt 2008).

7. Notice that after a single bolus injection, a limited amount of dye(s) will distribute in both cells. Under this condition, fluorescence loss can occur by transmembrane leaking (e.g., from the donor cell puncture) and/or by UV exposure (photobleaching). To minimize the effect of these factors on the calculated diffusion rate constant, only experiments where the combined fluorescence loss is <10% from point to point should be used.

5.5.6.3.1 Hints

- Excessive dye loss at start can only be remedied by perfecting the injection technique.
- Dye equilibrium between the cells renders the later points irrelevant, as exponential functions cannot fit the lack of change from point to point. Often, eliminating very early and late recorded points of the dye intercellular diffusion allows for a reasonable fitting of the remaining data points.

5.5.6.4 Limitations and Other Issues

- In dual dye injections, there is no measurement of g_j to normalize the diffusion rates against the estimated channels present. Instead, the diffusion rates are compared with one another.
- Only a fraction of attempted sharp electrode injections will be amenable to analysis and produce reliable data.
- Nevertheless, provided that the junctions are permeable to the selected dyes, the relative simplicity and brevity of these experiments makes them highly valuable to study gap junction channel function.
- The results using this method indicate that the junctions formed by Cx43 are size selective, but not charge selective.

5.5.7 RECORDING AND ANALYSIS OF JUNCTIONAL DYE/ELECTRICAL PERMSELECTIVITY

5.5.7.1 Overview

Dye injection with a patch pipette is required for assessing junctional permselectivity, a term we use to particularly designate the correlation of dye and electrical coupling when they are concurrently measured. The advantage of this approach is that the transjunctional dye diffusion is correlated with a direct estimate of the number of channels present at that particular junction.

5.5.7.2 Recording: Step by Step

1. For dye/g_j measurements, the whole patch clamp configuration must be achieved on each cell with speed and ease.
2. Prepare your cells as described earlier.
3. Directly fill one electrode of the pair with the dye mixture using a syringe and a needle (as mentioned earlier) and place in electrode holder.
4. Fill the second electrode with an internal solution devoid of dyes.

5. Lower both electrodes toward the cells (as explained earlier) while keeping a small positive pressure on the tips (for instance, push ~50–100 uL volume of air from a 1 or 3 mL syringe into the pressure controlling tubing; stronger pressure could be used to quickly clean the tips, but not when getting closer to the cells).

6. Notice that, unlike with sharp electrodes, the membranes are not pierced by the blunt pipettes used in a patch clamp. Rather, cell access is gained by first making a segment of the membrane go into and stick to the interior of the glass pipette and then breaking the top patch of that membrane (a dome that has not stuck to the pipette).

7. For the dye coupling part of the experiment, attempt first to obtain a gigaohm seal with the dye-containing pipette. While applying repeated short, small pulses at a continuous rate (as described earlier), approach the cell (Figure 5.5) while gradually decreasing the speed of the movements in the micromanipulator controller until a change of pitch (in the sound monitor) and/or an increase of R_{EL} (in the electrical traces) and/or a dimple in the cell membrane surface occur.

8. Release the positive pressure, and (1) if you were close enough to the cell, the membrane will immediately contact the electrode tip, and a stronger change of pitch and increase of R_{EL} will be noticed. Then, apply a slight suction with the syringe (the equivalent of 100–200 uL volume) and allow the seal to complete. (2) If you are not close enough to the cell, there will be no clear changes. Reapply positive pressure and push the electrode tip closer to cell. (3) Another possibility is that the electrode tip is clogged and a seal cannot be achieved (see Section 5.5.7.2.1 Hints).

9. In CC mode, the voltage deflections will increase (up to 300 mV or more), and in VC mode, the current deflections will decrease (to almost 0 pA) as a gigaohm seal develops.

10. Once the membrane starts rapidly sticking to the (inside) electrode glass, releasing suction in small steps will further accelerate the formation of the seal. With a three-way valve between the syringe and the pressure system tubing (running to the pipette holder), a gradual suction release is achieved by first shunting the syringe and the exterior and then the syringe and the tubing; this will divide the original negative pressure, partially decreasing the stress on the cell membrane and allowing it to further stick to the glass.

11. In ideal conditions (including the absolute cleanliness of the system, as noted Section 5.5.7.2.1 Hints), a very fast resistance increase occurs at the very start; then, the sudden release of all suction will immediately collapse the membrane within the electrode against the glass surface, thoroughly completing a firm gigaohm seal in a few seconds.

12. Once a gigaohm seal has formed, break into the cell with increasing suction or with the smallest effective zap/buzz. A combination of light suction and buzz is often effective.

13. When the dye starts diffusing from the pipette to the (donor) cell, immediately start taking timely fluorescent images (as indicated earlier) and store for future analysis.

Model to fit data for dye spread from one cell to a second cell
TWS, May 2003　　Version using measured donor cell data
With RMS deviation computed. Minimize by varying k2.

Parameters
$V2$ = 1
$k2$ = 0.029683

ek 130130 time (min)	Experiment donor	recipient	Model recipient	Square deviation
0	24.88811	1.972025	1.972025	
0.083333	35.05827	2.032634	2.041203	7.34E-05
0.166667	46.73427	2.335668	2.137195	0.039391
0.25	58.86014	2.149182	2.262352	0.012808
0.333333	74.17016	2.717944	2.42109	0.088122
0.416667	88.51981	2.622377	2.616074	3.97E-05
0.5	102.8555	2.703965	2.846009	0.020176
0.583333	121.8648	2.9021	3.116566	0.045996
0.666667	139.2587	3.648015	3.431423	0.046912
0.75	156.0326	3.73893	3.787708	0.002379
0.833333	170.8089	4.100233	4.182085	0.0067
0.916667	185.3823	4.258746	4.611742	0.124606
1	198.8485	4.885777	5.074975	0.035796
2	347.4872	12.13753	12.91638	0.606616
3	439.4545	23.24709	24.04717	0.64013
5	505.1772	50.39394	49.8919	0.252047
7	536.7762	74.84383	77.05208	4.876377
10	564.4476	118.965	117.4245	2.373306
			RMS	0.734506

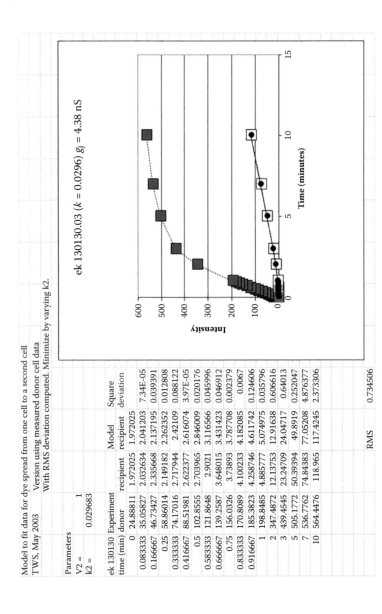

ek 130130.03 (k = 0.0296) g_j = 4.38 nS

FIGURE 5.7　Illustration of the fitting procedures to obtain a rate constant of transjunctional dye diffusion on (whole cell) patch-clamp style injection. The columns show time, donor and recipient fluorescent values (after background subtraction), recipient values predicted by the model, and squared deviation from the real values. The fit of the recipient fluorescent values was obtained with the Excel solver feature. The plot shows donor (dark squares), recipient (white squares), and model-predicted (black dots) fluorescent values with a rate constant (k) of −0.0296 for a cell pair with g_j = 4.38 nS.

14. Next, attempt a seal on the partner (recipient) cell with the second (no dye-containing) electrode when the fluorescence is equal in both cells, or after a reasonable period with little or no dye transfer to the recipient cell. Notice that for Cx43 gap junctions, the dye can equilibrate in a few seconds or after >20 minutes. Cx37 gap junctions generally show lower diffusion rates.

15. Take an image of Rhodex3000 signal after several minutes of the permeant dye diffusion or at the end of the recording.

16. The seals can be simultaneously initiated on the donor and the recipient, but notice that breaking into the recipient cell will lower the local concentration of dye, thus decreasing the apparent diffusion rate.

17. With complete access to both cells complete and in VC mode, the g_j can be measured for several minutes (see Chapter 4 for more specific details). In our laboratory, we only use small amplitude, short V_j pulses to obtain a measure of the maximum g_j (near $V_j = 0\,mV$), that is, unaffected by V_j-induced channel closure. For Cx43, for instance, 5-second long $V_j = \pm10\,mV$ pulses, separated by 1-second resting period, and applied every 20–30 seconds, work well. For gap junction channels that display a strong V_j dependence, shorter, smaller, and less frequent pulses may be appropriate.

18. With impeccable technique, the opportunity to uncouple the cell pair and explore the junctional channel activity (e.g., conductance, behavior under changing V_j values, open duration) may present.

19. Depending on the rate of success and any given treatment, one to three attempts can be made in a single coverslip; thus, several coverslips must be available for these experiments.

5.5.7.2.1 Hints

- For these procedures to work well, extreme cleanliness of the system (including internal solution, syringes, filters, tubing, holders, and electrode glass) is required.
- Many things can clog the electrode tips, including grease (e.g., Vaseline, hair and hand creams), dust and other particles (e.g., rubber from gaskets, Lucite from electrode holders, glass from electrodes, paper fibers) that are easy to accidentally pick up when the filling needle is left exposed; however, the more challenging problem are microorganisms (fungi, bacteria) that can easily grow in the humid, saline environment of an unattended tubing system. Bacteria can be seen as round or oval small droplets sticking inside or outside the electrode. Fungi are most easily recognized by their long filaments (hyphae) inside the electrode.
- Besides the obvious precautions of washing hands with soap and water and avoiding contact of the microfilling needle with any object, additional measures may help increase the likelihood of obtaining stable seals. For instance, degreasing the fingertips with ethanol-sprayed paper before pulling electrodes or using gloves may be advisable.
- Do not use the internal solution from previous days; instead, discard the solution at the end of the day and rinse as much of the system as possible with distilled water, particularly the filling needle.

- If it becomes necessary (as evidenced by persistently unstable, noisy, or short-lived recordings), wash the pressure system tubes and the associated components, including pipette holders, with alcohol (or sometimes 20–50% bleach) and abundant distilled water, and air dry them; alternatively, change the entire system.
- Some signs of pollution (i.e., contamination introduced by researchers into any part of the system—tubing, holders, needles, solutions) are excessive bubble formation in the electrode tips, difficulty to get rid of those bubbles, ineffectiveness of the pressure/suction to arrive to the electrode tip, need of strong suction to initiate the seals, persistent leak currents despite apparent good seal formation and good cell access, and ultimately, lack of seal formation.

5.5.7.3 Special Issue: Series Resistance

To get reliable patch clamp g_j measurements, several conditions must be met; obtaining a good seal, clean access to both cells and no significant current membrane (I_m) leak are basic. However, even in the best recording conditions, electrode series (or access) resistance must always be taken in account. To review the electrical concepts involved in voltage (patch) clamp, the reader is referred to specialized texts (including the *Axon Manual by Molecular Devices*). Nevertheless, a brief overview is offered next.

When a current (I) passes through a resistor (R), a voltage develops and can be measured between the two extremes (source and ground) of the resistor. This voltage increases with increasing values of the resistor and of the applied current, according to Ohm's Law ($V = RI$). If the applied current is kept constant, the value of the voltage between source and zero (ground) is directly proportional only to the resistor value (e.g., $1\,mA \times 1\Omega = 1\,mV$; $1\,mA \times 2\Omega = 2\,mV$; $1\,mA \times 5\Omega = 5\,mV$). In an electrical circuit that contains several resistances in series, the total voltage that is measured from the source point to the ground is distributed on every serial resistance proportionally to each resistance value: the higher the resistance, the higher the proportion of voltage drop. For instance, for two resistance in series A = 2Ω and B = 5Ω, the total resistance is 7Ω, and the total voltage developed by 1 mA along both is $7\,V = 1\,mA \times 7\Omega$; however, the voltage measured along the 2Ω and 5Ω resistors will be $2\,mV$ and $5\,mV$, respectively. In summary, it is expected that a proportion of any voltage from the amplifier to the cell will be lost (it will drop) at the electrode (series) resistance. In Axopatch amplifiers, the resistance of the junction should be at least 10-fold higher than that of the electrodes, for the voltage drop to occur mostly (at least 90%) at the junction.

Conductance is the reciprocal of resistance ($G = 1/R$). The conductance value of $10\,M\Omega$ electrodes is $100\,nS$, so they can reliable clamp a junction with g_j of ~$10\,nS$ (a 10-fold lower conductance). Likewise, $5\,M\Omega$ electrodes (conductance = $200\,nS$) can reliably clamp junctions with g_j ~$20\,nS$. The electrode resistance compensation by the amplifier decreases the calculated voltage loss at the electrode (due to its access resistance). However, further calculations for g_j >20 nS may be needed.

SEC0.5LX amplifiers obviate in good part the need for series resistance compensation, because the actual voltage at the tip of the electrode is simultaneously measured during the clamping (that being the nature of switching clamp amplifiers [cf. Ek-Vitorin and Burt 2005]). Nevertheless, when two similarly powerful amplifiers are trying to clamp the opposite sides of a highly conductive junction at different voltages, there is no absolute certainty that the voltage is not dropping between the electrode tip (the point of measurement) and the junction. In summary, the assessment of very high g_j values should be scrupulously considered and may require some adjustment.

5.5.7.4   Analysis

1. The fluorescent values from both cells, obtained by subtracting a noncellular area (background) from equivalent brightest areas on each cell, are plotted as a function of time. The recipient values are fit with a numerical method (stated immediately below (in the next paragraph) and illustrated in Figure 5.7) to calculate the diffusion rate constant (k) between the donor and the recipient.
2. The rate constants (k) are obtained by fitting each time (t) point with the equation: $C_2(t + \Delta t) = C_2(t) + k\Delta t/2[C_1(t) + C_1(t + \Delta t) - C_2(t) - C_2(t + \Delta t)]$, where Δt is the time between points, C_1 is the donor, and C_2 is the recipient fluorescence (cf. Ek-Vitorin and Burt 2005; Ek-Vitorin et al. 2006).
3. Using the solver feature in Excel (Figure 5.7), k is varied to minimize the RMS between the predicted and experimental values of $C_2(t)$. k from each junction is plotted as a function of the corresponding electrical conductance value, calculated as $g_j = I_j/V_j$.
4. The analysis of the transjunctional dye coupling with the whole cell patch clamp style injection is based on the assumptions that (1) the diffusion occurs between two equal compartments and is contingent upon the coupling level and the instantaneous dye gradient between them; (2) the pipette is a virtually inexhaustible source of dye for the duration of the recording; (3) the donor cell dye concentration plateaus toward that of the pipette, followed by the recipient with a delay dependent on the coupling level; (4) no significant changes in the channel function occur during the entire recording period, unless induced by the researcher (Ek-Vitorin and Burt 2005).
5. If permselectivity were constant, a plot of k versus g_j for multiple junctions would describe a linear relationship that represents the signature of a connexin type. At least for Cx43 that has not been the case, indicating that channels in different functional states (some more and some less permeable to the dyes compared to current-carrying ions) exist in the junction (Ek-Vitorin et al. 2006).

5.5.7.4.1   Hints

- As with the single bolus injection (with sharp electrode), in the patch-clamp style dye injection, the dye diffusion between donor and recipient cells is sometimes very fast, and no difference can be discerned between the fluorescence intensity of the cells from time point to time point, even as the fluorescence keeps increasing in both cells, until a maximum (indicating equilibrium with the pipette concentration) is observed. Although these experiments serve to confirm the high degree of metabolic coupling that can be achieved by junctional channels (at least for Cx43), they cannot be analyzed with the mathematical method described here, which requires a difference between the donor and the recipient. These experiments must be discarded.

- In contrast, dye diffusion may be very slow, and the fluorescence of the recipient cell remains below detectable (over the background) levels at least during the first 1–2 minutes of recording. These experiments may need to be prolonged over 15 minutes, and yet equilibrium between the donor and the recipient will not occur during the observation period. However, a reliable estimation of the diffusion rate can be obtained with the measurable recipient fluorescence during the latest time points. Interestingly, at least for Cx43, very low dye coupling is not always linked to a low electrical coupling.

- In our experience, whether the whole cell area or a fraction is used in the calculations, the rate constants are comparable. However, cell shrinking (due to excessive pipette suction) or swelling (due to osmotic pressure), as well as slight cell drifts on the field of view, will affect the calculations.

5.5.7.5 Limitations and Other Issues

- As with the sharp electrode injections, extremely fast or slow transjunctional diffusions rarely produce usable data.

- In contrast to injections with sharp electrodes, no dye leak should occur from cells once a gigaohm seal and eventually the whole cell mode is achieved. However, the dye will escape from the electrode tip while approaching the cell; the resulting increase on the background brightness can be decreased by exchanging the solution (with superfusion) right before breaking into the donor cell.

- Despite the more difficult and time-consuming process of performing patch-clamp style dye transfer experiments, their analysis is appealingly simpler if certain conditions are met.

- The equipment and the experimentalist dexterity to quickly achieve the whole cell configuration must be dependable, for the shortest delay between the dye and the electrical coupling assessment.

- Those pairs of healthy and well-maintained cells (with minimum debris) that appear very similar in shape and size and completely separated from any other should be used, to fulfill the prerequisite of diffusion between two equal compartments.

- Recordings should be performed within ≤2.5 hours of taking the cells out of the incubator, and refreshing the external solution should be easy to accomplish while the chamber is in place.

5.5.8 Final Notes

Recording junctional permselectivity (particularly the micromanipulation and long hours looking through a microscope) can be a tiring task. The reader is strongly advised not to start an experiment until a reasonable high level of comfort for a good body posture and freedom of movement is felt. "Reasonable comfort" may vary from person to person. However, following some rules can minimize the physical stress, if only to achieve a good productivity while avoiding excessive fatigue. Thus,

1. Use a good swivel chair (where you can sit stably and comfortably and turn around as needed to manipulate all devices), but *not* wheeled, as the instability caused by unexpected sliding may interfere with micromanipulator handling; a lockable wheel system can be useful;
2. Have everything you need at a reachable distance from your seating position (Figure 5.1), and the armrests and the microscope oculars at a height that allows a correct seating;
3. Do not have on your armrests and other work surfaces anything that is not absolutely essential to perform your work (Figure 5.1 and 5.3); exclude empty bottles, paper, pipettes, tubes, wires, or any other equipment/tool that is not under immediate use;
4. Be sure that the whole rig, including software, computer, and illumination devices are in working condition before starting an experiment. Good luck.

ACKNOWLEDGMENTS

The author is deeply grateful to Dr. Janis M. Burt for providing a superb work environment that allowed for the refinement of these techniques, to Tasha Pontifex for her excellent work (Cell and Molecular techniques) and support, and to the innumerable students that have offered me the opportunity of continuous learning.

REFERENCES

Aavula BR, Ahad Ali M, Mash EA, Bednarczyk D, and Wright SH. 2006. Synthesis and fluorescence of N, N, N-Trimethyl-2-[methyl(7-nitrobenzo[c][1,2,5]oxadiazol-4-yl) amino] ethanaminium Iodide: A pH-insensitive reporter of organic cation transport. *Synth Commun* 36 (6): 701–705.

Beblo DA, and Veenstra RD. 1997. Monovalent cation permeation through the connexin40 gap junction channel Cs, Rb, K, Na, Li, TEA, TMA, TBA, and effects of anions Br, Cl, F, acetate, aspartate, glutamate, and NO_3. *J Gen Physiol* 109 (4): 509–522.

Bruzzone R, White TW, and Paul DL. 1996, May 15. Connections with connexins: The molecular basis of direct intercellular signaling. *Eur J Biochem* 238 (1): 1–27.

Bukauskas FF, Bukauskiene A, and Verselis VK. 2002. Conductance and permeability of the residual state of connexin43 gap junction channels. *J Gen Physiol* 119 (2): 171–186.

Burt JM, Nelson TK, Simon AM, and Fang JS. 2008. Connexin 37 profoundly slows cell cycle progression in rat insulinoma cells. *Am J Physiol Cell Physiol* 295 (5): C1103–C1112.

Eckert R. 2006. Gap-junctional single channel permeability for fluorescent tracers in mammalian cell cultures. *Biophys J* 91: 565–579.

Ek-Vitorin JF, and Burt JM. 2005. Quantification of gap junction selectivity. *Am J Physiol Cell Physiol* 289 (6): C1535–C1546.

Ek-Vitorin JF, and Burt JM. 2013. Structural basis for the selective permeability of channels made of communicating junction proteins. *Biochim Biophys Acta* 1828 (1): 51–68.

Ek-Vitorin JF, King TJ, Heyman NS, Lampe PD, and Burt JM. 2006. Selectivity of connexin 43 channels is regulated through protein kinase C-dependent phosphorylation. *Circ Res* 98 (12): 1498–1505.

Ek-Vitorin JF, Pontifex TK, and Burt JM. 2016, Jan 5. Determinants of Cx43 channel gating and permeation: The amino terminus. *Biophys J* 110 (1): 127–140.

Elfgang C, Eckert R, Lichtenberg-Frate H, Butterweck A, Traub O, Klein RA, Hulser D, and Willecke K. 1995. Specific permeability and selective formation of gap junction channels in connexin-transfected HeLa cells. *J Cell Biol* 129 (3): 805–817.

Gazdar AF, Chick WL, Oie HK, Sims HL, King DL, Weir GC, and Lauris V. 1980. Continuous, clonal, insulin- and somatostatin-secreting cell lines established from a transplatable rat islet cell tumor. *Cell Biol* 77 (6): 3519–3523.

Goldberg GS, Lampe PD, and Nicholson BJ. 1999. Selective transfer of endogenous metabolites through gap junctions composed of different connexins. *Nat Cell Biol* 1 (7): 457–459.

Gutstein DE, Morley GE, Tamaddon H, Vaidya D, Schneider MD, Chen J, Chien KR, Stuhlmann H, and Fishman GI. 2001, Feb 16. Conduction slowing and sudden arrhythmic death in mice with cardiac-restricted inactivation of connexin43. *Circ Res* 88 (3): 333–339.

Heyman NS, and Burt JM. 2008. Hindered diffusion through an aqueous pore describes invariant dye selectivity of Cx43 junctions. *Biophys J* 94 (3): 840–854.

Heyman NS, Kurjiaka DT, Ek Vitorin JF, and Burt JM. 2009. Regulation of gap junctional charge selectivity in cells coexpressing connexin 40 and connexin 43. *Am J Physiol Heart Circ Physiol* 297 (1): H450–H459.

Hille B. 2001. *Ion Channels of Excitable Membranes*. Sunderland, MA: Sinauer Associates.

Kanaporis G, Mese G, Valiuniene L, White TW, Brink PR, and Valiunas V. 2008. Gap junction channels exhibit connexin-specific permeability to cyclic nucleotides. *J Gen Physiol* 131 (4): 293–305.

Kanno S, Kovacs A, Yamada KA, and Saffitz JE. 2003. Connexin43 as a determinant of myocardial infarct size following coronary occlusion in mice. *J Am Coll Cardiol* 41 (4): 681–686.

Kelsell DP, Dunlop J, Stevens HP, Lench NJ, Liang JN, Parry G, Mueller, RF, and Leigh IM. 1997. Connexin 26 mutations in hereditary non-syndromic sensorineural deafness. *Nature* 387 (6628): 80–83.

Keung EC, Keung CS, and Aronson RS. 1982. Passive electrical properties of normal and hypertrophied rat myocardium. *Am J Physiol* 243 (6): H917–H926.

Kleber AG, and Riegger CB. 1987. Electrical constants of arterially perfused rabbit papillary muscle. *J Physiol* 385: 307–324.

Kovbasnjuk ON, and Spring KR. 2000. The apical membrane glycocalyx of MDCK cells. *J Membr Biol* 176 (1): 19–29.

Mesnil M, Piccoli C, and Yamasaki H. 1997. A tumor suppressor gene, Cx26, also mediates the bystander effect in HeLa cells. *Cancer Res* 57: 2929–2932.

Morley GE, Vaidya D, and Jalife J. 2000, March. Characterization of conduction in the ventricles of normal and heterozygous Cx43 knockout mice using optical mapping. *J Cardiovasc Electrophysiol* 11 (3): 375–377.

Rohr S, Kucera JP, and Kleber AG. 1998. Slow conduction in cardiac tissue: I: Effects of a reduction of excitability versus a reduction of electrical coupling on microconduction. *Circ Res* 83: 781–794.

Tamaddon HS, Vaidya D, Simon AM, Paul DL, Jalife J, and Morley GE. 2000. High-resolution optical mapping of the right bundle branch in connexin40 knockout mice reveals slow conduction in the specialized conduction system. *Circ Res* 87 (10): 929–936.

Trosko JE, Chang CC, Madhukar BV, and Klaunig JE. 1990. Chemical, oncogene and growth factor inhibition of gap junctional intercellular communication: An integrative hypothesis of carcinogenesis. *Pathobiology* 58: 265–278.

Vaidya D, Tamaddon HS, Lo CW, Taffet SM, Delmar M, Morley GE, and Jalife J. 2001. Null mutation of connexin43 causes slow propagation of ventricular activation in the late stages of mouse embryonic development. *Circ Res* 88 (11): 1196–1202.

Valiunas V, Beyer EC, and Brink PR. 2002. Cardiac gap junction channels show quantitative differences in selectivity. *Circ Res* 91 (2): 104–111.

Valiunas V, Polosina YY, Miller H, Potapova IA, Valiuniene L, Doronin S, Mathias RT et al. 2005. Connexin-specific cell-to-cell transfer of short interfering RNA by gap junctions. *J Physiol* 568 (Pt 2): 459–468.

Veenstra RD, Wang HZ, Beblo DA, Chilton MG, Harris AL, Beyer EC, and Brink PR. 1995. Selectivity of connexin-specific gap junctions does not correlate with channel conductance. *Circ Res* 77 (6): 1156–1165.

Vozzi C, Ullrich S, Charollais A, Philippe J, Orci L, and Meda P. 1995. Adequate connexin-mediated coupling is required for proper insulin production. *J Cell Biol* 131 (6 PT 1): 1561–1572.

Wang H-Z, and Veenstra RD. 1997. Monovalent ion selectivity sequences of the rat connexin43 gap junction channel. *J Gen Physiol* 109: 491–507.

Weber PA, Chang HC, Spaeth KE, Nitsche JM, and Nicholson BJ. 2004. The permeability of gap junction channels to probes of different size is dependent on connexin composition and permeant-pore affinities. *Biophys J* 87 (2): 958–973.

Weidmann S. 1952. The electrical constants of Purkinje fibres. *J Physiol* 118 (3): 348–360.

Weidmann S. 1970. Electrical constants of trabecular muscle from mammalian heart. *J Physiol* 210: 1041–1054.

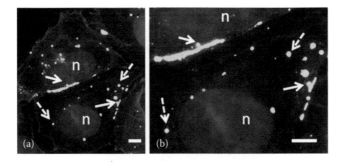

FIGURE 1.2   Immunofluorescence of SW-13 human adrenal cortical tumor cells (a, b).

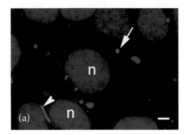

FIGURE 1.4   Immunolabeling Cx43 with (a) immunofluorescence and (b, c) quantum dot streptavidin-biotin conjugate methods.

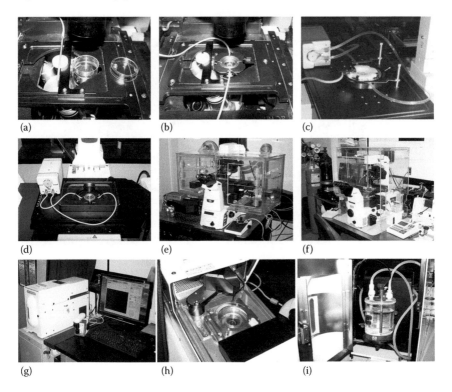

FIGURE 2.1   Live-cell microscope equipment.

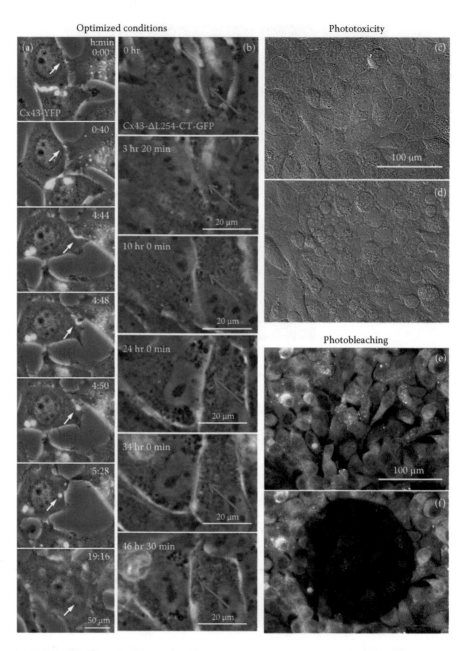

FIGURE 2.2 Optimized imaging conditions and negative effects of excitation light.

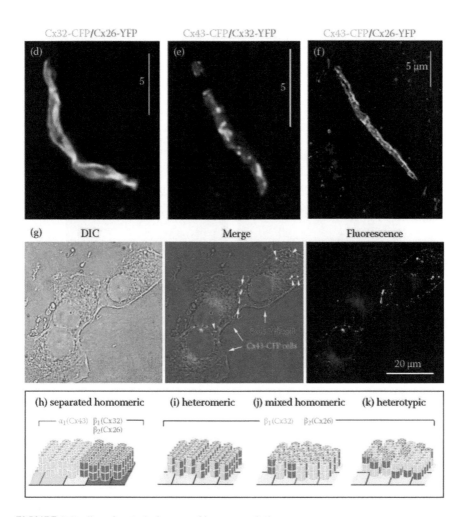

FIGURE 2.3 Imaging techniques and image resolution.

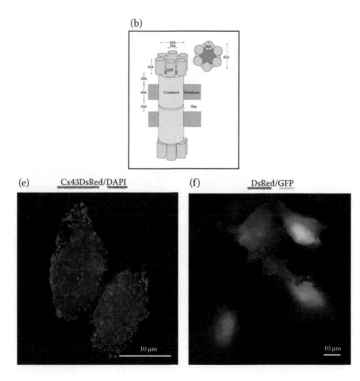

(b)

(e) Cx43DsRed/DAPI   (f) DsRed/GFP

10 μm   10 μm

FIGURE 2.5 Fluorescent probes.

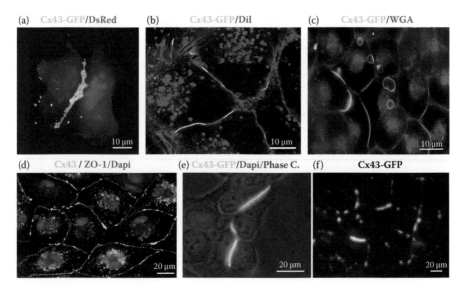

(a) Cx43-GFP/DsRed   (b) Cx43-GFP/DiI   (c) Cx43-GFP/WGA

10 μm   10 μm   10 μm

(d) Cx43 / ZO-1/Dapi   (e) Cx43-GFP/Dapi/Phase C.   (f) **Cx43-GFP**

20 μm   20 μm   20 μm

FIGURE 2.6 Identification of gap junction plaques.

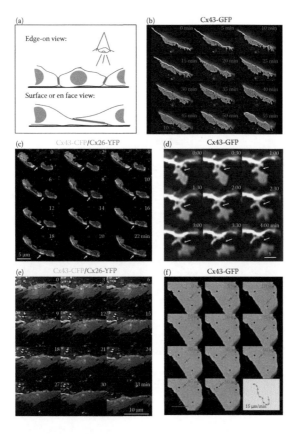

FIGURE 2.7 Gap junction plaque dynamics.

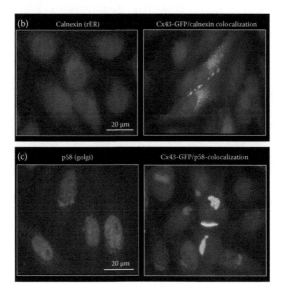

FIGURE 2.8 Connexins trafficking along the secretory pathway: Colocalization with relevant compartment markers.

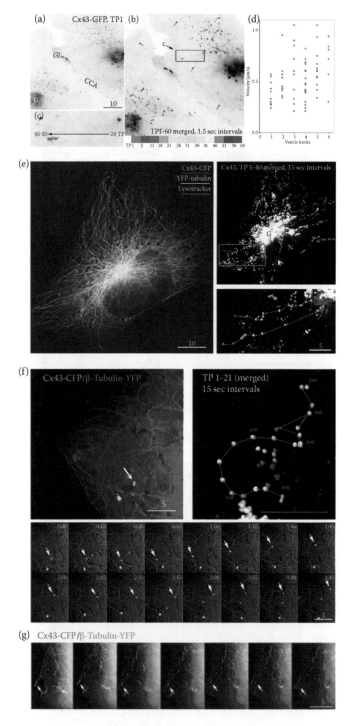

FIGURE 2.9 Connexins trafficking along the secretory pathway: Trafficking along the microtubules.

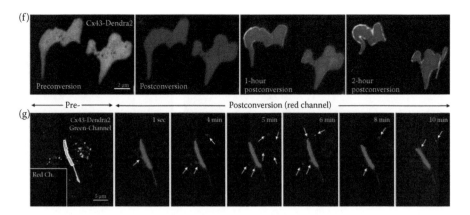

FIGURE 2.10 Plasma membrane dynamics of connexons and gap junctions.

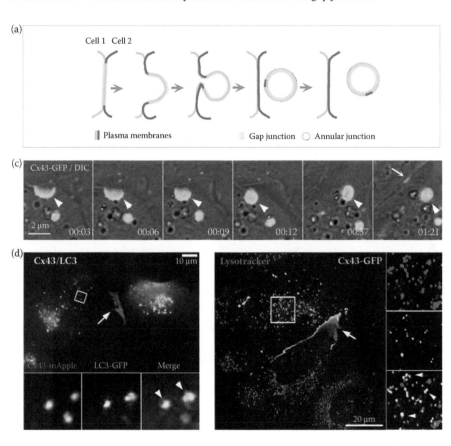

FIGURE 2.11 Internalization and degradation of gap junctions.

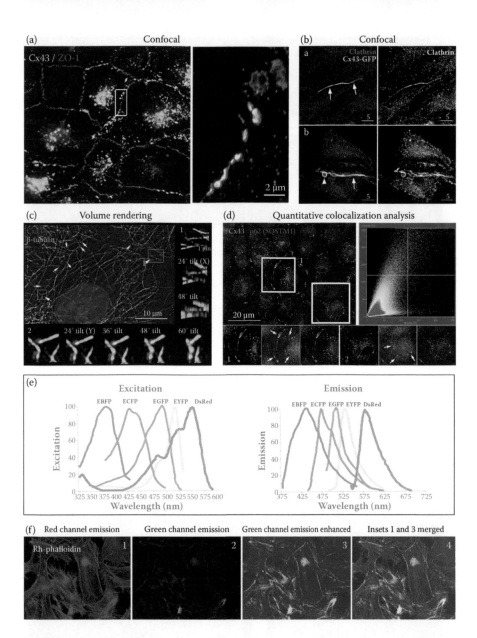

FIGURE 2.12 Colocalization: Tools and techniques and potential false positives due to emission light bleed-through.

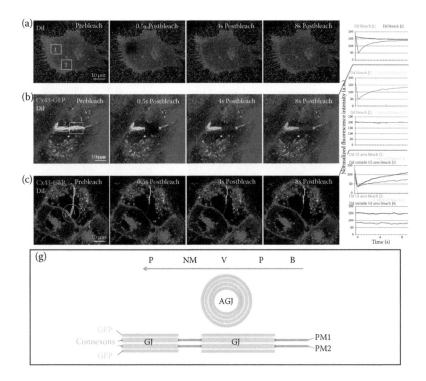

FIGURE 2.13 Quantitative fluorescence analyses.

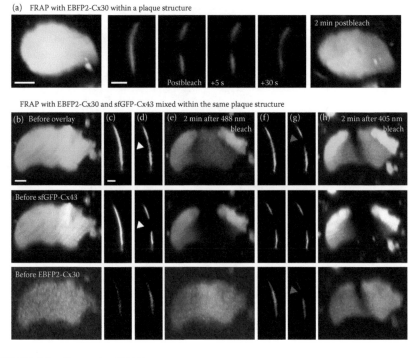

FIGURE 3.3 Two-color FRAP.

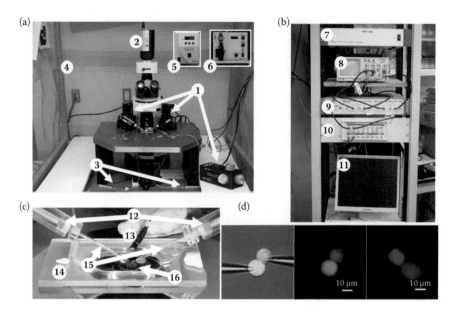

FIGURE 4.3 Equipment required for dual patch clamp technique.

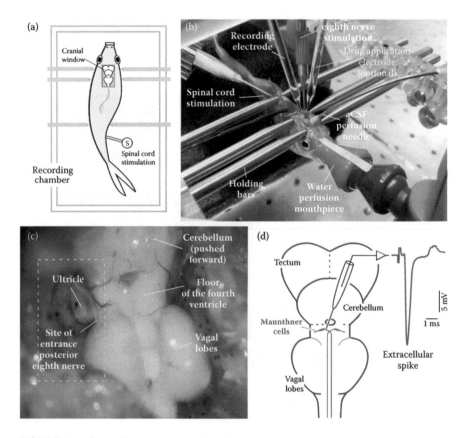

FIGURE 6.1 Recording apparatus and brain exposure.

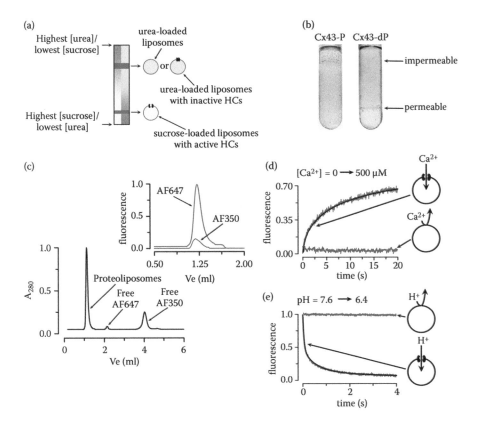

FIGURE 10.2 Transport assays in purified and reconstituted hemichannels.

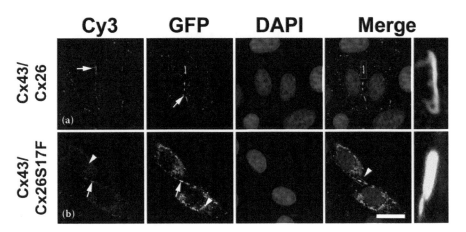

FIGURE 11.1 Syndromic mutation Cx26S17F colocalizes with Cx43.

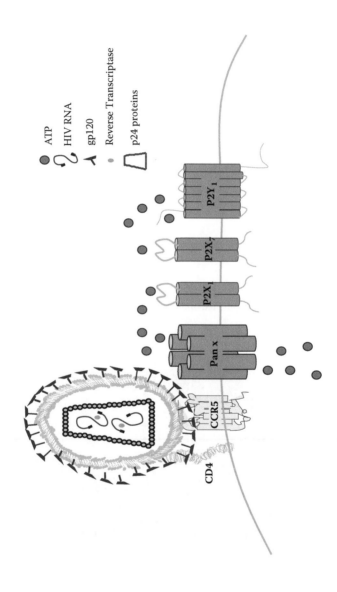

FIGURE 12.1 Illustration of HIV binding to the cell surface of the host cells and facilitating pannexin1 channel (Panx) opening and purinergic receptor activation.

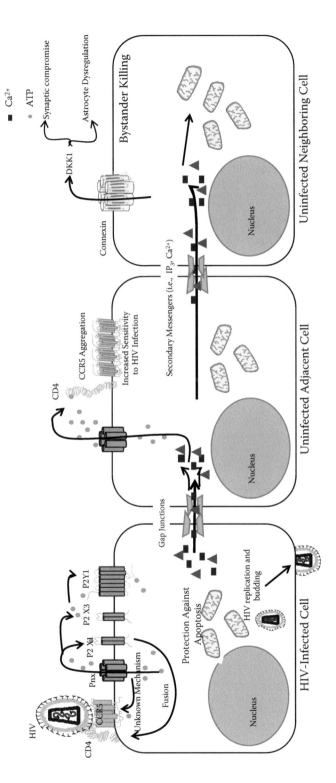

FIGURE 12.2 HIV infection and mechanism of bystander apoptosis in the host cells involving pannexin1 channels and connexin containing channels, as well as secondary messengers including IP₃, calcium ions, and ATP.

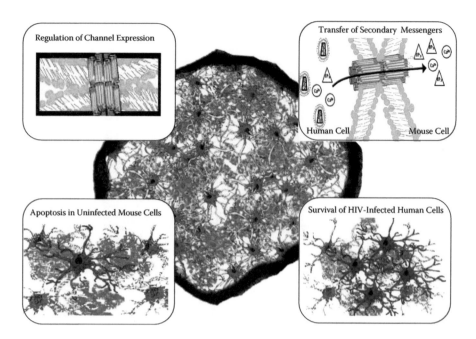

FIGURE 12.3 A dual species model of HIV infection in the CNS allows for the characterization and the analysis of HIV gap junctions and bystander apoptosis.

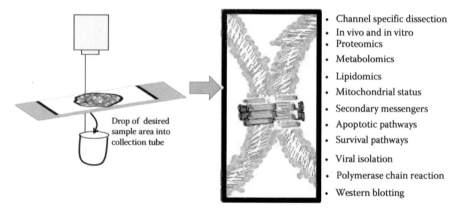

FIGURE 12.4 A schematic of microdissection as a method for isolating the channels for precise downstream analysis.

6 Recording Gap Junction-Mediated Synaptic Transmission In Vivo at Mixed Synapses on the Goldfish Mauthner Cells

Roger Cachope, Sebastian Curti,
and Alberto E. Pereda

CONTENTS

6.1   INTRODUCTION

The protocol describes the necessary steps required to record electrical coupling between a special class of primary auditory afferents and the goldfish Mauthner cells, which are interconnected by gap junctions. The anatomical features and electrophysiological properties of these neurons make it possible to monitor gap junction-mediated electrical communication and its variations in-vivo, at which anatomical connections and normal physiological conditions are preserved.

Gap junctions mediate the electrical communication between neurons by providing a pathway of low resistance for the spread of the electrical currents between coupled cells. Monitoring electrical communication between neurons usually requires performing simultaneous recordings from two adjacent neurons (Galarreta and Hestrin, 1999; Gibson et al., 1999). Gap junctional communication is generally tested under these conditions by changing the membrane potential of one of the recorded cells with brief pulses of current while monitoring the corresponding voltage responses in the second (coupled) cell. These recordings are generally obtained in slices of brain or neuronal cultures using the whole cell technique, under either CC or VC configurations. The strength of electrical transmission is described under CC by calculating the coupling coefficient, a ratio obtained by dividing the amplitude of the depolarization recorded in the coupled cell (coupling potential) over the amplitude of the depolarization recorded in the cell where the current was injected, or by just directly recording the junctional current under VC configuration (Pereda et al., 2013). We describe here the necessary steps that make possible the monitoring of electrical transmission in living fish. This in vivo approach provides the opportunity of exploring the properties of neuronal gap junctions in a more natural context under which the anatomical connections and the higher regulatory systems are preserved.

The Mauthner cells are a pair of unusually large reticulospinal neurons located in the medulla of fish (Figure 6.1). Because of the spinal connectivity of their axons, these neurons mediate sensory-evoked tail-flip escape responses in these animals (Faber and Pereda, 2011). Although the Mauthner cells can also be activated by visual or somesthetic inputs, they are commonly recruited by sound and vibratory stimuli (Faber and Pereda, 2011; Pereda and Faber, 2011). The auditory afferents that are likely to provide this information originate in the sacculus (the auditory organ in fish) and terminate on the lateral dendrite of the Mauthner cells as single large myelinated club endings, or club endings (Bartelmez and Hoerr, 1933; Lin and Faber, 1988a). These single terminals contain both gap junctions and specializations for chemical transmission, and therefore support both electrical and chemical transmissions (Robertson et al., 1963; Furshpan, 1964; Tuttle et al., 1986; Lin and Faber, 1988a; Pereda et al., 2004). They are therefore considered mixed synapses.

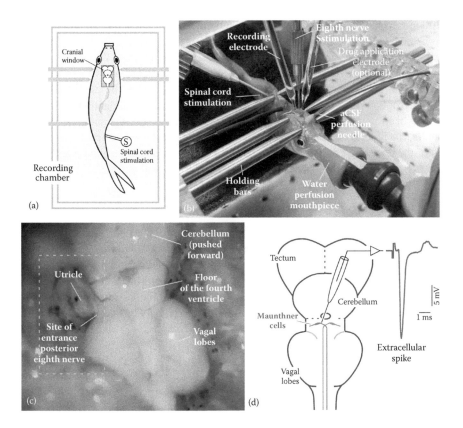

FIGURE 6.1 **(See color insert.)** Recording apparatus and brain exposure. (a) The cartoon illustrates the recording chamber and the cranial window performed to expose the fourth ventricle of the fish (Pereda and Faber, 2011). (Adapted from Pereda AE, Faber DS, *Physiology of the Mauthner Cell: Discovery and Properties*, Elsevier Inc., Amsterdam, 2011. With permission.) (b) The picture illustrates a fish placed in the recording chamber and the placement of the recording and stimulating electrodes. (c) Goldfish brain after dissection. The asterisk marks the usual point of entrance of the microelectrode into the brainstem, to record the left Mauthner cell. (d) The identification of the Mauthner cell is based on the large extracellular action potential generated in the axon hillock, following the stimulation of the axon in the spinal cord (Pereda and Faber, 2011). (Adapted from Pereda AE, Faber DS, *Physiology of the Mauthner Cell: Discovery and Properties*, Elsevier Inc., Amsterdam, 2011. With permission.)

The morphological characteristics of both the auditory afferents and the Mauthner cells make the use of the whole-cell voltage clamp technique impractical to estimate the junctional conductance. These characteristics include the fact that auditory afferents are heavily myelinated, and both the afferents and the Mauthner cell are exceptionally large, making it impossible to obtain adequate spatial control of their membrane voltage. Instead, the electrical transmission can be assessed with sharp electrodes in CC by recording the electrical synaptic potentials in both directions (coupling of action potentials or "spikelets"). Moreover, the unusually fast membrane time constant of both cells (~400 microseconds for the Mauthner cell and

~200 microseconds for the afferents) (Fukami et al., 1965; Curti et al., 2008) allows the use of action potentials to measure the strength of the electrical transmission, as they are not become filtered by the postsynaptic membrane. Thus, the coupling coefficients can be obtained during the simultaneous recordings of saccular club ending afferents and Mauthner cell lateral dendrite by using the amplitude of the coupling potential (or spikelet) divided by the amplitude of the action potential or spike evoked in the activated cell (either the Mauthner cell or the auditory afferent). Finally, the junctional conductance can be estimated at these contacts using the values of the coupling coefficients and the input resistances of the coupled cells (Rash et al., 2013). For detailed information, please see Rash et al. (2013).

6.2 EXPERIMENTAL ARRANGEMENT

The following sections describe the procedures required to perform (1) intradendritic recordings from the Mauthner cell to monitor the mixed synaptic response evoked by stimulation of saccular afferents, (2) intracellular recordings from saccular afferents and terminals to monitor the antidromic electrical coupling (from the dendrite to the auditory afferent), and finally (3) simultaneous intracellular recordings to monitor bidirectional electrical communication between a single club ending and the Mauthner cell lateral dendrite.

6.2.1 RECORDING MIXED SYNAPTIC RESPONSES FROM THE MAUTHNER CELL

The experiment consists in performing intradendritic recordings from the lateral dendrite of the Mauthner cell in vivo, while monitoring the mixed synaptic response evoked by stimulation of the posterior eighth nerve, where the saccular auditory afferents terminating as club endings run. This procedure is facilitated by (1) the large diameter and orientation of the Mauthner cell lateral dendrite and (2) the electrophysiological properties of the Mauthner cells, whose extracellular spike can be easily identified following the antidromic stimulation of its axon (Figures 6.1d and 6.2a) (Furshpan and Furukawa, 1962). We strongly encourage the reader to read the article by Furshpan and Furukawa (1962) in order to become familiar with the unique electrophysiological properties of the Mauthner cells, whose detailed description exceeds the scope of this article. The full understanding of these properties is essential for the successful completion of the experiments described in the sections described below.

6.2.1.1 Materials
- Subjects: Adult *Carassius auratus* (~4–5 inches head to the insertion of the tail)
- Artificial cerebrospinal fluid (aCSF) (see recipe in Section 6.2.1.2)
- Intracellular sharp microelectrodes
- Bipolar stimulating electrode for stimulation of the spinal cord
- Small bipolar stimulating electrode for the stimulation of the posterior branch of the eighth nerve
- Microelectrode filling solution (see recipe in Section 6.2.1.3)

- Surgical tools: small scissors, small rongeur (3–4 mm), teethed forceps, serrated forceps, fine-tip forceps (#4 or 5), and scalpel
- Consumables: tricaine methanesulfonate (MS-222), lidocaine gel or cream, lidocaine solution 1%, saline solution NaCl 0.9%, small cotton-tipped applicators, tissue paper (Kimber wipes type), gauze
- D-tubocurarine dissolved in physiological solution (NaCl 0.9%) for intramuscular administration
- Electrophysiology setup, equipped with
  - Antivibrating table
  - Dissecting microscope (i.e., Zeiss OPMI Pico)
  - Recording chamber (the chamber consists of an acrylic box equipped with a number of metal rods with conical ends on each side to secure the fish's head and body on a fixed position (Figure 6.1a and b) and an outlet connection for recirculating water (see Section 6.2.1.5)
  - Recirculating water system (it is required to perfuse the fish's gills with water and maintain the animal in an oxygenated state during the experiment. This system consists of a 3–4 L water reservoir (4 L bucket) positioned below the recording table, a submersible pump, which pushes water through a piece tubing connected to a mounted mouthpiece, which is inserted in the fish's mouth. The water that leaves the gills collects in the recording chamber, passes through the outlet mentioned earlier, and finally travels back through a second piece of tubing to the water reservoir.)
  - aCSF gravity-driven brain irrigation system
  - Micromanipulator for intracellular electrodes (Narishige Canberra or Sutter MP-285)
  - Three-axis positioner manipulator for the eighth nerve-stimulating electrode (i.e., Newport).
  - Amplifier for current clamp recordings (i.e., Axoclamp 2A; Axoprobe)
  - Isolated stimulators (i.e., Digitimer DS2A [battery-operated constant voltage isolation unit], or DS3 [battery-operated constant current isolation unit])
  - Computer equipped with an analog-to-digital (AD) converter and acquisition software.
  - Oscilloscope; Master 8 or similar timing control unit; Brownlee signal conditioner amplifier equipped with filters or similar

6.2.1.2 Artificial Cerebrospinal Fluid

The aCSF consists of (in mM): 124 NaCl, 5.1 KCl, 3.0 $NaH_2PO_4-H_2O$, 0.9 $MgSO_4$, 5.6 dextrose, 1.6 $CaCl_2-H_2O$, and 20 HEPES, and pH 7.2–7.4.

6.2.1.3 Microelectrode Filling Solution

The intracellular electrode solution consists of either 5 M KAc or 2.5 M KCl, in 10 mM HEPES at pH 7.2. KAc is recommended for intradendritic recordings, as the use of KCl-containing electrodes could result in an increase of the intracellular concentration of Cl^-, which results in the reversal of spontaneous inhibitory synaptic potentials that interfere with the detection of the excitatory synaptic response. These spontaneous inhibitory responses are quite numerous but are normally not detected as a variations in the membrane potential, as the resting potential of the Mauthner cell lies at the equilibrium potential of Cl−.

6.2.1.4   Fabrication of Sharp Microelectrodes

The procedure consists of two steps: (1) fabricating a glass microelectrode with the help of a microelectrode puller and then (2) breaking its tip using a standard light microscope. This method allows obtaining a low-resistance (ideal for the identification of the extracellular field produced by the spike of the Mauthner cell) and a sharper electrode (required for good penetration of the cell). The glass used to make electrodes can be either 2 or 3 mm in diameter, and should be cut so that electrodes are between 100–120 mm in length. The micropipette puller settings, including heat and speed, should be set so that a conical tip length of about 10–12 μm is obtained. To break the tip of the electrode, the pulled capillary is placed on a glass slide on its side where it is kept stationary with a piece of clay. The slide is then placed on the stage of a light microscope with low magnification objectives (×2–40), which has a second glass capillary (~120 mm in length) situated in the field of view, attached with clay to the static portion of the stage, and oriented perpendicular to the electrode tip to be broken. It is important to note that the soon-to-be-broken tip should be in the same z plane as the middle of the perpendicular electrode. Using ×160 total magnification or similar, slowly move the slide containing the pulled microelectrode, so it just touches the fixed glass capillary in a kiss-and-run fashion. If correctly performed, the now-broken tip should be barely visible through the scope. Tips that look obviously broken are always too large and will damage the Mauthner cell. Alternative methods for breaking the electrode tip include gently touching the tip of the microelectrode with a small piece of cotton or a small strip of aluminum foil. Once the procedure is completed, using a syringe equipped with a 0.22 μm syringe filter and a quartz flexible needle (i.e., WPI Microfil), fill the electrode with the desired recording solution (either KAc of KCl) and store it in a jar or a petri dish (usually affixed with clay to avoid any contact with the dish). The microelectrode impedance after this procedure should range between 3 and 6 MΩ.

6.2.1.5   Anesthesia and Positioning in the Recording Chamber

The anesthetic and the surgical procedures described here and throughout this chapter do not constitute a detailed description of an animal use protocol (which should be obtained to perform these experiments) and only highlights the main steps required to obtain these recordings. Fish are initially immobilized and anesthetized by immersion in chilled water containing tricaine methanesulfonate (MS-222; 75–125 mg/L). Mix about one-third of ice with two-thirds water and the MS-222. For this purpose, the goldfish in a bucket containing this solution and monitor its movements. Once anesthesia is reached, quickly transfer the fish to the recording chamber. The recording chamber must include a circulating water system running a maintenance solution of tricaine methanesulfonate (MS-222; 50–75 mg/L) in cold water. The water flow should be adjustable so that the small mouthpiece provides a slow but steady flow of water, which can be readily positioned into the fish's mouth once the fish is anesthetized and positioned into the chamber.

6.2.1.6 Surgery

The surgery is performed in the recording chamber and consists of a limited craniotomy to expose the floor of the fourth ventricle (brainstem), where the Mauthner cells are located (Figure 6.1). The fish is situated in the recording chamber on a sponge, and the skull and the body are held in place by metal rods and the mouthpiece. In addition, a small incision is made near the tail to expose the spinal cord in order to position a stimulating electrode that allows antidromic activation of the Mauthner cell for its identification.

6.2.1.7 Placement of the Fish in the Recording Chamber

Apply lidocaine gel (2.5%) to the regions of the fish's skull and body where the metal rods will be placed. With the mouthpiece inserted, tilt the fish laterally to the right, between 15–20° before securing the head with two rods per side (Figure 6.1a and b). The tilting of the animal will facilitate the dissection and the exposure of the posterior branch of the eighth nerve (see Section 6.2.1.8). Two additional rods (one per side) are required to secure the dorsum of the fish's body, at the level of the dorsal fin.

6.2.1.8 Cranial Dissection

Topical application of lidocaine gel (2.5%) is also used during the removal of the skin that covers the cranium. Then, using either a small spatula or similar, remove the skin covering the dorsal aspect of the cranium. Remove a band of 2–3 mm of scales, skin, and muscle from both sides immediately caudal to the cranium maintaining the integrity of the subjacent spinal cord. Using a 3 mm rongeur, open a 4–5 mm wide rectangular window in the skull. Rostrally continue using the rongeur, carefully removing small pieces of bone while avoiding touching the brain and blood vessels. Using a gentle suction needle gently remove the fat and membranes that cover the brain. Ultimately, a window should be created that extends rostrocaudally from the position of the fish eyes to the most caudal end of the cranium (Figure 6.1a). Once the fat is removed, this window should allow visualization of the optic lobes, the cerebellum, the vagal lobes, and the initial portion of the spinal cord (Figure 6.1c and d).

Build a small cotton-tipped applicator with a fine-tip pair of forceps and cotton, and then gently push the cerebellum forward in the rostral direction, until it lies on the optic lobes. Push gently on it, so that it adapts to its new position, and place a 2 mm strip of tissue paper on it extending out of the cranial window to hold it in place (Figure 6.1b). This movement will expose the brainstem, whose dorsal surface constitutes the floor of the fourth ventricle (the cerebellum is considered part of the roof of the fourth ventricle). In the rostral left corner of the cranial window, place the perfusion needle providing aCSF at a slow flow of about 5–10 mL/h (Figure 6.1b). A small drain channel should be created by removing the bone and the surrounding muscle at the caudal right corner of the cranial window, which provides an easy outlet for the perfusate.

Finally, the craniotomy is extended on the left side to expose the posterior eighth nerve (containing the saccular club ending afferents). This will reveal a compartment between the fish's operculum and a cartilaginous lateral wall that protects the brain;

this cavity contains the muscles that control the movement of the operculum. With gentle suction, remove some of the muscle that insert in this cartilaginous lateral wall, and once isolated, remove it with the help of a pair of forceps. The entrance of various cranial nerves to the brainstem should now be visible after removing some membranes and blood clots. Please see Furukawa and Ishii (1967) for a detailed description of the anatomy of the cranial nerves (Furukawa and Ishii, 1967). The anterior eighth nerve innervating the utricle should be clearly visible (Figure 6.1c). The posterior eighth nerve is the root immediately posterior to it and meets the anterior branch as it enters the brain. Because the sacculus is located underneath the brainstem, most of the trajectory of the posterior eighth nerve is hidden, and only a small portion of it is visible at its entrance into the brainstem. Therefore, the visualization of the nerve is greatly facilitated by tilting the animal, as described previously (Figure 6.1c).

6.2.1.9   Positioning the Eighth Nerve-Stimulating Electrode

The saccular afferents terminating on the Mauthner cell enter the brain in the posterior branch of the eighth nerve. As mentioned earlier, tilting the fish during its positioning in the recording chamber greatly helps its visualization. Stimulation is obtained with small metallic bipolar electrodes (FHC Inc.; spacing of the tips: $70\,\mu m$). Only the very fine tips of the electrode prongs are free from insulation, which limits the spread of the current to neighboring nerve fibers. The stimulating electrode should be brought closer to the nerve and, under visual guidance, placed in contact with the posterior branch of the eighth nerve, barely touching it to avoid nerve damage.

6.2.1.10   Spinal Cord Dissection

On the right side of the body at the level of the dorsal fin, remove a band of scales of about 5 mm (dorsoventral) by 1 cm (rostrocaudal), then cut a flap of skin/muscle to expose the vertebral column containing the spinal cord. Place a metallic bipolar stimulating electrode (parallel, 2 mm separation between prongs) in contact with the spinal cord.

6.2.1.11   Immobilization

The fish needs to be immobilized during the experiment, since the stimulation of the Mauthner cell will otherwise result in a tail flip. For this purpose, apply D-tubocurarine ($1–3\,\mu g/g$) is administered intramuscularly. It is important to note that immobilization is only done after surgery, making sure the animal remains anesthetized. Pain can be assessed by direct observation of the fish's eye, gill, and body movements in response to increasing levels of mechanical stimulation of the skin (from initial light touch to the early stages of surgery).

6.2.1.12   Locating the Mauthner Cell with Electrophysiology

The recording microelectrode is positioned near the floor of the fourth ventricle near the cerebellar peduncle (Figure 6.1c), and the Mauthner cells are identified by the extracellular field potential evoked by antidromic activation (100 microsecond in duration, 4–6 V intensity, 5 Hz) of its axon with the spinal cord stimulating electrode. Please note that the entry position of the electrode in the brain depends on the angles of the manipulator, usually 22°. The functional properties of the M-cell axon hillock and its surrounding structures (only the axon and its initial segment are excitable, while the soma and dendrites are unexcitable) allow for the Mauthner cell action potential to generate a large negative extracellular field potential that spreads over several tens of micrometers from its site of generation (Furshpan and Furukawa, 1962). (The axon hillock of the Mauthner cell is encapsulated by an anatomical specialization with high resistivity, the "axon cap" which contributes to generate a large negative extracellular field.) An extracellular negative potential of amplitude larger than 12–15 mV is indicative of a close proximity to the axon hillock (Furshpan and Furukawa, 1962). Identifying the position of the axon hillock electrophysiologically usually requires probing multiple positions, generally at 50 μm intervals. The maximum depth attempted during each probe should not exceed 2 mm in order to avoid breaking the microelectrode tip against the bone. A 12–15 mV negative field is usually a good indication of the proximity of the recording electrode to the axon hillock (Figure 6.2a).

6.2.1.13   Intradendritic Recording in the Mauthner Cell Lateral Dendrite

Once the position of the Mauthner cell axon hillock has been physiologically identified the electrode is moved to the dendrite in a stereotaxic fashion. Withdraw the electrode from the tissue near the hillock, and move it about 300 μm lateral and 100 μm caudal. The lateral dendrite should be identifiable in this position by the small positive extracellular aciton potential (1–2 mV) in response to the spinal cord stimulation (Figure 6.2a). The extracellular action potential in the dendrite is positive because it represents the movement of the positive charges from the dendrite (source) to the axon hillock of the cell, at which the action potential was generated (sink). Be cautious and slowly advance the electrode while gently tapping on the micromanipulator until the final depth is reached. The amplitude of the positive extracellular potential should increase as the recording electrode advances closer to the source. An extracellular action potential amplitude of about 2–3 mV should be taken as an indication of immediate proximity to the Mauthner cell lateral dendrite. At this point, it is advisable to wait for a few minutes, so that the microelectrode finds a stable position with respect to the Mauthner cell dendrite. After this, gently advance the microelectrode while tapping on the manipulator to penetrate the Mauthner cell dendrite. The presence of the Mauthner cell action potential (which is only ~10 mV because it passively spreads from the axon hillock; Figure 6.3c) and a resting membrane potential of about −80 mV will indicate that the microelectrode is now intracellularly placed. A similar procedure but without inserting the electrode intradentrically can be used for placing a second extracellular electrode for the local application of pharmacological agents (Figure 6.1b).

6.2.1.14   Evoking the Mixed Synaptic Response

Once the intradendritic recordings have been established, stimulation of the posterior eighth nerve (50–100-microsecond pulses of 20–100 μA) will evoke a mixed synaptic response (Figure 6.2b) comprised of (1) an early, brief, electrical postsynaptic potential, representing the coupling via gap junctions of the action potentials at the presynaptic auditory afferents, and (2) a delayed, longer-lasting response

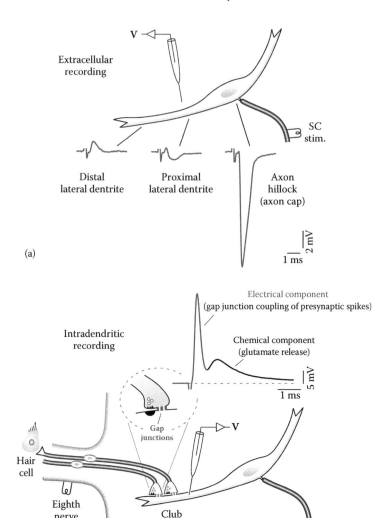

FIGURE 6.2 Extracellular and intracellular recordings of the Mauthner cell. (a) Different portions of the Mauthner cell can be identified during extracellular recordings based on the polarity and the amplitude of its extracellular action potential. The extracellular action potential is large and negative at the axon hillock, which is encapsulated by an anatomical specialization with high resistivity (axon cap), thus generating a large negative extracellular field. The extracellular action potential is otherwise small and biphasic in the proximal lateral dendrite and small and positive at the distal lateral dendrite (the soma and the dendrites of the Mauthner cell are unexcitable). (b) The stimulation of the posterior eighth nerve (50–100-microsecond pulses of 20–100 µA) evokes a mixed synaptic response made by an early, brief, electrical component representing the coupling of the action potential of the presynaptic auditory afferents, and a delayed, longer-lasting response representing the release of glutamate from the chemically mediated regions of the terminals (Pereda and Faber, 2011). (Adapted from Pereda AE, Faber DS, *Physiology of the Mauthner Cell: Discovery and Properties*, Elsevier Inc., Amsterdam, 2011. With permission.)

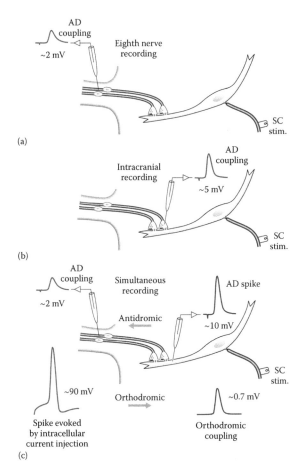

FIGURE 6.3 Intracellular recordings from the auditory afferents terminating as club endings on the lateral dendrite of the Mauthner cell in three different configurations. The panels show schematic drawings of saccular club ending afferents terminating at the distal portion of the Mauthner cell lateral dendrite. Also depicted are the positions of the intracellular recording electrodes and the extracellular stimulating electrodes for the antidromic activation of the Mauthner cell (Pereda and Faber, 2011). The traces are drawings representing characteristic responses of typical amplitude and time course. (a) Intracellular recording from an afferent at the root of the posterior eighth nerve. The antidromic activation of the Mauthner cell with an extracellular stimulating electrodes in the spinal cord (SC stim.) produces an action potential at the axon hillock that passively propagates along the lateral dendrite. This dendritic depolarization retrogradely spreads through the gap junctions toward the presynaptic auditory afferents producing an antidromic coupling potential at these afferents (AD coupling). (b) The antidromic coupling potential intracranially recorded close to or at the synaptic terminals is larger in amplitude and has a faster time course. (c) Simultaneous intracellular recordings of one saccular club ending afferent at the eighth nerve and the lateral dendrite of the Mauthner cell. Above: the antidromic activation of the Mauthner cell evokes a dendritic action potential (AD spike) and a corresponding antidromic coupling potential (AD coupling) in the afferent. Below: the direct activation with a depolarizing current pulse evokes a presynaptic spike in the afferent, which in turn generates an orthodromic coupling potential (spikelet) in the lateral dendrite of the Mauthner cell.

representing the release of glutamate from the chemically mediated regions of the terminals (Furshpan, 1964; Lin and Faber, 1988a; Wolszon et al., 1997; Pereda et al., 2004). The short time constant of the M-cell membrane (~400 microsecond) allows recording the electrical and glutamatergic components without attenuation, making it ideal for monitoring electrical (and chemical) transmission. The amplitude of the mixed synaptic response varies with the intensity of the stimulation of the posterior branch of the eighth nerve, due to the variations in the number of afferents activated (see Figure 6.2b) (Pereda et al., 2004). The polarity and the intensity of the stimulus can be adjusted by monitoring the mixed postsynaptic response according to the requirements of the experiment. Because the saccular club ending afferents are unusually large in diameter, these afferents can be activated in isolation using weak extracellular stimulation. The number of activated fibers can be estimated by dividing the amplitude of the electrical component by 0.74 mV, the average value of the electrical component of an individual club ending (Lin and Faber, 1988a; Smith and Pereda, 2003).

The antidromic action potential that passively propagates through the somatodendritc membrane of the Mauthner cell is a convenient tool to assess the cell's input resistance (Furukawa and Furshpan, 1963). Assuming that the amplitude of the action potentials triggered at the axon hillock does not change between trials the amplitude of the action potential recorded at the distal dendrite is greatly altered during its passive spread through the somatodendritic membrane, and therefore, the changes in its amplitude might reflect changes in membrane conductance. Monitoring the amplitude of the antidromic action potential therefore provides a convenient way to distinguish the changes in the synaptic strength from modifications in the amplitude of synaptic responses due to changes in the cell's input resistance (Oda et al., 1995).

6.2.2   RECORDING ANTIDROMIC ELECTRICAL TRANSMISSION
AT SACCULAR CLUB ENDING AFFERENTS

This section describes the general procedure to obtain intracellular recordings from saccular afferents terminating as club endings in the lateral dendrite of the Mauthner cell. The large diameter and the advantageous experimental accessibility of these saccular afferents make them amenable for in vivo intracellular recordings. Recordings from eighth nerve fibers can be performed either at the root of the eighth nerve or intracranially, at the club ending terminal near the dendrite.

6.2.2.1   Recording of Saccular Afferents in the Posterior
Eighth Nerve Root

Because the saccular club ending afferents are densely myelinated, it is not possible to apply the whole cell method, which requires the surface membranes to be relatively clean and devoid of cellular debris. Instead, the intracellular recordings can be obtained using sharp microelectrodes. These microelectrodes are characterized by a long tapered neck (shank) and smaller tip diameters, and have resistances (despite the use of high-conductive ICSs) 10 times greater than those found using patch electrodes.

6.2.2.1.1   Microelectrode Preparation

Sharp microelectrodes are pulled from borosilicate capillary glasses with a micro-filament (Outside Diameter: 2.0 mm, ID: 1.16 mm). Heat, pull, speed, and other parameters of the puller should be set to obtain an electrode shank ~10–12 mm long. The microelectrodes are then backfilled using a syringe with a syringe filter (0.22 μm) and then attached to a specially designed pipette holder mounted on a high-precision micromanipulator (Sutter MP-285). The microelectrode impedance should range between 30 and 45 MΩ; the recommended intracellular electrode solution is 2.5 M KCl, 10 mM HEPES, pH: 7.2 (see Section 6.2.1.3).

6.2.2.1.2   Advancing the Electrode through the Tissue and Obtaining Intracellular Recordings

The microelectrode is lowered into the aCSF and its impedance monitored on a computer screen or an oscilloscope by applying 1 nA current pulses. With the aid of a dissecting microscope, the microelectrode is then placed on the root of the posterior eighth nerve, just before it joins the brainstem. The microelectrode is then advanced through the nerve using short steps of about 5–10 μm while gently applying light taps to the micropipette holder to facilitate its penetration in the afferent. The sudden presence of a −65 to −75 mV membrane resting potential in the recording will be indicative of afferent penetration.

6.2.2.1.3   Identification of Club Ending Afferents

Once an intracellular recording is obtained, the club ending afferents can be recognized based on the following physiological properties:

- The presence of a coupling potential produced by the antidromic activation of the Mauthner cell (Figure 6.3a)
- A characteristic lack of spontaneous activity
- High threshold for acoustic stimulation (clapping, voice)
- Higher threshold for intracellular stimulation with the recording electrode (2–3 nA)

The intracellular dye fills of cells exhibiting these properties have invariably resulted in the labeling of afferents which, because of their size, myelinization, saccular origin, and distribution along the Mauthner cell lateral dendrite, unambiguously corresponded to club endings afferents (Lin and Faber, 1988b; Smith and Pereda, 2003).

6.2.2.1.4   Recording of Antidromic Coupling Potentials

Once the intracellular recording from a saccular afferent has been established, the antidromic stimulation of an action potential in the Mauthner cell will evoke a depolarization that is due to the passive spread of the current through the gap junctions (identification criteria as well). This coupling potential typically has an amplitude of about 2 mV when recorded at resting membrane potential (Figure 6.3a).

6.2.2.2   Intracranial Recordings of Club Endings

The coupling potentials recorded at the club ending afferents in the eighth nerve are greatly attenuated due to electrotonic decay (over a distance of 1–2 mm from the synapse). A better assessment of the strength of the antidromic coupling can be obtained by performing intracellular recordings at the club ending–Mauthner cell synapse or in close proximity (Figure 6.3b). The procedure should be performed in two steps:

Step 1: Localization of the Mauthner cell axon hillock and the lateral dendrite with a low-resistance electrode See Section 6.2.1.12 for a description of the necessary steps.

Step 2: Recording of the club ending terminal with a high-resistance electrode
- Replace the low-resistance microelectrode with a high-resistance microelectrode suitable for intracellular recording from club ending afferents (30–45 MΩ filled with 2.5 KCl).
- Place the electrode in the area where the axon hillock was found with the low–resistance electrode and move to the lateral dendrite (250–300 μm lateral, 100 μm posterior) and identify the extracellular field potential corresponding to the M-cell lateral dendrite (~2 mV positive going potential) (Figure 6.2a).
- Remove the electrode from the brain; move 50 μm laterally from the place where the extracellular Mauthner cell lateral dendrite was located and initiate another recording track. Make the electrode tracks ~20 μm apart in the rostrocaudal direction until an intracellular recording from a club ending afferent is obtained.
- Intracellular recordings from club ending afferents can be identified by the presence of the antidromic coupling potential. The antidromic coupling potentials recorded at these terminals are of higher amplitude (up to 4 mV at resting potential) and possess faster onset and decay times compared to those recorded more distally on peripheral eighth nerve sites (Figure 6.3b). If dual recordings from the proximal nerve and the distal Mauthner cell are performed, one will notice that it is sometimes difficult to distinguish proximal nerve and distal dendrite recordings, as both the dendritic action potential and the coupling potential have similar amplitudes and kinetics. However, they can be distinguished because the depolarization of the nerve membrane will trigger an action potential at the terminal, while the depolarization of the electrically passive lateral dendrite will not.

6.2.3   SIMULTANEOUS RECORDINGS OF CLUB ENDING AFFERENTS AND M-CELL LATERAL DENDRITE

In order to monitor the bidirectional electrical communication between the auditory afferents and the M-cell, it is necessary to perform simultaneous intracellular recordings from both cells. Simultaneous intracellular recordings from club ending afferents at the eighth nerve root and the M-cell lateral dendrite provide a good approach to study bidirectional electrical communication at single club endings (Figure 6.3c). (A better assessment of antidromic coupling would although require simultaneously recording from the club ending terminals (intracranial recordings) and the Mauthner cell lateral dendrite. However, although feasible (Pereda et al., 2003), this configuration is highly unstable, making it difficult to obtain lasting recordings to monitor the changes in the junctional conductance in response to experimental manipulations.) The procedure combines two protocols already described: intradendritic Mauthner cell recording and intracellular recording from club ending afferents in the root of the eighth nerve. In order to obtain simultaneous recordings, these procedures should be performed in the following sequence:

- Prepare and position the two microelectrodes, a low-resistance one for the Mauthner cell intradendritic recording (KAC) and a second higher-resistance electrode for afferent recording (KCl). In our experimental setup, the micromanipulator for recording the left Mauther cell is on the right side of the recording chamber, and the micromanipulator for recording the afferents from the left eighth nerve is on the left side.
- Following the steps described in Section 6.2.1.12, obtain first the intradendritic recording from the Mauthner cell lateral dendrite.
- Following the steps described in Section 6.2.2.1, then obtain an intracellular recording from a saccular club ending afferent to obtain a simultaneous recording. The presynaptic electrode can be removed and used in another afferent to obtain multiple simultaneous recordings in the same animal (Smith and Pereda, 2003).

Electrical coupling in the orthodromic direction (from the afferent to the Mauthner cell) can be monitored by evoking an action potential in the presynaptic afferent with the recording electrode (depolarizing current pulse) and then monitoring the responses in the lateral dendrite of the Mauthner cell (Figure 6.3c). Characteristically, the chemical component of the mixed response is observed in only 15% of the cases following the activation of a single club ending afferent. This is due to the low probability of release at the club ending terminals, which has been shown to greatly increase during the coactivation of multiple afferents (Lin and Faber, 1988b; Pereda et al., 2004). As described in Section 6.2.2.1.4, the electrical transmission from the Mauthner cell to the eighth nerve afferent can be assessed by monitoring the coupling potential produced in the nerve fibers by passive the spread of the current from the Mauthner cell due to antidromic activation of the action potential at the spinal cord.

6.3 SUMMARY

The steps described here are meant to assist those interested in recording the electrical transmission at mixed synapses between auditory afferents and Mauthner cells in the goldfish. A full understanding of this protocol requires an understanding of the anatomical and electrophysiological properties of the Mauthner cell and the club ending terminals. Suggested reading is therefore included throughout the chapter, which is intended to provide the reader with the additional necessary information. Thus, rather than an exhaustive description of the properties of these cells, this protocol is meant to provide a general description of the elements and conditions required to perform these recordings.

ACKNOWLEDGMENTS

This chapter was supported by the National Institutes of Health (NIH) grants DC03186, DC011099, and NS0552827 (to AEP).

REFERENCES

Bartelmez GW, Hoerr NL (1933) The vestibular club endings in ameiurus: Further evidence on the morphology of the synapse. *J Comp Neurol* 57:401–428.

Curti S, Gómez L, Budelli R, Pereda AE (2008) Subthreshold sodium current underlies essential functional specializations at primary auditory afferents. *J Neurophysiol* 99:1683–1699.

Faber DS, Pereda AE (2011) *Physiology of the Mauthner Cell: Function.* Elsevier Inc., Amsterdam.

Fukami Y, Furukawa T, Asada Y (1965) Excitability changes of the Mauthner cell during collateral inhibition. *J Gen Physiol* 48:581–600.

Furshpan EJ (1964) Electrical transmission at an excitatory synapse in a vertebrate brain. *Science* 144:878–880.

Furshpan EJ, Furukawa T (1962) Intracellular and extracellular responses of the several regions of the Mauthner cell of the goldfish. *J Neurophysiol* 25:732–771.

Furukawa T, Fushapan EJ (1963) Two inhibitory mechanisms in the Mauthner neurons of goldfish. *J Neurophysiol* 26:140–176.

Furukawa T, Ishii Y (1967) Neurophysiological studies on hearing in goldfish. *J Neurophysiol* 30:1377–1403.

Galarreta M, Hestrin S (1999) A network of fast-spiking cells in the neocortex connected by electrical synapses. *Nature* 402:72–75.

Gibson JR, Beierlein M, Connors BW (1999) Two networks of electrically coupled inhibitory neurons in neocortex. *Nature* 402:75–79.

Lin JW, Faber DS (1988a) Synaptic transmission mediated by single club endings on the goldfish Mauthner cell: I: Characteristics of electrotonic and chemical postsynaptic potentials. *J Neurosci* 8:1302–1312.

Lin JW, Faber DS (1988b) Synaptic transmission mediated by single club endings on the goldfish Mauthner cell: II: Plasticity of excitatory postsynaptic potentials. *J Neurosci* 8:1313–1325.

Oda Y, Charpier S, Murayama Y, Suma C, Korn H (1995) Long-term potentiation of glycinergic inhibitory synaptic transmission. *J Neurophysiol* 74:1056–1074.

Pereda AE, Faber DS (2011) *Physiology of the Mauthner Cell: Discovery and Properties.* Elsevier Inc., Amsterdam.

Pereda AE, Curti S, Hoge G, Cachope R, Flores CE, Rash JE (2013) Gap junction-mediated electrical transmission: Regulatory mechanisms and plasticity. *Biochim Biophys Acta* 1828:134–146.

Pereda A, O'Brien J, Nagy JI, Bukauskas F, Davidson KG V, Kamasawa N, Yasumura T, Rash JE (2003) Connexin35 mediates electrical transmission at mixed synapses on Mauthner cells. *J Neurosci* 23:7489–7503.

Pereda AE, Rash JE, Nagy JI, Bennett MVL (2004) Dynamics of electrical transmission at club endings on the Mauthner cells. *Brain Res Brain Res Rev* 47:227–244.

Rash JE, Curti S, Vanderpool KG, Kamasawa N, Nannapaneni S, Palacios-Prado N, Flores CE et al. (2013) Molecular and functional asymmetry at a vertebrate electrical synapse. *Neuron* 79:957–969.

Robertson JD, Bodenheimer TS, Stage DE (1963) The ultrastructure of Mauthner cell synapses and nodes in goldfish brains. *J Cell Biol* 19:159–199.

Smith M, Pereda AE (2003) Chemical synaptic activity modulates nearby electrical synapses. *Proc Natl Acad Sci U S A* 100:4849–4854.

Tuttle R, Masuko S, Nakajima Y (1986) Freeze-fracture study of the large myelinated club ending synapse on the goldfish Mauthner cell: Special reference to the quantitative analysis of gap junctions. *J Comp Neurol* 246:202–211.

Wolszon LR, Pereda AE, Faber DS (1997) A fast synaptic potential mediated by NMDA and non-NMDA receptors. *J Neurophysiol* 78:2693–2706.

7 Assessing Connexin Hemichannel Function during Ischemic Injury and Reperfusion

Yeri Kim and Colin R. Green

CONTENTS

In this chapter we give an overview of connexin hemichannels and their potential roles under normal and pathological conditions, and discuss common methods used for assessing the functional state of connexin hemichannels. We then introduce hypoxic acidic ion shifted ringer (HAIR) solution as an alternative means to assess hemichannel opening during ischemia injury *in vitro*, and separately during post-ischemia reperfusion, by quantifying ATP release. As pannexin channels are also proposed to open under injury conditions, we also evaluate the contribution of pannexin channels during ischemic injury and reperfusion. The HAIR model indicates that the amount of the ATP released from connexin hemichannel opening during ischemia is over one and half times that released through pannexin channels. Following reperfusion, connexin hemichannel opening alone may account for most of the ATP released from those two channel types.

7.1 INTRODUCTION

7.1.1 CONNEXIN HEMICHANNELS

Gap junction channels are constructed of paired connexin hemichannels; each is composed of six protein subunits and occurs at areas of close plasma membrane apposition. Hemichannels are continuously recruited from surrounding unopposed plasma membrane areas and subsequently docked head to head with partners from adjacent cells at the edge of the existing gap junction plaques (Bai and Wang 2014; Goodenough and Paul 2009; Laird 2006). However, hemichannels may also be present in unopposed membranes, allowing ionic and molecular exchanges, and thus offering a pathway for autocrine and paracrine signaling between the intra- and extracellular environments (Evans, Bultynck, and Leybaert 2012; Wang, De Bock, et al. 2013a). It has been shown that the cytoplasmic Ca^{2+} changes observed within the physiological range can trigger the hemichannel opening (De Vuyst et al. 2006, 2009), and the cells exposed to endogenous ligands associated with cell proliferation or differentiation have been reported to show increased hemichannel activity (Belliveau et al. 2006; Schalper et al. 2008). In another example, the transmembrane passage of the neuroactive substances adenosine triphosphate (ATP) and glutamate may be involved in multiple functions across the central nervous system, including neurotransmission and neuromodulation, memory, and learning (for review, see Cheung, Chever, and Rouach 2014). A hemichannel role has been shown for ephaptic signaling in the retina (for review, see Klaassen et al. 2011 and Vroman, Klaassen, and Kamermans 2013).

Open hemichannels, however, constitute a large, relatively nonspecific membrane channel permeable to numerous low-molecular weight molecules, and hemichannels are considered to have a low open probability under resting conditions (Contreras et al. 2003; Giaume et al. 2013). Most reports of hemichannel opening are associated with cellular stress or related to pathological situations (Cheung, Chever, and Rouach 2014; Giaume et al. 2013; Orellana, Martinez, and Retamal 2013), including membrane potential depolarization (Contreras et al. 2003), mechanical stimulation (Gomes et al. 2005), ischemia or oxidative stress (Contreras et al. 2002; Ramachandran et al. 2007), proinflammatory cytokines (Retamal et al. 2007), or changes in Ca^{2+} concentration (an increase in cytoplasmic Ca^{2+} concentration or a decrease in extracellular Ca^{2+} concentration [De Vuyst et al. 2009; Gomez-Hernandez et al. 2003; Thimm et al. 2005]). Connexin expression is usually upregulated after injury, in many cases two- to three-fold within four to eight hours (Danesh-Meyer et al. 2012; Guo et al. 2014, O'Carroll et al. 2013a), and furthermore, the gating of connexin channels is often inversely regulated, with increased hemichannel opening under pathological conditions but decreased cell-to-cell coupling. The factors oppositely influencing the channel opening in this manner include kinase-activating stimuli (lipopolysaccharide and fibroblast growth factor-1 [FGF-1]), proinflammatory cytokines (tumor necrosis factor alpha and IL-1 beta), ischemic conditions, or calcium concentration changes (Contreras et al. 2002; De Vuyst et al. 2006, 2007, 2009; Kielian 2008; Retamal et al. 2007; Torres et al. 2012). As large pores, hemichannels may themselves affect big changes in cytoplasmic Ca^{2+} (Sanchez et al. 2009).

The hemichannel opening contributes to lesion spread after an injury and clearly plays a significant role under pathological conditions. This has been extensively reviewed elsewhere (Bosch and Kielian 2014; Castellano and Eugenin 2014; Davidson et al. 2013a, 2014a; De Vuyst et al. 2011; Eugenin 2014; Evans, Bultynck, and Leybaert 2012; Kielian 2008; Mallard et al. 2014; O'Carroll et al. 2013b; Zhang et al. 2013). In brief, and as examples only, a number of studies have shown that under pathological conditions, the hemichannel opening contributes to (1) inability of cells to osmoregulate (Danesh-Meyer et al. 2012; Davidson et al. 2012; Quist et al. 2000; Rodriguez-Sinovas et al. 2012), (2) vascular hemorrhage (Danesh-Meyer et al. 2012; De Bock et al. 2011), (3) excitotoxic cell death (Froger et al. 2010), (4) neuronal cell death (Chen et al. 2014; Danesh-Meyer et al. 2012; Orellana et al. 2011), (5) onset of seizures, reduced electroencephalogram (EEG) activity, and loss of sleep cycling after perinatal brain ischemia (Davidson et al. 2012), (6) lesion spread, increased inflammation, and poor behavioral outcomes after spinal cord injury (O'Carroll et al. 2013a), (7) increased myocardial infarction (Hawat et al. 2010), and (8) reduced neuronal loss, EEG activity, and oligodendrocyte death after preterm brain asphyxia (Davidson et al. 2014b). In all cases, the modulation of the connexin hemichannels (either the prevention of protein translation or the hemichannel block) resulted in significantly improved outcomes.

7.1.2 PANNEXIN CHANNELS

A level of complexity became apparent with the discovery of another family of channel-forming proteins, the pannexins (Baranova et al. 2004; Penuela, Gehi, and Laird 2013). The pannexin family has three isoforms that are vertebrate homologues of the invertebrate innexin gap junction channel-forming family (Panchin et al. 2000). They are highly expressed in the central nervous system (Pannexin1 [Baranova et al. 2004]; Pannexin2 [Swayne, Sorbara, and Bennett 2010]), and also, for example, in synovial fibroblasts and osteoblasts (Pannexin3 [Baranova et al. 2004]). Pannexin1 is localized to the plasma membrane, whereas Pannexin2 is intracellularly located (Boassa et al. 2014; Wicki-Stordeur, Boyce, and Swayne 2013). Pannexin1 channels are well known as ATP release channels allowing for purinergic signaling in multiple cell types (Locovei, Bao, and Dahl 2006; Seminario-Vidal et al. 2011), including vascular smooth muscle and endothelial cells (Billaud et al. 2011; Godecke et al. 2012) and microglia and astrocytes (Dahl and Muller 2014; Iwabuchi and Kawahara 2011). There is some controversy with the latter, however, as the presence of Pannexin1 has not yet been demonstrated in those cell types *in vivo*. One major difference between connexin and pannexin channels is that pannexin channels do not form cell-to-cell channels, and it has been suggested that the highly glycosylated extracellular loops of the pannexin proteins interferes with the docking process (Penuela et al. 2007). As with connexin hemichannels, the pannexin channels are said to be activated by a number of factors, but also show some differences. Pannexin channels, for example, are insensitive to decreases in calcium ion concentration (Ma et al. 2009; Sandilos et al. 2012). Both connexin hemichannels and pannexin channels contribute to gluta-mate and ATP release though, and a major constraint in discriminating the roles that these respective channels play is the limited availability of suitable research tools.

7.1.3 CHANNEL BLOCKERS

Several pharmacological agents are available to inhibit connexin-based channels includ-ing the anaesthetic halothane, aliphatic alcohols octanol and heptanol, glycyrrhizic metabolites 18α-glycyrrhetinic acid and carbenoxolone (CBX), and fenamates niflumic acid and flufenamic acid (for recent review, see Bodendiek and Raman [2010]). However, it is only possible to indirectly infer the effects of these reagents on hemichannel activity; they close the gap junctions too, and they affect other membrane channels including the pannexin channels. CBX, for example, inhibits connexin and pannexin channels, although with a higher affinity for pannexin channels ($EC_{50} = \sim 5\,\mu M$) com-pared to Connexin43 hemichannels ($10–100\,\mu M$) (Ye et al. 2003). Another tool avail-able for the regulation of pannexin channels is Probenecid, which has been used for the treatment of gout (Silverman et al. 2008), an inflammatory condition that leads to a buildup of urate crystals and the activation of the NOD-like receptor (NLR) family, pyrin domain-containing protein 3 (NLRP3) inflammasome (also known as NALP3 or cryopyrin) (Martinon et al. 2006). Probenecid inhibits the urate reuptake in the kidney, and may also attenuate the inflammatory signals by the possible inactivation of the inflammasome (Silverman et al. 2009). Probenecid inhibits Pannexin1 channels and does not affect the connexin channels (Silverman, Locovei, and Dahl 2008).

Hence, more specific approaches such as the genetic ablation of connexins have been widely used. Such approaches, however, again lack the ability to target hemi-channel activity independently of the cell-to-cell gap junction coupling. Mimetic peptides derived from the connexin protein sequences have been shown to act as selective inhibitors of connexin channels (De Vuyst et al. 2011; Evans, Bultynck, and Leybaert 2012; Iyyathurai et al. 2013; O'Carroll et al. 2013b). Peptides targeting the extracellular loops that inhibit the hemichannels only or the hemichannels and the gap junctions, based on a time- or concentration-dependent manner, are available for the connexins: Connexin26, Connexin32, Connexin37, Connexin40 and Connexin43 (Evans, Bultynck, and Leybaert 2012; O'Carroll et al. 2008; Ponsaerts et al. 2010; Wang et al. 2012a). Other peptides identical to the sequences on the intracellular parts of the Connexin43 protein have been exploited to specifically inhibit the hemichannels without influencing the gap junctions (Ponsaerts et al. 2010; Wang et al. 2013b). Based on this approach too, a synthetic mimetic peptide has also been developed to block the Pannexin1 channels, termed [10]Panx1 (Pelegrin and Surprenant 2006). Based on the sequence dissimilarity between connexin and pannexin proteins, mimetic peptides therefore offer more specific tools to interfere with these channel functions.

7.1.4 MODELS FOR ASSESSING HEMICHANNEL FUNCTION

A second research constraint is the availability of suitable models to assess the hemichannel blockers and to enable the differentiation between connexin hemichan-nel and pannexin channel roles. The three major methods for examining connexin and pannexin channel activities *in vitro* include fluorescent dye uptake or release, electrophysiological readouts, and ATP quantification. Small synthetic fluorescent dyes have been routinely used as they provide a fast, simple, and low-cost means for measuring hemichannel activity, and have also been useful for examining the gap junction activity (for review, see Saez and Leybaert 2014). The *in vitro* dye assays are dependent on the molecular size, charge, and permeability of the dyes and an appropriate stimulus for opening connexin hemichannels or pannexin channels.

A divalent-free cation-free ECS is widely used to stimulate the hemichannel opening (Braet et al. 2003; DeVries and Schwartz 1992). Under normal conditions, the extracellular Ca^{2+} ($[Ca^{2+}]_o$) undergoes rhythmic fluctuations of $1–2\,mM$ that is under tight feedback regulation in order to keep it within a narrow range (for review, see Thomas 2000). Lowering the $[Ca^{2+}]_o$ was first shown to activate the hemichannels in the catfish retina (DeVries and Schwartz 1992), and $[Ca^{2+}]_o$ has been proposed to affect the voltage gate that induces a reversible conformation change in the hemichannel. For example, a study in the Connexin46 hemichannels demonstrated that extracellular calcium ions bind to the extracellular voltage-dependent gate to lock the channel into a long-lived closed state (Verselis and Srinivas 2008).

The relevance of the dye uptake to the permeability of the physiological molecules has recently been debated in a Xenopus oocyte expression system (Hansen et al. 2014). Connexin30 and Connexin43 hemichannels exhibited highly selective permeability to physiologically relevant molecules that did not equate to the uptake of synthetic fluorescent dyes (Hansen et al. 2014). However, a fluorescent glucose derivative, 2-(N-(7-nitrobenz-2-oxa-1, 3-diazol-4-yl) amino)-deoxyglucose (2-NBDG) has been used for a biologically relevant comparison of uptake mechanisms via the hemichannels (Retamal et al. 2007). Electrophysiology has been the gold standard for assessing the functional state of connexin hemichannels due to distinct unitary channel conductance of the connexin hemichannels and pannexin channels. Connexin43 hemichannels have been well characterized with a commonly reported conductance of ~220 pS (Contreras et al. 2003; Wang et al. 2012a), although the conductances of 165 pS (Kang et al. 2008) and 753 pS (500 mM KCl buffer) (Brokamp et al. 2012) have also been reported. Pannexin1 channels have been suggested to display a conductance of ~550 pS (in 150 mM KCl) (Bao, Locovei, and Dahl 2004). Connexin hemichannels are stimulated to open at depolarized potentials ($\geq$+60 mV) (Contreras et al. 2003; Wang et al. 2012a). In contrast, the pannexin channels favor potentials of -20 to $+20$ mV range (Bao, Locovei, and Dahl 2004; Bruzzone et al. 2003). The evidence for the permeation of ATP through Connexin43 hemichannels has also been demonstrated using a combination of bioluminescence imaging and single-channel recording (Kang et al. 2008). The permeability of the pannexin channels to the ATP has been demonstrated in excised patches following mechanical stress (Bao, Locovei, and Dahl 2004). A limitation of electrophysiology, however, is that the flow of the current does not equate to the flux of the physiological molecules. For further review of electrophysiology approaches see Patel, Zhang, and Veenstra (2014).

Table 7.1 summarizes some of the approaches used and lists their advantages and disadvantages. These approaches include the use of tracers permeable to the hemichannels, the quantification of released molecules, the electrophysiological recordings of hemichannel currents, and the genetic manipulation of connexin isoforms.

TABLE 7.1

Overview of the Techniques Used to Study Connexin Hemichannel Function

Technique	Description	Advantages	Disadvantages
Dye uptake or efflux	Hemichannel activity is quantified as a direct function of the permeability to the fluorescence reporter dyes in vitro and in situ.	Simple technique, low cost, high-contrast fluorescence, wide range of reporter dyes with differences in net charge and molecular weight, can be high throughput	Influx/efflux of dye ≠ flux of physiological molecules, irreversible reaction, low temporal resolution, nonspecific (i.e., permeation through other channels, and some dyes are also markers for dead cells, such as propidium iodide [PI])
Electrophysiology	Hemichannel (HC) kinetics and gating properties are determined through patch-clamp recordings of the hemichannel currents in whole-cell, cell-attached, or excised patch methods	Distinct conductance signature, high temporal resolution, reversible reaction (i.e., drug wash-out)	Technically demanding, expensive equipment, low throughput, flow of current ≠ flux of physiological molecule, invasive, nonspecific (differentiating channels with similar conductance and/or the number of channels)
Knockout animal models	Targeted deletion of genes for understanding the contribution of specific connexin isoforms to function	Stable, highly specific to connexin isoform, single mechanism, a stable population may be obtained	Specialist equipment required, suitable for a limited number of species, expensive set up and maintenance costs, phenotypes may be lethal, perturbations in one area may mask others, internal controls are not possible, nonspecific (hemichannels or gap junctions)
Knockdown animal models (antisense oligodeoxynucleotides, RNAi, morpholinos)	As for knockout but targeted reduction, not deletion	Highly specific to connexin isoform, single mechanism, transient, not species constrained, dose controllable, low cost, can be topically targeted (spatial specificity), compensation by related gene family members less likely	Good consequence design can be complex, nonspecific (hemichannels or gap junctions)

These techniques have greatly contributed to the identification of many unique features of connexin hemichannels. These models, however, have limitations and have given variable results between laboratories. In the following sections, therefore, we outline in some detail a simple, *in vitro* model for differentiating connexin hemichannels and pannexin channel activities during ischemic injury and during reperfusion. We outline the results using the connexin mimetic peptide hemichannel blocker Peptide5 and the Pannexin1 channel blocker Probenecid, with this model enabling the separation of connexin hemichannels and pannexin channel roles in a reproducible manner. A standard protocol has been described here, but this model may be suitable for testing the pharmacology of other connexin hemichannel and pannexin channel inhibitors during ischemic injury and reperfusion.

7.2 METHODS

7.2.1 HYPOXIC ACIDIC ION-SHIFTED RINGER PROTOCOL FOR ASSESSING HEMICHANNEL ROLES DURING ISCHEMIA AND REPERFUSION

Dramatic ionic shifts are a principle feature of brain injury with the $[Ca^{2+}]_o$ falling to 90% of the normal value (Hansen 1985). At the same time, the injured brain is also subject to acidosis (Rosner and Becker 1984). Similar changes in the $[Ca^{2+}]_o$ have been reported for spinal cord injury (<0.1 mM) (Young, Yen, and Blight 1982) that is also accompanied by a decline in both extracellular sodium and chloride ion concentrations (Hansen 1985). Furthermore, the injured site is subject to glucose and oxygen deprivation that leads to energy failure and oxidative stress (reviewed in the study by Werner and Engelhard 2007). Therefore, in our *in vitro* model of the hemichannel assay, we used a hypoxic acidic ion-shifted ringer (HAIR) solution (Bondarenko and Chesler 2001), as it most closely resembles the complex changes of an acute insult that extends to the reperfusion injury *in vivo*. Our method extends the commonly used oxygen and glucose deprivation strategies to model central nervous system ischemia in vitro (for examples, see Plesnila et al. 2001, Strasser and Fischer 1995, Xu et al. 2000, Tasca, Dal-Cim, and Cimarosti 2015) by accounting for the ionic and acid shifts that are also present in ischemia and reperfusion injuries. Our rationale is further supported by hemichannel-gating mechanisms that are sensitive to pH (Saez et al. 2005; Wang et al. 2012b) and the ionic shifts (Lopez et al. 2014; Verselis and Srinivas 2008). Since Connexin43 hemichannels have been reported to have a high affinity for ATP (Goldberg, Lampe, and Nicholson 1999), the following *in vitro* model of ischemia–reperfusion injury has been combined with a highly sensitive bioluminescence reaction to quantify the ATP release as a function of Connexin43 hemichannel activity.

7.2.2 EQUIPMENT

1. Collagen I, rat tail, 3 mg/mL (Invitrogen, A10483–01)
2. 0.02 M acetic acid, sterile
3. 12-well, flat-bottom, clear plates, sterile (Falcon)
4. Reagent reservoirs, sterile

5. Membrane 0.22 μm filters, sterile (Millex-GS)
6. 50 mL sterile syringes, sterile (Terumo)
7. 96-well, flat, black bottom (Corning)
8. Magnetic stir bar and plate
9. Cell culture incubator (37°C, 95% O_2, and 5% CO_2)
10. Cell culture hood (i.e., laminar-flow hood)
11. Water bath
12. pH meter
13. Multichannel pipette, 10–100 μL (Eppendorf)
14. Luminescence reader (VICTOR X, Perkin Elmer #2030–0010)

7.2.3 CELLS AND REAGENTS

1. Human cerebral microvascular endothelial cells (hCMVEC)
2. Culture medium (M199 media, Invitrogen; 10% heat-inactivated fetal calf serum, 1 μg/mL hydrocortisone, 3 ng/mL human fibroblast growth factor (FGF) (Peprotech), 1 ng/mL human epidermal growth factor (EGF) (Peprotech), 10 μg/mL heparin (Sigma) 80 μM dibutyryl cyclic adenosine monophosphate (cAMP) (Sigma-D0627)
3. Tryple™ Express (Invitrogen)
4. 1× PBS (P4417, Sigma)
5. Connexin peptidomimetic Peptide5 (sequence VDCFLSRPTEKT, >95% purity dissolved in dH_2O, 10 mM stock, Auspep)
6. Probenecid ($C_{13}H_{19}NO_4S$, Sigma P8761, 50 mg/mL stock in 1 M NaOH)
7. Carbenoxolone ($C_{34}H_{48}O_7Na_2$, Sigma C4790, 100 mM stock in dH_2O)
8. 5 M HCl
9. Luciferin/luciferase bioluminescence reaction kit stored at ≤−20°C (ATP Determination Kit #A22066, Molecular Probes)
10. Standard reaction solution from the ATP determination kit consisting of 8.9 mL H_2O, 0.5 mL 20× reaction buffer, 0.1 mL 0.1 M dithiothreitol, 0.5 mL of 10 mM D-luciferin, 2.5 μL of firefly luciferase, 5 mg/mL stock solution
11. A 100 μM ATP solution (500 μL)
12. Nitrogen gas (>99.9% purity)
13. The working concentration of normal physiological solution is (in mM): 124 NaCl, 3 KCl, 26 $NaHCO_3$, 1 NaH_2PO_4, and 1.5 $MgCl_2$ in dH_2O (Prepare a 10× stock of the normal physiological solution in dH_2O)
14. The working concentration of hypoxic, acidic, HAIR injury solution is (in mM): 34 NaCl, 13 $NaHCO_3$, 3 Na-gluconate, 1 NaH_2PO_4, and 1.5 $MgCl_2$ in dH_2O (Prepare a 10× stock of the HAIR injury solution in dH_2O)

7.2.4 EXPERIMENT PREPARATION

1. Prepare 12 mL of collagen I solution at 30 μg/mL in 0.02 M acetic acid. Place 1 mL of 30 μg/mL collagen I solution in each well of a 12-well plate for an hour. Remove collagen I solution and wash the wells at least three times with PBS.

2. Dissociate one confluent T25 flask of hCMVEC using 3 mL TrypleE solution. Resuspend and seed the cells at 0.025×10^6 cells per 12-well a day before the experiment.

3. Prepare a 1-in-10 dilution of the 10× normal physiological stock solution with dH_2O, and add 1.3 mM $CaCl_2$ and 10 mM glucose to the 1× solution on the day of the experiment. Adjust the pH of the 1× solution to 7.4 using 5 M HCl and filter sterilize in the cell culture hood.

4. Prepare a 1-in-10 dilution of the 10× HAIR injury stock solution with dH_2O, and add 0.13 mM $CaCl_2$, 65 mM K-gluconate, and 38 mM NMDG-Cl to the 1× solution on the day of the experiment. Adjust the pH to 6.6 using 5 M HCl. Bubble the 1× solution in N_2 gas for 5 minutes at 20 L/min, and then readjust the pH to 6.6 using 5 M HCl. Filter sterilize the 1× solution in the cell culture hood.

5. Heat the 1× normal and injury solutions in a water bath at 37.5°C for 15 minutes.

6. Dilute the connexin and/or pannexin channel inhibitors at a desired final concentration in the injury solution for the injury-only model.

7. Dilute the connexin and/or pannexin channel inhibitors at a desired final concentration in 1× normal physiological solution for the injury–reperfusion (IR) model.

8. Preheat the luminometer to 28°C.

7.2.4.1 Simulated Ischemic Injury

1. Remove the culture medium from each well and rinse the hCMVEC cells six times with the 1× normal physiological solution.

2. Add 500 μL of the 1× normal physiological solution to the negative control group.

3. Rinse the rest of the hCMVEC cells two more times with the 1× HAIR injury solution.

4. Add 500 μL of 1x HAIR injury solution to the positive control group.

5. For the treatment groups, add 500 μL 1× injury solution + connexin hemichannel and/or pannexin channel inhibitors.

6. Place the 12-well plate in the cell culture incubator for 2 hours.

7. Prepare the 100 μM ATP solution and the standard reaction solution kit about 30 minutes prior to the end of the 2-hour incubation, as these reagents can degrade over time. Do not vortex the standard reaction solution, as the firefly luciferase enzyme can easily denature. Protect the solution from light and keep on ice until use.

8. In a 96-well black-walled black flat-bottom plate, prepare a serial dilution of ATP standard (0–500 nM). Standard curves must be generated for each experiment. The ATP range from the test samples must span at least five to six points on the ATP standard curve. The total volume in each well should be no more than 200 μL (Figure 7.1).

9. Place 180 μL of standard reaction solution for the ATP standard samples as shown in Figure 7.1.

10. Place 180 μL of the standard reaction solution for the test samples as shown in Figure 7.1.
11. Following a 2-hour incubation, transfer 100 μL of the test samples into an empty column as shown in Figure 7.1. Use fresh pipettes to transfer each sample to avoid ATP contamination. Conduct the following steps immediately after the sample collection. The fluorescence background levels can increase over time, and this can decrease the sensitivity of the detection.
12. Using a multichannel pipette, simultaneously transfer 20 μL of standard ATP samples into the standard reaction solution.
13. Using a multichannel pipette, simultaneously transfer 20 μL of test samples into a new column of the standard reaction solution. The total volume of the experimental sample assay in each well should be equal to the volume of the ATP standard assay in each well. Repeat step 13 three times for obtaining triplicate measurements.
14. Place the plate in the luminometer and make 10 repeated measurement readings. The Luciferin will react with the ATP in the sample to produce light with an emission maximum of ~560 nm at pH 7.8 (Chittock et al. 1998):

$$Luciferin + ATP + O_2 \xrightarrow{\text{Luciferase}} Oxyluciferin + AMP + Pyriphosphatase + CO_2 + Light \tag{7.1}$$

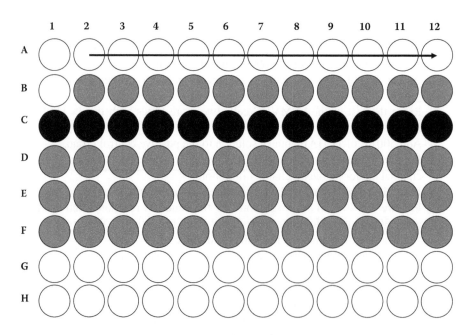

FIGURE 7.1 Standard layout of a 96-well plate for determining the ATP from samples. Serial dilutions are performed in column A starting on A2 (black arrow). The standard reaction solutions are shown in grey. The ATP samples (20 μL) are transferred from columns A to B. The test samples (20 μL) in column C (black) are transferred to D, E, and F for triplicate measurement.

7.2.4.2 Simulated Reperfusion (Post-Ischemia)

1. Remove the culture medium from each well, and rinse the hCMVEC cells six times with the 1× normal physiological solution.
2. Add 500 µL of the 1× normal physiological solution to the negative control.
3. Rinse the other remaining wells two more times with the 1× injury solution.
4. Add 500 µL of the 1× injury solution to all the remaining groups.
5. Place the 12-well plate in the cell culture incubator for 2 hours.
6. Remove solutions from the wells. Ensure that all the solution is removed from the well so that there is no residual ATP.
7. Add 500 µL of the 1× normal physiological solution to the negative control and positive control wells.
8. Add 500 µL of the 1× normal physiological solution + connexin hemichannel and/or pannexin channel inhibitors to the wells.
9. Place the 12-well plate in the cell culture incubator for 2 hours.
10. Repeat steps 7–14 in Section 7.2.4.1.

7.2.5 ANALYSIS

1. The background luminescence can be inferred from the blank sample in the standard curve. Average the background luminescence and subtract this value from all arbitrary bioluminescence units from the test samples.
2. The arbitrary bioluminescence units are converted into ATP concentration (nM) using an equation of best fit from the standard curve, $y = mx + c$, where x represents the unknown ATP concentration. The standard curve values with an R^2 value of ≥0.99 were used for further analysis (Figure 7.2).

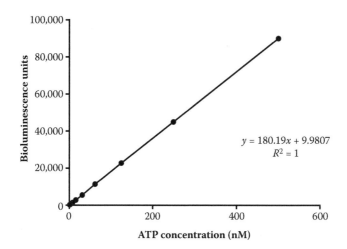

FIGURE 7.2 An example of a standard curve from ATP concentration samples of 0–500 nM. The arbitrary bioluminescence units are converted into ATP concentration (nM) using an equation of best fit from the standard curve, $y = mx + c$, where x represents the unknown ATP concentration. Standard curve values with an R^2 value of ≥0.99 were used for further analysis.

7.3 RESULTS

7.3.1 Injury (Ischemia)-Only Model

With the injury (ischemia)-only model, we aimed to differentiate Connexin43 hemichannel- and Pannexin1 channel-mediated ATP releases using the connexin hemichannel blocking mimetic Peptide5 at 100 μM concentration (O'Carroll et al. 2008) and 1 mM Probenecid (Silverman et al. 2009), respectively. Figure 7.3 summarizes the total ATP released from a subconfluent culture of hCMVEC following 2 hours in the in vitro injury model. When compared with the injury group (100 ± 2.2%), there was a basal level of 35 ± 2.6% ATP released from no-injury control that may reflect the normal cellular activity and/or the dead

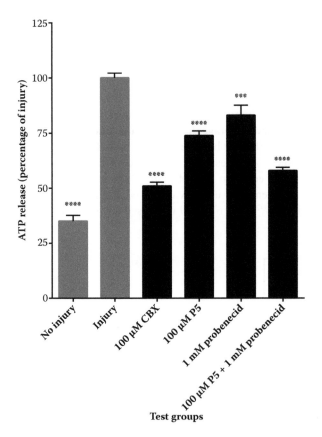

FIGURE 7.3 ATP released from hCMVEC cells following a 2-hour exposure to injury (ischemia)-only model *in vitro*. The quantification of the total extracellular ATP release is presented as a percentage of the injury control for each treatment group. A significant reduction in ATP was present across all treatment groups compared to injury control: 100 μM CBX, 100 μM Peptide5 (P5), 1 mM Probenecid, and 100 μM Peptide5 in combination with 1 mM Probenecid; the latter group now reducing the total ATP release to the same level as CBX. One-way analysis variance, Tukey's multiple comparison test. The values represent mean ± standard error. ***$p < 0.001$, ****$p < 0.0001$ against injury control.

cells prior to injury. A significant difference of 65 ± 3.7% in the ATP release was observed between no-injury and injury control ($p < 0.0001$). In comparison to injury, Peptide5 significantly reduced the ATP to 73.9 ± 2.1% ($p < 0.0001$), whereas Probenecid significantly decreased the ATP to 83.1 ± 4.5%, ($p < 0.001$). The nonspecific inhibitor of both connexin and pannexin channels CBX (Chekeni et al. 2010) significantly reduced the ATP release to 50.9 ± 1.7% of the injury control. Interestingly, a combined treatment of Peptide5 and Probenecid reduced the ATP to a level that was comparable to 100 μM CBX, with no significant difference observed between these two groups ($p = 0.28$).

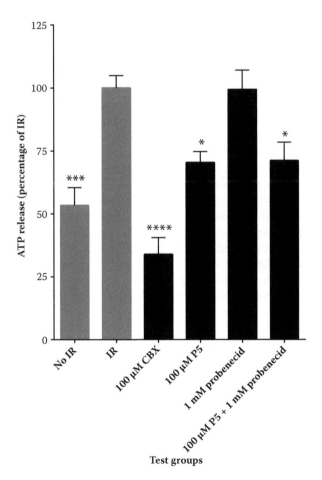

FIGURE 7.4 ATP released from subconfluent hCMVEC cells following a 2-hour exposure to injury, then a 2-hour exposure to reperfusion in vitro. The ATP is quantified as a percentage of the IR control. A significant reduction in ATP was observed with 100 μM CBX, 100 μM Peptide5 (P5), and 100 μM P5 and 1 mM Probenecid, but not with 1 mM Probenecid on its own. The values represent mean ± standard error. One-way ANOVA Tukey's multiple comparison test. *$p \leq 0.03$, ***$p \leq 0.006$, ****$p < 0.0001$ against IR control.

7.3.2 INJURY–REPERFUSION MODEL

We further aimed to characterize the role of Connexin43 hemichannel and pannexin channel during the reperfusion period postischemia (return to the normal media post two hours in HAIR solution). We observed a significant difference in ATP release between no IR and IR controls ($p = 0.0006$) (Figure 7.4). While the connexin mimetic Peptide5 significantly reduced the ATP compared to the IR control ($70.3 \pm 4.4\%$, $p = 0.02$), there was no significant difference to the control when Probenecid was used on its own. We observed a significant difference between $100\,\mu M$ CBX ($33.8 \pm 6.6\%$) and Peptide5 ($p = 0.0086$); the possible sources for this discrepancy are discussed in Section 7.4. However, in a combination treatment with Peptide5, Probenecid did not further reduce the ATP release levels below those obtained with Peptide5 on its own.

7.4 CONCLUSION

In this chapter, we have given an overview of the connexin hemichannels and their potential roles under normal and pathological conditions and summarized the common methods used for assessing the connexin hemichannel function. In particular, we have introduced the HAIR solution as a means to assess the hemichannel opening during the ischemia injury in vitro, and separately during postischemia reperfusion, by quantifying the amount of the ATP released into the extracellular space. As Pannexin1 channels permeate the ATP in endothelial cells (Godecke et al. 2012), and are also proposed to open under injury conditions (Weilinger, Tang, and Thompson 2012), we also examined the contribution of the pannexin channels to injury and reperfusion. We used the connexin peptidomimetic Peptide5 and Probenecid to modulate the connexin hemichannel and pannexin channel functions, respectively, but this model can be adapted for other channel inhibitors. The limitations of this method include possible leak of ATP currents from other maxi-anion plasma membrane channels, such as voltage-dependent anion channels and nonconnexin or nonpannexin mechanosensitive or volume-sensitive channels (Bader and Weingart 2006; Rutledge, Mongin, and Kimelberg 1999). The contribution of these channels to the total ATP is likely to be low relative to the connexin hemichannels, although during the reperfusion phase, there was a higher level of ATP leak which was not associated with either connexin hemichannel or pannexin channel activity, but which was nonetheless ameliorated with CBX. These leak currents could be further controlled by using additional blockers.

The timing of pannexin or connexin hemichannel opening as a result of ischemia and their relative roles in subsequent pathology, remain unclear because much of the literature does not distinguish between those events occurring during ischemia, as opposed to those occurring after ischemia. This may be significant as the in vivo upregulation of connexin and the associated hemichannel-mediated events appear to be important in reperfusion injury but may turn out to be limited (or have relatively little therapeutic impact) during the ischemia itself (Danesh-Meyer et al. 2012; Davidson et al. 2012, 2013b; Mallard et al. 2014). The results using the HAIR model suggest that the amount of the ATP leak resulting from the connexin hemichannel opening during ischemia is over one and half times of that lost through the pannexin channels. Combined pannexin channel block with Probenecid and hemichannel block with Peptide5 reduces the ATP release to the same level as the nonspecific channel blocker CBX used as a positive control during injury. Following the reperfusion, it appears that the pannexin channels may close, and the connexin hemichannel opening appears to account for most of the ATP leak through those two channel types. In vivo, reperfusion may not be so uniform though, and it remains to be seen to what extent this model reflects pathological conditions. The results go some way, however, to explaining reports that both pannexin channel and connexin hemichannel blocks can contribute to the improved outcomes after a ischemic injury and the neuronal sparing effect of the hemichannel blockers delivered during the postinjury reperfusion period.

ACKNOWLEDGMENTS

This work was supported by the University of Auckland Doctoral Scholarship to Yeri Kim. Colin Green is grateful for the support of Wendy and Bruce Hadden. We acknowledge Jarred M. Griffin for his technical contribution toward the ATP Determination.

REFERENCES

Bader, P., and R. Weingart. 2006. Pitfalls when examining gap junction hemichannels: Interference from volume-regulated anion channels. *Pflugers Arch* 452 (4):396–406.

Bai, D., and A. H. Wang. 2014. Extracellular domains play different roles in gap junction formation and docking compatibility. *Biochem J* 458 (1):1–10.

Bao, L., S. Locovei, and G. Dahl. 2004. Pannexin membrane channels are mechanosensitive conduits for ATP. *FEBS Lett* 572 (1–3):65–8.

Baranova, A., D. Ivanov, N. Petrash, A. Pestova, M. Skoblov, I. Kelmanson, D. Shagin et al. 2004. The mammalian pannexin family is homologous to the invertebrate innexin gap junction proteins. *Genomics* 83 (4):706–16.

Belliveau, D. J., M. Bani-Yaghoub, B. McGirr, C. C. Naus, and W. J. Rushlow. 2006. Enhanced neurite outgrowth in PC12 cells mediated by connexin hemichannels and ATP. *J Biol Chem* 281 (30):20920–31.

Billaud, M., A. W. Lohman, A. C. Straub, R. Looft-Wilson, S. R. Johnstone, C. A. Araj, A. K. Best et al. 2011. Pannexin1 regulates alpha1-adrenergic receptor- mediated vasoconstriction. *Circ Res* 109 (1):80–5.

Boassa, D., P. Nguyen, J. Hu, M. H. Ellisman, and G. E. Sosinsky. 2014. Pannexin2 oligomers localize in the membranes of endosomal vesicles in mammalian cells while Pannexin1 channels traffic to the plasma membrane. *Front Cell Neurosci* 8:468.

Bodendiek, S. B., and G. Raman. 2010. Connexin modulators and their potential targets under the magnifying glass. *Curr Med Chem* 17 (34):4191–230.

Bondarenko, A., and M. Chesler. 2001. Rapid astrocyte death induced by transient hypoxia, acidosis, and extracellular ion shifts. *Glia* 34 (2):134–42.

Bosch, M., and T. Kielian. 2014. Hemichannels in neurodegenerative diseases: Is there a link to pathology? *Front Cell Neurosci* 8:242.

Braet, K., S. Aspeslagh, W. Vandamme, K. Willecke, P. E. Martin, W. H. Evans, and L. Leybaert. 2003. Pharmacological sensitivity of ATP release triggered by photoliberation of inositol-1,4,5-trisphosphate and zero extracellular calcium in brain endothelial cells. *J Cell Physiol* 197 (2):205–13.

Brokamp, C., J. Todd, C. Montemagno, and D. Wendell. 2012. Electrophysiology of single and aggregate Cx43 hemichannels. *PLoS One* 7 (10):e47775.

Bruzzone, R., S. G. Hormuzdi, M. T. Barbe, A. Herb, and H. Monyer. 2003. Pannexins, a family of gap junction proteins expressed in brain. *Proc Natl Acad Sci U S A* 100 (23):13644–9.

Castellano, P., and E. A. Eugenin. 2014. Regulation of gap junction channels by infectious agents and inflammation in the CNS. *Front Cell Neurosci* 8:122.

Chekeni, F. B., M. R. Elliott, J. K. Sandilos, S. F. Walk, J. M. Kinchen, E. R. Lazarowski, A. J. Armstrong et al. 2010. Pannexin 1 channels mediate "find-me" signal release and membrane permeability during apoptosis. *Nature* 467 (7317):863–7.

Chen, Y., C. R. Green, K. Wang, H. V. Danesh-Meyer, and I. D. Rupenthal. 2014. Sustained intravitreal delivery of connexin43 mimetic peptide by poly(d,l-lactide-co-glycolide) acid micro- and nanoparticles—Closing the gap in retinal ischaemia. *Eur J Pharm Biopharm* 95:378–86.

Cheung, G., O. Chever, and N. Rouach. 2014. Connexons and pannexons: Newcomers in neurophysiology. *Front Cell Neurosci* 8:348.

Chittock, R. S., J. M. Hawronskyj, J. Holah, and C. W. Wharton. 1998. Kinetic aspects of ATP amplification reactions. *Anal Biochem* 255 (1):120–6.

Contreras, J. E., J. C. Saez, F. F. Bukauskas, and M. V. Bennett. 2003. Gating and regulation of connexin 43 (Cx43) hemichannels. *Proc Natl Acad Sci U S A* 100 (20):11388–93.

Contreras, J. E., H. A. Sanchez, E. A. Eugenin, D. Speidel, M. Theis, K. Willecke, F. F. Bukauskas, M. V. Bennett, and J. C. Saez. 2002. Metabolic inhibition induces opening of unapposed connexin 43 gap junction hemichannels and reduces gap junctional communication in cortical astrocytes in culture. *Proc Natl Acad Sci U S A* 99 (1):495–500.

Dahl, G., and K. J. Muller. 2014. Innexin and pannexin channels and their signaling. *FEBS Lett* 588 (8):1396–402.

Danesh-Meyer, H. V., N. M. Kerr, J. Zhang, E. K. Eady, S. J. O'Carroll, L. F. Nicholson, C. S. Johnson, and C. R. Green. 2012. Connexin43 mimetic peptide reduces vascular leak and retinal ganglion cell death following retinal ischaemia. *Brain* 135 (Pt 2):506–20.

Davidson, J. O., P. P. Drury, C. R. Green, L. F. Nicholson, L. Bennet, and A. J. Gunn. 2014b. Connexin hemichannel blockade is neuroprotective after asphyxia in preterm fetal sheep. *PLoS One* 9 (5):e96558.

Davidson, J. O., C. R. Green, L. Bennet, and A. J. Gunn. 2014a. Battle of the hemichannels— Connexins and pannexins in ischemic brain injury. *Int J Dev Neurosci* 45:66–74.

Davidson, J. O., C. R. Green, L. Bennet, L. F. Nicholson, H. Danesh-Meyer, S. J. O'Carroll, and A. J. Gunn. 2013a. A key role for connexin hemichannels in spreading ischemic brain injury. *Curr Drug Targets* 14 (1):36–46.

Davidson, J. O., C. R. Green, L. F. Nicholson, L. Bennet, and A. J. Gunn. 2013b. Connexin hemichannel blockade is neuroprotective after, but not during, global cerebral ischemia in near-term fetal sheep. *Exp Neurol* 248:301–8.

Davidson, J. O., C. R. Green, L. F. Nicholson, S. J. O'Carroll, M. Fraser, L. Bennet, and A. J. Gunn. 2012. Connexin hemichannel blockade improves outcomes in a model of fetal ischemia. *Ann Neurol* 71 (1):121–32.

De Bock, M., M. Culot, N. Wang, M. Bol, E. Decrock, E. De Vuyst, A. da Costa et al. 2011. Connexin channels provide a target to manipulate brain endothelial calcium dynamics and blood-brain barrier permeability. *J Cereb Blood Flow Metab* 31 (9):1942–57.

De Vuyst, E., K. Boengler, G. Antoons, K. R. Sipido, R. Schulz, and L. Leybaert. 2011. Pharmacological modulation of connexin-formed channels in cardiac pathophysiology. *Br J Pharmacol* 163 (3):469–83.

De Vuyst, E., E. Decrock, L. Cabooter, G. R. Dubyak, C. C. Naus, W. H. Evans, and L. Leybaert. 2006. Intracellular calcium changes trigger connexin 32 hemichannel opening. *EMBO J* 25 (1):34–44.

De Vuyst, E., E. Decrock, M. De Bock, H. Yamasaki, C. C. Naus, W. H. Evans, and L. Leybaert. 2007. Connexin hemichannels and gap junction channels are differentially influenced by lipopolysaccharide and basic fibroblast growth factor. *Mol Biol Cell* 18 (1):34–46.

De Vuyst, E., N. Wang, E. Decrock, M. De Bock, M. Vinken, M. Van Moorhem, C. Lai et al. 2009. Ca(2+) regulation of connexin 43 hemichannels in C6 glioma and glial cells. *Cell Calcium* 46 (3):176–87.

DeVries, S. H., and E. A. Schwartz. 1992. Hemi-gap-junction channels in solitary horizontal cells of the catfish retina. *J Physiol* 445:201–30.

Eugenin, E. A. 2014. Role of connexin/pannexin containing channels in infectious diseases. *FEBS Lett* 588 (8):1389–95.

Evans, W. H., G. Bultynck, and L. Leybaert. 2012. Manipulating connexin communication channels: Use of peptidomimetics and the translational outputs. *J Membr Biol* 245 (8):437–49.

Froger, N., J. A. Orellana, C. F. Calvo, E. Amigou, M. G. Kozoriz, C. C. Naus, J. C. Saez, and C. Giaume. 2010. Inhibition of cytokine-induced connexin43 hemichannel activity in astrocytes is neuroprotective. *Mol Cell Neurosci* 45 (1):37–46.

Giaume, C., L. Leybaert, C. C. Naus, and J. C. Saez. 2013. Connexin and pannexin hemichannels in brain glial cells: Properties, pharmacology, and roles. *Front Pharmacol* 4:88.

Godecke, S., C. Roderigo, C. R. Rose, B. H. Rauch, A. Godecke, and J. Schrader. 2012. Thrombin-induced ATP release from human umbilical vein endothelial cells. *Am J Physiol Cell Physiol* 302 (6):C915–23.

Goldberg, G. S., P. D. Lampe, and B. J. Nicholson. 1999. Selective transfer of endogenous metabolites through gap junctions composed of different connexins. *Nat Cell Biol* 1 (7):457–9.

Gomes, P., S. P. Srinivas, W. Van Driessche, J. Vereecke, and B. Himpens. 2005. ATP release through connexin hemichannels in corneal endothelial cells. *Invest Ophthalmol Vis Sci* 46 (4):1208–18.

Gomez-Hernandez, J. M., M. de Miguel, B. Larrosa, D. Gonzalez, and L. C. Barrio. 2003. Molecular basis of calcium regulation in connexin-32 hemichannels. *Proc Natl Acad Sci U S A* 100 (26):16030–5.

Goodenough, D. A., and D. L. Paul. 2009. Gap junctions. *Cold Spring Harb Perspect Biol* 1 (1):a002576.

Guo, C. X., H. Tran, C. R. Green, H. V. Danesh-Meyer, and M. L. Acosta. 2014. Gap junction proteins in the light-damaged albino rat. *Mol Vis* 20:670–82.

Hansen, A. J. 1985. Effect of anoxia on ion distribution in the brain. *Physiol Rev* 65 (1):101–48.

Hansen, D. B., T. H. Braunstein, M. S. Nielsen, and N. MacAulay. 2014. Distinct permeation profiles of the connexin 30 and 43 hemichannels. *FEBS Lett* 588 (8):1446–57.

Hawat, G., M. Benderdour, G. Rousseau, and G. Baroudi. 2010. Connexin 43 mimetic peptide Gap26 confers protection to intact heart against myocardial ischemia injury. *Pflugers Arch* 460 (3):583–92.

Iwabuchi, S., and K. Kawahara. 2011. Functional significance of the negative-feedback regulation of ATP release via pannexin-1 hemichannels under ischemic stress in astrocytes. *Neurochem Int* 58 (3):376–84.

Iyyathurai, J., C. D'Hondt, N. Wang, M. De Bock, B. Himpens, M. A. Retamal, J. Stehberg, L. Leybaert, and G. Bultynck. 2013. Peptides and peptide-derived molecules targeting the intracellular domains of Cx43: Gap junctions versus hemichannels. *Neuropharmacology* 75:491–505.

Kang, J., N. Kang, D. Lovatt, A. Torres, Z. Zhao, J. Lin, and M. Nedergaard. 2008. Connexin 43 hemichannels are permeable to ATP. *J Neurosci* 28 (18):4702–11.

Kielian, T. 2008. Glial connexins and gap junctions in CNS inflammation and disease. *J Neurochem* 106 (3):1000–16.

Klaassen, L. J., Z. Sun, M. N. Steijaert, P. Bolte, I. Fahrenfort, T. Sjoerdsma, J. Klooster et al. 2011. Synaptic transmission from horizontal cells to cones is impaired by loss of connexin hemichannels. *PLoS Biol* 9 (7):e1001107.

Laird, D. W. 2006. Life cycle of connexins in health and disease. *Biochem J* 394 (Pt 3):527–43.

Locovei, S., L. Bao, and G. Dahl. 2006. Pannexin 1 in erythrocytes: Function without a gap. *Proc Natl Acad Sci U S A* 103 (20):7655–9.

Lopez, W., Y. Liu, A. L. Harris, and J. E. Contreras. 2014. Divalent regulation and intersubunit interactions of human connexin26 (Cx26) hemichannels. *Channels (Austin)* 8 (1):1–4.

Ma, W., H. Hui, P. Pelegrin, and A. Surprenant. 2009. Pharmacological characterization of pannexin-1 currents expressed in mammalian cells. *J Pharmacol Exp Ther* 328 (2):409–18.

Mallard, C., J. O. Davidson, S. Tan, C. R. Green, L. Bennet, N. J. Robertson, and A. J. Gunn. 2014. Astrocytes and microglia in acute cerebral injury underlying cerebral palsy associated with preterm birth. *Pediatr Res* 75 (1–2):234–40.

Martinon, F., V. Petrilli, A. Mayor, A. Tardivel, and J. Tschopp. 2006. Gout-associated uric acid crystals activate the NALP3 inflammasome. *Nature* 440, 237–241.

O'Carroll, S. J., M. Alkadhi, L. F. Nicholson, and C. R. Green. 2008. Connexin 43 mimetic peptides reduce swelling, astrogliosis, and neuronal cell death after spinal cord injury. *Cell Commun Adhes* 15 (1):27–42.

O'Carroll, S. J., D. L. Becker, J. O. Davidson, A. J. Gunn, L. F. Nicholson, and C. R. Green. 2013b. The use of connexin-based therapeutic approaches to target inflammatory diseases. *Methods Mol Biol* 1037:519–46.

O'Carroll, S. J., C. A. Gorrie, S. Velamoor, C. R. Green, and L. F. Nicholson. 2013a. Connexin43 mimetic peptide is neuroprotective and improves function following spinal cord injury. *Neurosci Res* 75 (3):256–67.

Orellana, J. A., N. Froger, P. Ezan, J. X. Jiang, M. V. Bennett, C. C. Naus, C. Giaume, and J. C. Saez. 2011. ATP and glutamate released via astroglial connexin 43 hemichannels mediate neuronal death through activation of pannexin 1 hemichannels. *J Neurochem* 118 (5):826–40.

Orellana, J. A., A. D. Martinez, and M. A. Retamal. 2013. Gap junction channels and hemichannels in the CNS: Regulation by signaling molecules. *Neuropharmacology* 75:567–82.

Panchin, Y., I. Kelmanson, M. Matz, K. Lukyanov, N. Usman, and S. Lukyanov. 2000. A ubiquitous family of putative gap junction molecules. *Curr Biol* 10 (13):R473–4.

Patel, D., X. Zhang, and R. D. Veenstra. 2014. Connexin hemichannel and pannexin channel electrophysiology: How do they differ? *FEBS Lett* 588 (8):1372–8.

Pelegrin, P., and A. Surprenant. 2006. Pannexin-1 mediates large pore formation and interleukin-1beta release by the ATP-gated P2X7 receptor. *EMBO J* 25 (21):5071–82.

Penuela, S., R. Bhalla, X. Q. Gong, K. N. Cowan, S. J. Celetti, B. J. Cowan, D. Bai, Q. Shao, and D. W. Laird. 2007. Pannexin 1 and pannexin 3 are glycoproteins that exhibit many distinct characteristics from the connexin family of gap junction proteins. *J Cell Sci* 120 (Pt 21):3772–83.

Penuela, S., R. Gehi, and D. W. Laird. 2013. The biochemistry and function of pannexin channels. *Biochim Biophys Acta* 1828 (1):15–22.

Plesnila, N., S. Zinkel, D. A. Le, S. Amin-Hanjani, Y. Wu, J. Qiu, A. Chiarugi et al. 2001. BID mediates neuronal cell death after oxygen/glucose deprivation and focal cerebral ischemia. *Proc Natl Acad Sci U S A* 98 (26):15318–23.

Ponsaerts, R., E. De Vuyst, M. Retamal, C. D'Hondt, D. Vermeire, N. Wang, H. De Smedt et al. 2010. Intramolecular loop/tail interactions are essential for connexin 43-hemichannel activity. *FASEB J* 24 (11):4378–95.

Quist, A. P., S. K. Rhee, H. Lin, and R. Lal. 2000. Physiological role of gap-junctional hemichannels: Extracellular calcium-dependent isosmotic volume regulation. *J Cell Biol* 148 (5):1063–74.

Ramachandran, S., L. H. Xie, S. A. John, S. Subramaniam, and R. Lal. 2007. A novel role for connexin hemichannel in oxidative stress and smoking-induced cell injury. *PLoS One* 2 (8):e712.

Retamal, M. A., N. Froger, N. Palacios-Prado, P. Ezan, P. J. Saez, J. C. Saez, and C. Giaume. 2007. Cx43 hemichannels and gap junction channels in astrocytes are regulated oppositely by proinflammatory cytokines released from activated microglia. *J Neurosci* 27 (50):13781–92.

Rodriguez-Sinovas, A., J. A. Sanchez, C. Fernandez-Sanz, M. Ruiz-Meana, and D. Garcia-Dorado. 2012. Connexin and pannexin as modulators of myocardial injury. *Biochim Biophys Acta* 1818 (8):1962–70.

Rosner, M. J., and D. P. Becker. 1984. Experimental brain injury: Successful therapy with the weak base, tromethamine: With an overview of CNS acidosis. *J Neurosurg* 60 (5):961–71.

Rutledge, E. M., A. A. Mongin, and H. K. Kimelberg. 1999. Intracellular ATP depletion inhibits swelling-induced D-[3H]aspartate release from primary astrocyte cultures. *Brain Res* 842 (1):39–45.

Saez, J. C., and L. Leybaert. 2014. Hunting for connexin hemichannels. *FEBS Lett* 588 (8):1205–11.

Saez, J. C., M. A. Retamal, D. Basilio, F. F. Bukauskas, and M. V. Bennett. 2005. Connexin-based gap junction hemichannels: Gating mechanisms. *Biochim Biophys Acta* 1711 (2):215–24.

Sanchez, H. A., J. A. Orellana, V. K. Verselis, and J. C. Saez. 2009. Metabolic inhibition increases activity of connexin-32 hemichannels permeable to Ca2+ in transfected HeLa cells. *Am J Physiol Cell Physiol* 297 (3):C665–78.

Sandilos, J. K., Y. H. Chiu, F. B. Chekeni, A. J. Armstrong, S. F. Walk, K. S. Ravichandran, and D. A. Bayliss. 2012. Pannexin 1, an ATP release channel, is activated by caspase cleavage of its pore-associated C-terminal autoinhibitory region. *J Biol Chem* 287 (14):11303–11.

Schalper, K. A., N. Palacios-Prado, M. A. Retamal, K. F. Shoji, A. D. Martinez, and J. C. Saez. 2008. Connexin hemichannel composition determines the FGF-1-induced membrane permeability and free [Ca2+]i responses. *Mol Biol Cell* 19 (8):3501–13.

Seminario-Vidal, L., S. F. Okada, J. I. Sesma, S. M. Kreda, C. A. van Heusden, Y. Zhu, L. C. Jones et al. 2011. Rho signaling regulates pannexin 1-mediated ATP release from airway epithelia. *J Biol Chem* 286 (30):26277–86.

Silverman, W., S. Locovei, and G. Dahl. 2008. Probenecid, a gout remedy, inhibits pannexin 1 channels. *Am J Physiol Cell Physiol* 295 (3):C761–7.

Silverman, W. R., J. P. de Rivero Vaccari, S. Locovei, F. Qiu, S. K. Carlsson, E. Scemes, R. W. Keane, and G. Dahl. 2009. The pannexin 1 channel activates the inflammasome in neurons and astrocytes. *J Biol Chem* 284 (27):18143–51.

Strasser, U., and G. Fischer. 1995. Quantitative measurement of neuronal degeneration in organotypic hippocampal cultures after combined oxygen/glucose deprivation. *J Neurosci Methods* 57 (2):177–86.

Swayne, L. A., C. D. Sorbara, and S. A. Bennett. 2010. Pannexin 2 is expressed by postnatal hippocampal neural progenitors and modulates neuronal commitment. *J Biol Chem* 285 (32):24977–86.

Tasca, C. I., T. Dal-Cim, and H. Cimarosti. 2015. In vitro oxygen-glucose deprivation to study ischemic cell death. *Methods Mol Biol* 1254:197–210.

Thimm, J., A. Mechler, H. Lin, S. Rhee, and R. Lal. 2005. Calcium-dependent open/closed conformations and interfacial energy maps of reconstituted hemichannels. *J Biol Chem* 280 (11):10646–54.

Thomas, A. P. 2000. Sharing calcium opens new avenues of signalling. *Nat Cell Biol* 2 (7):E126–7.

Torres, A., F. Wang, Q. Xu, T. Fujita, R. Dobrowolski, K. Willecke, T. Takano, and M. Nedergaard. 2012. Extracellular Ca(2)(+) acts as a mediator of communication from neurons to glia. *Sci Signal* 5 (208):ra8.

Verselis, V. K., and M. Srinivas. 2008. Divalent cations regulate connexin hemichannels by modulating interfacial voltage-dependent gating. *J Gen Physiol* 132 (3):315–27.

Vroman, R., L. J. Klaassen, and M. Kamermans. 2013. Ephaptic communication in the vertebrate retina. *Front Hum Neurosci* 7:612.

Wang, N., M. De Bock, G. Antoons, A. K. Gadicherla, M. Bol, E. Decrock, W. H. Evans, K. R. Sipido, F. F. Bukauskas, and L. Leybaert. 2012a. Connexin mimetic peptides inhibit Cx43 hemichannel opening triggered by voltage and intracellular Ca2+ elevation. *Basic Res Cardiol* 107 (6):304.

Wang, N., M. De Bock, E. Decrock, M. Bol, A. Gadicherla, M. Vinken, V. Rogiers, F. F. Bukauskas, G. Bultynck, and L. Leybaert. 2013a. Paracrine signaling through plasma membrane hemichannels. *Biochim Biophys Acta* 1828 (1):35–50.

Wang, N., E. De Vuyst, R. Ponsaerts, K. Boengler, N. Palacios-Prado, J. Wauman, C. P. Lai et al. 2013b. Selective inhibition of Cx43 hemichannels by Gap19 and its impact on myocardial ischemia/reperfusion injury. *Basic Res Cardiol* 108 (1):309.

Wang, X., X. Xu, M. Ma, W. Zhou, Y. Wang, and L. Yang. 2012b. pH-dependent channel gating in connexin26 hemichannels involves conformational changes in N-terminus. *Biochim Biophys Acta* 1818 (5):1148–57.

Weilinger, N. L., P. L. Tang, and R. J. Thompson. 2012. Anoxia-induced NMDA receptor activation opens pannexin channels via Src family kinases. *J Neurosci* 32 (36):12579–88.

Werner, C., and K. Engelhard. 2007. Pathophysiology of traumatic brain injury. *Br J Anaesth* 99 (1):4–9.

Wicki-Stordeur, L. E., A. K. Boyce, and L. A. Swayne. 2013. Analysis of a pannexin 2-pannexin 1 chimeric protein supports divergent roles for pannexin C-termini in cellular localization. *Cell Commun Adhes* 20 (3–4):73–9.

Xu, J., L. He, S. H. Ahmed, S. W. Chen, M. P. Goldberg, J. S. Beckman, and C. Y. Hsu. 2000. Oxygen-glucose deprivation induces inducible nitric oxide synthase and nitrotyrosine expression in cerebral endothelial cells. *Stroke* 31 (7):1744–51.

Ye, Z. C., M. S. Wyeth, S. Baltan-Tekkok, and B. R. Ransom. 2003. Functional hemichannels in astrocytes: A novel mechanism of glutamate release. *J Neurosci* 23 (9):3588–96.

Young, W., V. Yen, and A. Blight. 1982. Extracellular calcium ionic activity in experimental spinal cord contusion. *Brain Res* 253 (1–2):105–13.

Zhang, J., S. J. O'Carroll, H. V. Danesh-Meyer, H. C. Van der Heyde, D. L. Becker, Nicholson L. F. B., and C. R. Green. 2013. Connexin-based therapeutic approaches to inflammation in the central nervous system. In *Connexin Cell Communication Channels: Roles in The Immune System and Immunopathology*, Oviedo-Orta, E., B. R. Kwak, and W. H. Evans (eds.), 273–305. Boca Raton, FL: CRC Press.

8 Whole-Cell Patch Clamp Recordings of Unitary Connexin 43 Hemichannel Currents

Nan Wang, Alessio Lissoni, Maarten De Smet, Karin R. Sipido, and Luc Leybaert

CONTENTS

8.1 INTRODUCTION

Connexins are tetraspan transmembrane proteins engaging in the formation of gap junctions that facilitate electrical and metabolic couplings between neighboring cells (Goodenough and Paul, 2009). Prior to intracellular trafficking to the plasma membrane, the connexins synthesized in the ER assemble into a hexameric hemichannel, also known as connexon, in the trans-Golgi network (Laird, 2006). There is ample evidence showing that more than the gap junction precursors en route to dock with their counterparts, the connexin hemichannels retained at the nonjunctional site of the plasma membrane are functional and are regulated channels in their own right, forming an indispensable autocrine/paracrine signaling pathway that tightly controls cellular responses in physiology and pathology (Wang et al., 2013b; Evans et al., 2006). Of all the connexin isoforms, Cx43 is the most widely expressed member in the human body, including the heart and the brain (Schulz et al., 2015). Typically, unapposed Cx43 hemichannels remain closed under resting conditions but can be activated by electrical and chemical stimuli including membrane depolarization (Contreras et al., 2003), mechanical stimulation, altered redox potential, reduced extracellular Ca^{2+} concentration, and increased intracellular Ca^{2+} ($[Ca^{2+}]_i$) (Wang et al., 2013b). An open Cx43 hemichannel constitutes a large-conductance and nonselective transmembrane conduit allowing the passive diffusion of ions and substances with molecular weight of 1–2 kDa such as ATP and nicotinamide adenine dinucleotide, giving rise to various cellular processes including cell volume regulation (Quist et al., 2000), dynamic Ca^{2+} signals (De Bock et al., 2012a,b), release of neurotransmitter (Stehberg et al., 2012; Romanov et al., 2007), inflammatory responses (Calder et al., 2015; Lu et al., 2012), and cell death (Decrock et al., 2009; Kalvelyte et al., 2003). Studying the functions of Cx43 hemichannels mainly relies on three types of methods: (1) knock-down or knock-out approach which nevertheless depletes both gap junctions and hemichannels, (2) indirect measures making use of the cellular uptake of fluorescent probes with a molecular weight below 1 kDa or the release of endogenous ATP, and (3) electrophysiological recording of hemichannel currents. This chapter focuses on the third method that provides the most direct measure of hemichannel behavior at the single-channel level. The technique is unique in that the single-channel conductance serves as the signature of the channel being recorded from, and the voltage–conductance profile furthermore helps in identifying the hemichannel type. Here, we give particular attention to the technical aspects of performing high-quality recordings using an overexpression system as well as acutely dissociated adult cardiomyocytes expressing endogenous Cx43. Importantly, we use a whole-cell recording approach to resolve single-hemichannel currents. This approach is possible because of the rather low open probability of the hemichannels, with maximally about 3–4 channels being active in a single isolated ventricular cardiomyocyte. Yet because of the large conductance of the Cx43 hemichannels, such low activity can have serious consequences on the electrical excitability of these cells. This is also the case in neurons, where it has been demonstrated that the opening of a single hemichannel may in fact trigger action potentials (Moore et al., 2014).

8.2 RECODING OF UNITARY CONNEXIN 43 HEMICHANNEL CURRENTS IN AN OVEREXPRESSION SYSTEM

The functional expression of Cx43 hemichannels in heterologous systems provides the easiest way for addressing the biophysical fingerprint of hemichannels. Through electrophysiological recordings that allow detailed analysis at the highest resolution, our understanding of electrically and chemically driven Cx43 hemichannel gatings have substantially advanced, which in turn has made it possible to prosper the designs of pharmacological tools that target Cx43 hemichannels with enhanced specificity and efficacy.

8.2.1   Host Cells and Plasmid DNA

Communication-deficient cell lines including HeLa and human embryonic kidney (HEK) 293 cells have been used for electrophysiological recordings of Cx43 hemichannels through the introduction of cDNA encoding Cx43 given their connexin-free background and low expression level of endogenous ion channels (Contreras et al., 2003; Wang et al., 2012; John et al., 1999). Although *Xenopus* oocyte has been widely used for functional studies of cloned Cx26, Cx37, and Cx46, the injection of Cx43 mRNA yields dysfunctional Cx43 hemichannels due to impaired interdomain interaction between the intracellular loop and the CT (Ponsaerts et al., 2010). We have made ample use of HeLa cells stably transfected with Cx43 (kindly provided by Dr. Willecke [Elfgang et al., 1995]) to investigate the basic properties of Cx43 hemichannels. To establish the cell line, mouse Cx43 gene was cloned into the expression vector PMJgree that contains the cytomegalovirus (CMV) promoter and a puromycin N-acetyl-transferase gene-encoding region. Cx43-expressing plasmid may include a fluorescent protein such as a GFP tagged to the Cx43CT for identifying successfully transfected cells. It is important to note that the gating behaviors of C-terminally tagged Cx43 hemichannels slightly deviate from those of wild-type channels, while adding a bulk molecule to the N-terminus fails to form functional hemichannels despite of the abundance of Cx43 at the plasma membrane (Contreras et al., 2003). Here, we describe a protocol for whole-cell recordings of untagged Cx43 hemichannels.

8.2.2   Solution

Standard ECS contains (in mM) 140 NaCl, 4 KCl, 2 $CaCl_2$, 2 $MgCl_2$, 2 pyruvic acid, 5 glucose, and 5 HEPES, and pH 7.4. Modifications including the substitution of KCl by the equimolar of CsCl and the addition of $BaCl_2$ (1 mM) minimize Na^+-K^+-ATPase pump activities and K^+ currents.

The compositions of the pipette solution must be amenable to the nature of the channels of interest while concomitantly eliminating intrinsic channel activities in the cell line of choice. The pipette solutions for measurements in HeLa cells contain (in mM) 130 CsCl, 10 Na-aspartate (NaAsp), 0.26 $CaCl_2$, 1 $MgCl_2$, 2 ethylene glycol tetraacetic acid (EGTA), 5 tetraethylammonium TEACl, and 5 HEPES; the pH is adjusted to 7.2 with CsOH. Cs^+, the charge carrier of Cx43 hemichannel-mediated outward currents, does not pass through most of the K^+ channels, thereby suppressing the K^+ channel currents. TEACl, a K^+ channel blocker, further abolishes the K^+ current contamination. Cytoplasmic $[Ca^{2+}]_i$ is maintained at the physiological levels of ~50–100 nM by the inclusion of a Ca^{2+} chelator such as EGTA or 1,2-bis(o-aminophenoxy)ethane-N,N,N′,N′-tetraacetic acid (BAPTA) with much higher binding affinity for Ca^{2+} than Mg^{2+} (intracellular Mg^{2+} level is typically 10,000-fold higher than $[Ca^{2+}]_i$). The EGTA forms more stable complex with Ca^{2+} compared to BAPTA (Tsien, 1980), while BAPTA exhibits a higher rate of Ca^{2+}-binding and, more importantly, little pH dependence of its Ca^{2+}-binding affinity. Thus, one must select appropriate Ca^{2+}-chelating reagents in accordance with the aim of the experiments and calculate the free Ca^{2+} concentration at different pH levels and ionic strengths using online softwares such as Webmax Standard and MaxChelator. The osmolarity of the pipette solution lower than that of the ECS preferably keeps the cellular volume intact and promotes steady recordings. Therefore, the osmolarity of the ECS is adjusted so that it is ~10 mOsm higher than that of the pipette solution for HeLa-Cx43 cells.

8.2.3 EQUIPMENT

A patch-clamp amplifier consisting of a current-to-voltage convertor and a main control unit are commercially available from HEKA Elektronik, Axon Instruments, or others. An external ADDA convertor serves as a data-acquisition device communicating between most of the patch-clamp amplifiers and the computers, although some newer modules of amplifiers (such as HEKA EPC10) with a built-in ADDA convertor are operated solely through a software designated by the manufacture. WinWCP designed by Dr. J. Dempster (University of Strathclyde, UK) provides a third-party acquisition system featuring both recording and analysis functions.

8.2.4 METHODS TO OBTAIN UNITARY HEMICHANNEL CURRENTS

Mammalian cells stably transfected with Cx43 are a good starting point to become acquainted with recording Cx43 hemichannel currents (Wang et al., 2012). The cells are plated at a density of 50,000 cells/cm^2 on a glass coverslip 24 hours prior to patch-clamp experiments, and this guarantees a large population of solitary cells in culture on the day of measurements. Low-resistance gap junctions in a cell that is in contact with the neighboring cells may affect the input resistance, and therefore, communicating cells are not ideal for whole-cell recordings and should be avoided. The hemichannel currents can also be assessed with the use of transient transfection approaches including electroporation and cationic lipid that aids DNA delivery. In our laboratory, we have used Lipofectamine LTX and PLUS™ reagents (Sigma) to obtain high-transfection efficiency in HeLa cells cultured in wells of a 24-well plate. On the day of transfection, the cells grown to 50–60% confluency in each well were subject to 50 μL of DNA–lipid complex containing 0.5 μg of Cx43 plasmid DNA (e.g., pCMV6-XL5 vector from OriGene), 5% LTX, and 1% PLUS reagents. After a 48-hour incubation at 37°C in 5% CO$_2$, the cells with transfection efficiency of up to ~70% are ready for electrophysiological measurements.

To proceed with the whole-cell recording, a coverslip that contains cells stably or transiently expressing Cx43 is transferred into a recording chamber filled with a modified Krebs–Ringer external solution. A 3 M KCl agar bridge connects the recording chamber to a grounding compartment in order to eliminate liquid junction potentials. Ideally, the patch pipettes with a resistance of ~3–4 MΩ are used. A four-stage pull procedure in which the heat and the velocity of the initial three steps shape the taper of the pipette, and the final pull that tightly controls the pipette resistance is recommended to minimize the variability among pipette fabrication. A patch pipette is then loaded with a whole-cell recording pipette solution and positioned toward a solitary HeLa cell. To obtain a successful seal between the pipette and the plasma membrane, a crucial step consists of applying positive pressure to the patch pipette before dipping the tip of the pipette in the bath solution. A 10 mV test pulse at 200 Hz through the patch pipette is used to detect a fairly slight reduction of the pipette resistance once the patch pipette gently presses against the cell. This change can vary from one to few megaohms depending on the cell type and the recording conditions. An immediate negative pressure applied to the patch pipette brings the pipette resistance progressively above 1 GΩ at a holding potential of −70 mV (gigaseal formation). The plasma membrane is later ruptured through the repetitive injection of brief negative pressures to the patch pipette. The whole-cell configuration is evidenced by large transient capacitive currents arising from charging and discharging of the cell membrane. An access resistance of <10 MΩ should be sustained throughout the recording, allowing accurate clamping voltage and better cellular loading with the internal solution. After the whole-cell recording mode is formed, the membrane potential is held at −30 mV, which is a potential that resembles the resting condition of HeLa cells. The data are acquired at the sample frequency of 4–10 kHz and filtered at 1 kHz (seven-pole Bessel filter) with voltage commands following the protocols described:

1. Linear voltage ramps that provide a first approximation of voltage-dependent hemichannel openings (Figure 8.1)
2. Repetitive 30-second voltage steps from −30 mV to $V_m \geq +40$ mV that measure steady-state hemichannel currents (Figure 8.2) (Note that there is a

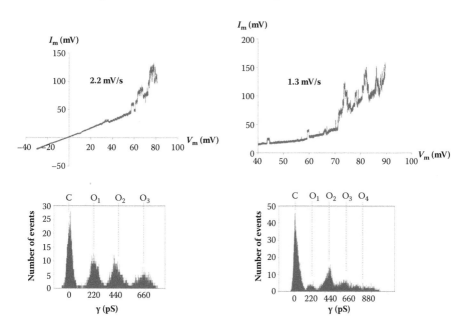

FIGURE 8.1 I_m–V_m plots illustrating voltage ramp experiments in Cx43-expressing HeLa cells. A fast ramp from −30 to +80 mV (2.2 mV/s) revealed the appearance of unitary hemichannel events from +60 mV onward. A slow ramp (1.3 mV/s) exemplifies the detailed hemichannel activation over a range of positive V_m. All-point histograms determined from the corresponding traces earlier show ~220 pS unitary conductance at the voltages being applied.

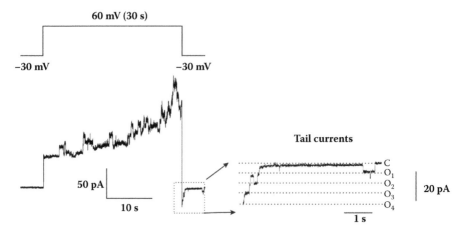

FIGURE 8.2 An example trace showing unitary current events elicited by stepping V_m to +60 mV (30 seconds). The expanded trace highlights the stepwise closing of the hemichannels upon switching back to −30 mV.

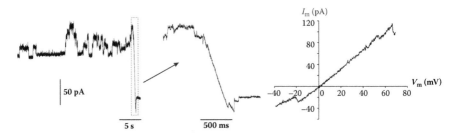

FIGURE 8.3 Reversal potential of open hemichannels revealed by a 500-millisecond ramp. V_m was stepped to +70 mV (30 seconds) to activate the steady-state hemichannel currents. This was followed by a fast ramp repolarizing to −40 mV, applied to investigate the reversal potential of the open hemichannels. The expanded trace depicts the current evolution during the fast ramp. The $I–V$ plot indicates a reversal potential of ~0 for an open hemichannel.

delay [often of the order of seconds] before the unitary hemichannel currents appear, indicating relatively slow activation kinetics [~22 seconds] [Wang et al., 2013a]. The tail currents arising from the hemichannel deactivation appears upon switching V_m back to −30 mV.)

3. A 500-millisecond ramp from +70 mV to −40 mV that precisely marks the reversal potential of the open hemichannels (Figure 8.3)

Cx43 hemichannel activities feature a limited open probably (single-channel events in whole-cell recording), a reversal potential of ~0 mV and a unitary conductance of ~220 pS. A typical V_m-dependent stimulation of unitary currents with slow activation kinetics as well as the stepwise deactivation upon repolarization serve as additional hallmarks for Cx43 hemichannels.

8.2.5 APPLICATIONS

Connexin hemichannel inhibitors including the connexin mimetic peptides Gap26, Gap27, and Gap19 that correspond to specific sequences on the first extracellular loop, the second extracellular loop, and the cytoplasmic loop (CL) of Cx43, respectively, have been proposed as experimental and possibly also therapeutic tools, because they have better selectivity than general inhibitors like CBX. It is only recently that their actions have been investigated at the single-hemichannel level, bringing up interesting information on how they interact with the channel domain and how they affect Cx43 hemichannel gating. More importantly, electrophysiological data have demonstrated that, unlike Gap26 and Gap27 that impair both the gap junctions and the hemichannels, Gap19 peptide selectively affects the Cx43 hemichannels without inhibiting the junctional coupling. Interestingly, all three peptides inhibit Cx43 hemichannel openings triggered by voltage and $[Ca^{2+}]_i$ increase (steady-state Ca^{2+} increase to 200–500 nM via whole-cell recording pipette). $[Ca^{2+}]_i$ elevation reduces the threshold voltage for hemichannel activation, whereas the peptides counteracted this effect. Collectively, single-channel data obtained from HeLa-Cx43 cells highlight an intriguing network of voltage and Ca^{2+} effects on the Cx43 hemichannels and its modulation by connexin-targeting peptides most likely through a common target.

8.3 RECORDING OF CONNEXIN 43 HEMICHANNEL CURRENTS IN ACUTELY ISOLATED VENTRICULAR CARDIOMYOCYTES

To date, the presence of three principle connexins—Cx40, Cx43, and Cx45—has been well established in the adult mammalian heart, each characterized by a site-specific distribution pattern (reviewed by Jansen et al. [2010] and Severs et al. [2008]). Cx43 is the most abundant cardiac connexin that is primarily expressed in the working myocardium of the ventricles. Cx40 is a major atrial connexin, while Cx45 mostly constitutes the conduction system. Cx30.2 is a fourth connexin that is present in the atrioventricular node of mouse hearts (Kreuzberg et al., 2006); its orthologous protein Cx31.9 is not found in humans (Kreuzberg et al., 2009). Previous studies have demonstrated that Cx43 in the intercalated disk segregates into a nexus composed of gap junctions and a perinexus where a pool of unapposed hemichannels surrounds the junctional plaques (Rhett et al., 2013; Rhett and Gourdie, 2012). In line with this, we have demonstrated the presence of single-channel plasma membrane currents in ventricular cardiomyocytes (acutely isolated from adult pig) with biophysical behaviors that are typical for Cx43 hemichannels. Further work also showed a strong promotion of hemichannel currents under simulated ischemia, and in vivo work in a myocardial ischemia/reperfusion mouse model demonstrated that Gap19 inhibition of Cx43 hemichannels resulted in a reduction of the infarct size (Wang et al., 2013c). Here, we describe the cardiomyocyte patch-clamp recording protocol employed in this study, which allows clear isolation of unitary Cx43 hemichannel activities. Cx43 distribution seems to stay unaffected after the enzymatic digestion as reflected by the observation of the intact Cx43 structure at the intercalated disks of dissociated cardiomyocytes (Severs et al., 1989; Mazet et al., 1985). Prior to patch-clamp experiments, the isolated cells are suspended in a low-Ca^{2+} Tyrode solution (0.18 mM Ca^{2+}), in which the Ca^{2+} level is gradually restored to 1.8 mM at increments of 0.8 mM in two steps. Ca^{2+}-tolerant cells are stored at room temperature and used within 12 hours after isolation (cardiomyocyte isolation protocol previously described in the study by Stankovicova et al. [2000]).

8.3.1 SOLUTION

The Tyrode external solution used for bulk bath perfusion contains (in mM) 137 NaCl, 5.4 KCl, 1.8 $CaCl_2$, 0.5 $MgCl_2$, 11.8 HEPES, and 10 glucose, and pH adjusted to 7.4 with NaOH. The pipette solution is composed of (in mM) 120 CsCl, 5 NaCl, 1 $CaCl_2$, 1 $MgCl_2$, 10 EGTA, 10 TEACl, 2 MgATP, 10 HEPES, and pH adjusted to 7.2 with CsOH. The free Ca^{2+} concentration of the pipette solution is estimated to be ~50 nM as calculated with WEBMAX. The inclusion of a relatively high concentration of Mg^{2+} in the pipette suppresses TRPM6 and TRPM7 channels (Gwanyanya et al., 2004). When recording unitary hemichannel currents, the perforated cell is locally perfused with a Cs-Tyrode-based solution consisting of (in mM) 130 NaCl, 10 CsCl, 1.8 $CaCl_2$, 0.5 $MgCl_2$, 10 glucose, 11.8 HEPES, and pH 7.4. The cells exposed to this solution before being voltage clamped are depolarized due to the blockade of Na^+-K^+-ATPase pumps. Of note, eliminating unwanted channel currents with the use of pharmacological reagents may have implications on Cx43 hemichannel activities due to their nonspecific actions (Verselis and Srinivas, 2013). For example, 5-nitro-2-(3-phenylpropylamino)benzoic acid and flufenamic acid widely adopted as Cl^- channel antagonists are also potent Cx43 hemichannel inhibitors (Eskandari et al., 2002; Ye et al., 2009). Ionic substitution is an alternative approach that readily reduces the background membrane currents without affecting the hemichannel activities. Replacing NaCl in the Cs-based Tyrode solution with NaAsp brings the reversal potential of Cl^- up to +40 mV, at which the Cx43 hemichannel currents thrive, thereby minimizing outward currents conducted through the Cl^- channels. Finally, the inclusion of the selective hemichannel blocker Gap19 peptide in the whole-cell recording pipette solution allows the unequivocal identification of Cx43 hemichannels currents. This 9-mer synthetic peptide inhibiting Cx43 hemichannel currents with an IC_{50} of 6.5 μM is derived from a sequence located at the CL of Cx43 protein (K128–K136 in the L2 domain) and binds to the last 10 amino acids of the Cx43CT, resulting in perturbed interdomain CT–CL interaction that locks the Cx43 hemichannels in a closed state (Wang et al., 2013c). A mutant peptide containing the first isoleucine mutated to alanine fails to affect the Cx43 hemichannel currents due to impaired hydrogen bond formation with its counterpart in the CL, and it can therefore be used as a control peptide.

8.3.2 METHODS

Isolated pig ventricular cardiomyocytes are transferred to a recording bath filled with Tyrode external solution. It takes 10 minutes for the suspended cells to settle and adhere to the glass surface of the bath. Following the start of the bulk perfusion,

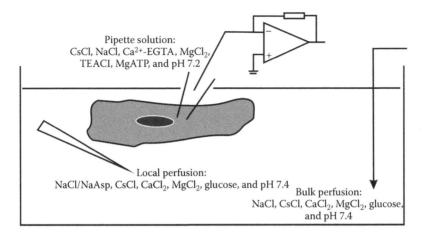

FIGURE 8.4 Recording conditions applied to identify unitary Cx43 hemichannel currents in ventricular cardiomyocytes acutely isolated from a pig.

a rod-shaped, striated, and quiescent cardiomyocyte is selected for electrophysiolgical recording. The fine outlet of the local perfusion system is then positioned adjacent to the cell of choice (Figure 8.4). The ideal resistance of a patch pipettes for recording from an adult ventricular cardiomyocyte is ~2–3 MΩ. Although lower pipette resistance makes the gigaseal formation more challenging, the large size of the pipette tip helps to attain a low-access resistance and improved dialysis of the cell interior.

Once the fast pipette capacitance is compensated upon the gigaseal formation, a whole-cell recording configuration can be obtained by applying a strong suction to the pipette at a holding potential of −70 mV. After the break-in, brief injections of slightly positive pressure to the patch pipette are often necessary to improve and stabilize the access resistance that is not more than 7 MΩ. Locally perfusing the cells with Cs-based Tyrode solution (K$^+$ replaced by Cs$^+$) imposes an immediate increase in the holding currents. The typical hemichannel activities are recorded by repetitively stepping the cell from the holding potential of −70 mV to $+V_m > +10$ mV. $I_m - V_m$ plots established by open hemichannels at positive V_m and tail currents at negative V_m provide information on two biophysical parameters of the channel: a reversal potential of ~0 mV and a ~200 pS unitary conductance. Similar hemichannel currents can be recorded when the Cs-based local perfusion solution contains NaAsp instead of NaCl (Figure 8.5), excluding the contaminant of any Cl$^-$ channel currents.

Cx43 hemichannels elicited by positive V_m are mechanistically distinct from voltage-gated Na$^+$, K$^+$, and L-type Ca^{2+} channels; these channels exhibit (1) voltage-dependent activation kinetics in the range of tens of seconds, (2) no apparent inactivation at positive V_m, (3) sporadic activities due to low open probability, and (4) large unitary conductance of ~200 pS and lack of ionic selectivity. With the typical values for membrane capacitance of a ventricular cardiomyocyte (90–150 pF), the access resistance (<7 MΩ), and the membrane resistance (>300 MΩ), the membrane time constant falls into the several hundred microseconds to one millisecond range, which

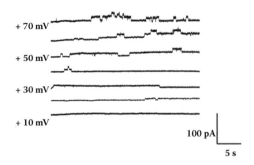
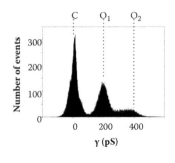

FIGURE 8.5 Biophysical properties of Cx43 hemichannels in pig ventricular cardiomy-ocytes. Whole-cell recording of single Cx43 hemichannels in ventricular cardiomyocytes demonstrated a V_m-dependent activation of the Cx43 hemichannels (30-second voltage step). The extracellular Cl⁻ ions were mostly replaced by Asp⁻ to elevate the reversal potential of Cl⁻ up to +40 mV. The all-point histogram illustrated below the example traces confirmed a ~200 pS unitary conductance.

is too rapid to influence the slow hemichannel currents. The limited contribution of the hemichannel openings to membrane resistance similarly makes the error in the steady-state value of V_m negligible (on the order of < 1/50).

8.4   CONCLUDING REMARKS

Cx43 hemichannels have emerged as active contributors to membrane conductance and cellular responses, with a broad spectrum of functional implications. However, unraveling their physiological/pathological significance demands direct measures and specific tools unequivocally distinguishing the Cx43 hemichannels from other molecular release/uptake pathways. In this chapter, we have described whole-cell patch clamp recordings in an overexpression system that establishes the signature of Cx43 hemichannels, which consists of their unitary conductance, their reversal potential, their V_m-sensitivity, and their activation kinetics. We further validated the protocol making use of acutely isolated adult cardiomyocyte endogenously expressing a complex ion channel network, confirming the presence of unitary current events that resemble the biophysical fingerprints of Cx43 hemichannels. However, even though a single-channel recording technique provides the most accurate acquisition for the subsequent interpretation of the experimental data, it is not always possible to perform these recordings in intact tissues. The reason for this is the contamination by cell–cell connecting gap junction channels that may compromise the space clamp quality and the limitations in the possibility of locally perfusing the patched cell. Therefore, a continuous effort should be made to develop alternative approaches to explore the hemichannel functions in vivo. For example, mutants with a "gain of function" of hemichannels and a "loss of function" of gap junctions have been characterized (Dobrowolski et al., 2008), allowing better distinction between hemichannel and gap junctional functions in knock-in animal studies.

ACKNOWLEDGMENTS

This work was supported by the Fund for Scientific Research Flanders, Belgium (Fonds Wetenschappelijk Onderzoek; grant numbers G0354.07, G.0140.08, G.0134.09, and G.0298.11 to LL) and by The Interuniversity Attraction Poles Program (Belgian Science Policy, project P7/10 to KRS and LL). Special thanks to Dr. E. Dries and K. Vermeulen for cardiomyocyte isolation.

REFERENCES

Calder, B. W., R. J. Matthew, H. Bainbridge, S. A. Fann, R. G. Gourdie, and M. J. Yost. 2015. Inhibition of connexin 43 hemichannel-mediated ATP release attenuates early inflammation during the foreign body response. *Tissue Eng. Part A.* 21:1752–1762.

Contreras, J. E., J. C. Saez, F. F. Bukauskas, and M. V. Bennett. 2003. Gating and regulation of connexin 43 (Cx43) hemichannels. *Proc. Natl. Acad. Sci. U. S. A.* 100:11388–11393.

De Bock, M., M. Culot, N. Wang, A. da Costa, E. Decrock, M. Bol, G. Bultynck, R. Cecchelli, and L. Leybaert. 2012a. Low extracellular Ca2+ conditions induce an increase in brain endothelial permeability that involves intercellular Ca2+ waves. *Brain Res.* 1487:78–87.

De Bock, M., N. Wang, M. Bol, E. Decrock, R. Ponsaerts, G. Bultynck, G. Dupont, and L. Leybaert. 2012b. Connexin 43 hemichannels contribute to cytoplasmic Ca2+ oscillations by providing a bimodal Ca2+-dependent Ca2+ entry pathway. *J. Biol. Chem.* 287:12250–12266.

Decrock, E., E. De Vuyst, M. Vinken, M. Van Moorhem, K. Vranckx, N. Wang, L. L. Van et al. 2009. Connexin 43 hemichannels contribute to the propagation of apoptotic cell death in a rat C6 glioma cell model. *Cell Death. Differ.* 16:151–163.

Dobrowolski, R., P. Sasse, J. W. Schrickel, M. Watkins, J. S. Kim, M. Rackauskas, C. Troatz et al. 2008. The conditional connexin43G138R mouse mutant represents a new model of hereditary oculodentodigital dysplasia in humans. *Hum. Mol. Genet.* 17:539–554.

Elfgang, C., R. Eckert, H. Lichtenberg-Frate, A. Butterweck, O. Traub, R. A. Klein, D. F. Hulser, and K. Willecke. 1995. Specific permeability and selective formation of gap junction channels in connexin-transfected HeLa cells. *J. Cell Bio.* 129:805–817.

Eskandari, S., G. A. Zampighi, D. W. Leung, E. M. Wright, and D. D. Loo. 2002. Inhibition of gap junction hemichannels by chloride channel blockers. *J. Membr. Biol.* 185:93–102.

Evans, W. H., E. De Vuyst, and L. Leybaert. 2006. The gap junction cellular internet: Connexin hemichannels enter the signalling limelight. *Biochem. J.* 397:1–14.

Goodenough, D. A., and D. L. Paul. 2009. Gap junctions. *Cold Spring Harb. Perspect. Biol.* 1:a002576.

Gwanyanya, A., B. Amuzescu, S. I. Zakharov, R. Macianskiene, K. R. Sipido, V. M. Bolotina, J. Vereecke, and K. Mubagwa. 2004. Magnesium-inhibited, TRPM6/7-like channel in cardiac myocytes: Permeation of divalent cations and pH-mediated regulation. *J. Physiol.* 559:761–776.

Jansen, J. A., T. A. van Veen, J. M. de Bakker, and H. V. van Rijen. 2010. Cardiac connexins and impulse propagation. *J. Mol. Cell Cardiol.* 48:76–82.

John, S. A., R. Kondo, S. Y. Wang, J. I. Goldhaber, and J. N. Weiss. 1999. Connexin-43 hemichannels opened by metabolic inhibition. *J. Biol. Chem.* 274:236–240.

Kalvelyte, A., A. Imbrasaite, A. Bukauskiene, V. K. Verselis, and F. F. Bukauskas. 2003. Connexins and apoptotic transformation. *Biochem. Pharmacol.* 66:1661–1672.

Kreuzberg, M. M., M. Liebermann, S. Segschneider, R. Dobrowolski, H. Dobrzynski, R. Kaba, G. Rowlinson, E. Dupont, N. J. Severs, and K. Willecke. 2009. Human connexin31.9, unlike its orthologous protein connexin30.2 in the mouse, is not detectable in the human cardiac conduction system. *J. Mol. Cell Cardiol.* 46:553–559.

Kreuzberg, M. M., K. Willecke, and F. F. Bukauskas. 2006. Connexin-mediated cardiac impulse propagation: Connexin 30.2 slows atrioventricular conduction in mouse heart. *Trends Cardiovasc. Med.* 16:266–272.

Laird, D. W. 2006. Life cycle of connexins in health and disease. *Biochem. J.* 394:527–543.

Lu, D., S. Soleymani, R. Madakshire, and P. A. Insel. 2012. ATP released from cardiac fibroblasts via connexin hemichannels activates profibrotic P2Y2 receptors. *FASEB J.* 26:2580–2591.

Mazet, F., B. A. Wittenberg, and D. C. Spray. 1985. Fate of intercellular junctions in isolated adult rat cardiac cells. *Circ. Res.* 56:195–204.

Moore, A. R., W. L. Zhou, C. L. Sirois, G. S. Belinsky, N. Zecevic, and S. D. Antic. 2014. Connexin hemichannels contribute to spontaneous electrical activity in the human fetal cortex. *Proc. Natl. Acad. Sci. U. S. A.* 111:E3919–E3928.

Ponsaerts, R., E. De Vuyst, M. Retamal, C. D'hondt, D. Vermeire, N. Wang, S. H. De Smedt et al. 2010. Intramolecular loop/tail interactions are essential for connexin 43-hemichannel activity. *FASEB J.* 24:4378–4395.

Quist, A. P., S. K. Rhee, H. Lin, and R. Lal. 2000. Physiological role of gap-junctional hemichannels: Extracellular calcium-dependent isosmotic volume regulation. *J. Cell Biol.* 148:1063–1074.

Rhett, J. M., and R. G. Gourdie. 2012. The perinexus: A new feature of Cx43 gap junction organization. *Heart Rhythm.* 9:619–623.

Rhett, J. M., R. Veeraraghavan, S. Poelzing, and R. G. Gourdie. 2013. The perinexus: Sign-post on the path to a new model of cardiac conduction? *Trends Cardiovasc. Med.* 23:222–228.

Romanov, R. A., O. A. Rogachevskaja, M. F. Bystrova, P. Jiang, R. F. Margolskee, and S. S. Kolesnikov. 2007. Afferent neurotransmission mediated by hemichannels in mammalian taste cells. *EMBO J.* 26:657–667.

Schulz, R., P. M. Gorge, A. Gorbe, P. Ferdinandy, P. D. Lampe, and L. Leybaert. 2015. Connexin 43 is an emerging therapeutic target in ischemia/reperfusion injury, cardioprotection and neuroprotection. *Pharmacol. Ther.* 153:90–106.

Severs, N. J., A. F. Bruce, E. Dupont, and S. Rothery. 2008. Remodelling of gap junctions and connexin expression in diseased myocardium. *Cardiovasc. Res.* 80:9–19.

Severs, N. J., K. S. Shovel, A. M. Slade, T. Powell, V. W. Twist, and C. R. Green. 1989. Fate of gap junctions in isolated adult mammalian cardiomyocytes. *Circ. Res.* 65:22–42.

Stankovicova, T., M. Szilard, I. De Scheerder, and K. R. Sipido. 2000. M cells and transmural heterogeneity of action potential configuration in myocytes from the left ventricular wall of the pig heart. *Cardiovasc. Res.* 45:952–960.

Stehberg, J., R. Moraga-Amaro, C. Salazar, A. Becerra, C. Echeverria, J. A. Orellana, G. Bultynck et al. 2012. Release of gliotransmitters through astroglial connexin 43 hemichannels is necessary for fear memory consolidation in the basolateral amygdala. *FASEB J.* 26:3649–3657.

Tsien, R. Y. 1980. New calcium indicators and buffers with high selectivity against magnesium and protons: Design, synthesis, and properties of prototype structures. *Biochemistry.* 19:2396–2404.

Verselis, V. K. and M. Srinivas. 2013. Connexin channel modulators and their mechanisms of action. *Neuropharmacology.* 75:517–524.

Wang, N., M. De Bock, G. Antoons, A. K. Gadicherla, M. Bol, E. Decrock, W. H. Evans, K. R. Sipido, F. F. Bukauskas, and L. Leybaert. 2012. Connexin mimetic peptides inhibit Cx43 hemichannel opening triggered by voltage and intracellular Ca2+ elevation. *Basic Res. Cardiol.* 107:304.

Wang, N., M. De Bock, E. Decrock, M. Bol, A. Gadicherla, G. Bultynck, and L. Leybaert. 2013a. Connexin targeting peptides as inhibitors of voltage- and intracellular Ca2+-triggered Cx43 hemichannel opening. *Neuropharmacology.* 75:506–516.

Wang, N., M. De Bock, E. Decrock, M. Bol, A. Gadicherla, M. Vinken, V. Rogiers, F. F. Bukauskas, G. Bultynck, and L. Leybaert. 2013b. Paracrine signaling through plasma membrane hemichannels. *Biochim. Biophys. Acta.* 1828:35–50.

Wang, N., E. De Vuyst, R. Ponsaerts, K. Boengler, N. Palacios-Prado, J. Wauman, C. P. Lai et al. 2013c. Selective inhibition of Cx43 hemichannels by Gap19 and its impact on myocardial ischemia/reperfusion injury. *Basic Res. Cardiol.* 108:309.

Ye, Z. C., N. Oberheim, H. Kettenmann, and B. R. Ransom. 2009. Pharmacological "cross-inhibition" of connexin hemichannels and swelling activated anion channels. *Glia.* 57:258–269.

9 Functional Characterization of Connexin Hemichannels Using *Xenopus* Oocytes and the Two-Electrode Voltage Clamp Technique

Juan Manuel Valdez Capuccino and
Jorge E. Contreras

CONTENTS

9.1 INTRODUCTION

Connexin hemichannels are nonselective channels that are permeable to atomic ions and small metabolites including transmitters such as ATP and glutamate. It is thought that the controlled opening of hemichannels plays physiological roles. However, the extensive opening of hemichannels (likely under pathological conditions) will lead to collapse of the ionic gradient, leak of small metabolites, and cell death. The former phenomena are readily observed with the several human connexin mutations that produce gain in hemichannel function when transfected into cell lines [1–3]. In nongenetic pathologies, such as ischemia, trauma, and inflammation, there is compelling evidence that cell damage is caused by exacerbated opening of the hemichannels [4–7]. Unfortunately, the poor understanding of the gating properties of connexin hemichannels and the lack of connexin-specific blockers make identifying the hemichannel currents in native systems a challenge. Gaining a better understanding of the fundamental principles of gating, permeation, and regulation of connexin hemichannels can help to elucidate the mechanisms by which cellular perturbations or disease-causing mutations modify connexin hemichannel function.

Thus far, the most direct way to evaluate the changes in the functional activity of connexin hemichannels is the use of electrophysiological techniques combined with pharmacological tools. Heterologous expression systems are the most suitable models to examine the biophysical properties (gating and permeation) of connexin channels. This chapter provides a short description of the advantages and the considerations for the use of *Xenopus* oocytes and the whole-cell two-electrode voltage clamp (TEVC) technique to examine the connexin hemichannel activity. *Xenopus* oocytes have been shown to be particularly appropriate for the characterization of connexin hemichannel gating kinetics because of their ability to tolerate and recover from the robust ion fluxes evoked by the hemichannel opening. This is partially due to their large cytoplasmic volume and various endogenous protein transporters that maintain the ionic electrochemical gradient. Many human connexin mutations that cause diseases have been functionally characterized using this methodology. However, as noted in the following sections, the system is not without drawbacks.

9.2 ASSESSING CONNEXIN HEMICHANNEL CURRENTS USING *XENOPUS* OOCYTES AND THE TWO-ELECTRODE VOLTAGE CLAMP TECHNIQUE

9.2.1 TWO-ELECTRODE VOLTAGE CLAMP

Electrophysiological techniques can be combined with an appropriate protein expression/assay system (cell lines, liposomes, *Xenopus* oocytes) to study the properties of gating and permeability of ion channels. Conventional whole-cell recording techniques have been long used to assay the currents from cell lines stably or transiently transfected with plasmid encoding for connexin proteins. Single- or multihemichannel openings are recorded under this configuration, but not large macroscopic currents like those observed for more conventional ion channels types (such as K^+, Na^+ or Ca^{2+} voltage-gated channels). Active hemichannels show long-lived mean open times, but they are a very small fraction of the total number of hemichannels present at the plasma membrane [8,9]. The specific signature of a maximal channel conductance helps to identify the connexin hemichannel activity; these are usually large conductance with values that are nearly twice that of their corresponding gap junction channels (as expected for hemichannels arranged in series) [10]. The open probability of hemichannels under physiological conditions is low, and nonphysiological stimuli such as large and prolonged changes in the voltage or low extracellular Ca^{2+} concentrations are required to increase the number of active hemichannels [8,11,12]. However, these conditions are stressful for mammalian cells, reducing the feasibility of obtaining suitable recordings. Conversely,

Xenopus oocytes are much larger and can better handle the harsh conditions that increase the hemichannel opening [13]. They are also appropriate for performing both single and macroscopic currents using patch-clamp techniques (cell attached or excised patches) or TEVC, respectively. In this chapter, we present only the general considerations involved in using electrophysiological approaches in *Xenopus* oocytes expressing connexin channels, as detailed procedures associated with the techniques have been extensively described by others [14,15].

The whole-cell TEVC uses two microelectrodes inserted into the oocyte, rather than using one patch electrode on the surface followed by rupturing the membrane, as is done in mammalian cells. One electrode is used to measure the internal potential of the oocyte, and the other electrode is used to inject current [14,15]. The large size of the oocyte (about 1 mm in diameter) makes this feasible and is both the major advantage and disadvantage of the system. One advantage is that the procedure is easy to learn and fast to perform. The large membrane capacitance of oocytes means that the voltage clamp must deliver large currents to generate voltage steps. Failure of the clamp to deliver these large currents will affect the quality of the recording, and it may not be noticed by the experimenter. To deliver these large currents, the current-delivering electrode needs to have as low a resistance as possible ($\sim$1 MΩ). This is accomplished by using a pipette solution with high ionic strength, usually 3 M KCl, and a large tip diameter of approximately 1–2 μm. The TEVC can manage high current levels, but for reliable measurements, no more than 20 μA is ideal. This is because the voltage drop across the current-delivering electrode can be considerable. It is imperative for the reliability of the recordings to always monitor the voltage output (reflecting the voltage changes at the plasma membrane of the oocyte) to be sure that the clamp is fast and accurate. This is especially important when the currents are large with fairly rapid kinetics. Connexin hemichannel currents are large, but they have slow activation and deactivation kinetics (in the range of seconds); thus, it is easier to correlate the voltage clamp changes with the macroscopic currents when monitoring the voltage outputs.

Many electrophysiology experiments using oocytes require changes in the bath solution; this becomes an important issue if the kinetic responses to the agonist/blockers are measured. The large size of the oocytes makes the rapid application of agonist/blockers very challenging. It is possible to overcome this in some instances by local application using an outlet valve located near the oocyte with an adjacent perfusion. Although the gating kinetics of the connexin hemichannels are in some cases slower than drug perfusion times, most work with agonists/blockers are done under steady-state conditions (e.g., when performing concentration–response curves upon changes in extracellular Ca^{2+} concentrations [16]). Bath perfusion is important to reduce the rundown of hemichannel currents. The reason is unknown, but it is possible that the opening of hemichannels triggers the release or the uptake of molecules that promote autocrine signaling that modulates the activity and/or the plasma membrane trafficking of the connexin hemichannels.

To record the hemichannel currents from *Xenopus* oocytes expressing heterologous connexin hemichannels, defolliculated oocytes are injected with mRNA encoding for a particular connexin protein. Different amounts of mRNA (0.2–1 μg/μL) per injection will lead to different expression levels of connexin proteins. The mRNA-injected oocytes are incubated in Ringer's or ND96 solutions supplemented with gentamicin (0.05 mg/mL) at 16°C before recording. In general, the hemichannel currents from heterologously expressed connexins are detected using TEVC 24–48 hours after the mRNA injection.

9.2.2 EQUIPMENT

To perform TEVC recordings, we utilize an OC-725C oocyte clamp amplifier (Warner Instrument). Left and right micromanipulators are magnetically attached to a solid steel plate that provides a stable platform for mounting the recording chamber. The currents from oocytes are sampled using a low-pass filter (LPF-8, Warner Instruments). The ADDA signal is converted using the Digidata 1440A (Axon CNS, Molecular Devices). The microelectrodes (with resistances of 0.1 and 1.2 MΩ when filled with 3 KCl) are made using a Flaming/Brown micropipette puller model P-97 (Sutter instrument Co.).

9.2.3 CONNEXIN HEMICHANNELS: GATING AND PERMEABILITY

The gating and permeability properties of connexin hemichannels are highly specific and tightly regulated. At the plasma membrane, the hemichannels are mostly closed, but they (or at least a fraction of them) are gated in response to the changes in membrane voltages and extracellular Ca^{2+} among other factors [17].

An important characteristic feature of connexin hemichannels, when compared to other ion channels, is their slow gating kinetics, which are in the range of seconds and can resemble those of electrogenic membrane transporters. These distinctive kinetic properties can be easily identified in electrophysiological recordings acquired by the TEVC technique. Figure 9.1a shows the current responses elicited by a depolarizing pulse from −80 to 0 mV, but with different pulse durations for two different connexins, hCx26 (human Cx26) and rCx46 (rat Cx46). The peak tail current magnitude as a function of the duration of the depolarizing pulse indicates that the maximal tail current activation is reached following the depolarizing pulses of ~40 seconds (Figure 9.1b). This suggests that channel activation by a depolarizing pulse is very slow and kinetically complicated. Channel deactivation kinetics are slower for hCx26 than for rCx46 and may be analyzed by fitting with single or double exponentials, respectively [16]. Figure 9.1c shows the differences between the currents activated by a depolarizing pulse from −80 to 0 mV in oocytes expressing hCx26 (black) or rCx46 (dark gray) compared to oocytes injected only with antisense against Cx38 (light gray), the endogenous connexin expressed by *Xenopus* oocytes. Note the prominent activation of outward connexin hemichannel currents and the corresponding tail currents only when the mRNA for either hCx26 or rCx46 is injected.

Unlike most ion channels, the relevant permeabilities of connexin hemichannels extend from the charge selectivity among atomic ions such as K^+ and Ca^{2+}, through size and charge selectivities among nonbiological tracer molecules, to highly specific selectivities among cytoplasmic molecules. Early notions were that connexin channels were nonspecifically permeable to molecules below a size cutoff. However, we

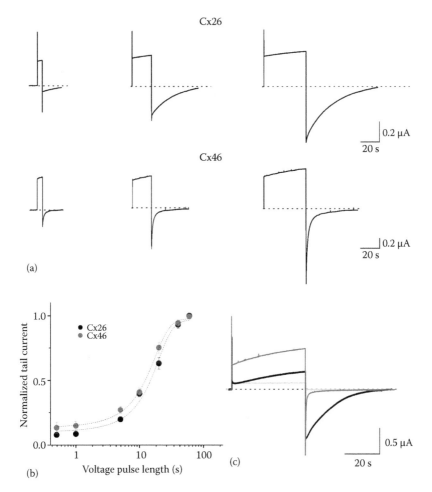

FIGURE 9.1 Slow gating kinetics of hCx26 and rCx46 hemichannel currents expressed in *Xenopus* oocytes. (a) Currents elicited in response to voltage steps from −80 mV to 0 mV from oocytes expressing mRNA hCx26 (top traces) or rCx46 (lower traces) in the presence of 1.8 mM external Ca^{2+}. Pulses of 5, 20, and 40 seconds show that the tail current magnitude is a function of the duration of the depolarizing pulse. (b) Normalized tail currents in response to voltage pulses from −80 to 0 mV of different durations. The data points represent mean ± SEM of at least three independent measurements. (c) Comparison of currents in the presence of 1.8 mM Ca^{2+} elicited by a voltage pulse from −80 to 0 mV from oocytes coinjected with hCx26 (black) or rCx46 (dark gray) mRNA plus Cx38 antisense or only with Cx38 antisense (light gray).

now know that their permeability properties are quite selective among molecules with similar size [18]. The selectivity depends on the type of connexin under study. The ionic selectivity properties of connexin hemichannels may be reliably assessed from single-channel recordings from *Xenopus* oocytes expressing connexin in the excised patch configuration (inside-out and outside-out patches). A complete description on how to perform these experiments and analysis can be found in several book chapters [19–21]. In particular, rCx46 has been shown to be amenable for these types of analyses, showing higher permeability to cations over anions [22]. Yet the permeability to large cations like tetramethylammonium (TMA$^+$) and TEA$^+$ is lower than that of atomic ions. For example, the permeability ratios relative to potassium are 0.34 and 0.2 for TMA and TEA, respectively.

Unfortunately, for many channels permeable to ions, including hemichannels formed by most connexin isoforms, the excised patch configuration is problematic due to the rundown and the instability over time during the course of the experiments. In this case, the estimation of the relative permeability properties of the connexin hemichannels may be assessed using the whole-oocyte currents if one assumes the cytoplasmic ionic concentrations of Na$^+$, K$^+$, and Cl$^-$ (e.g., 7, 98, 37 mM, respectively) [23]. As for excised patches, the permeability is estimated from the reversal potentials (E_{rev}) using the Goldman–Hodgkin–Katz constant field equation:

$$E_{rev} = \frac{RT}{F} \ln \frac{P_K[K^+]_o + P_{Na}[Na^+]_o + P_{Cl}[Cl^-]_i}{P_k[K^+]_i + P_{Na}[Na^+]_i + P_{Cl}[Cl^-]_o} \tag{9.1}$$

Here subscripts i and o denote intracellular and extracellular, respectively; R, T, and F have their usual meanings, and P_{Na}, P_K, and P_{Cl} are the membrane permeabilities to Na$^+$, K$^+$, and Cl$^-$, respectively.

In addition, a rough estimation of the pore diameter can be experimentally calculated for the molecules using the excluded volume model along with large permeable molecules (P_X), assuming that the pore is a cylinder formed by spherical amines. For these assumptions, the relationship between the permeability and ionic radius is examined using the following equation:

$$\frac{P_X}{P_{Na}} = A\left(1 - \frac{a}{r}\right)^2 \tag{9.2}$$

Here P_X/P_{Na} is the relative permeability, a is the radius of the amine compound, r is the radius of the pore, and A is a scaling factor.

9.2.4 ISOLATION OF CONNEXIN HEMICHANNEL CURRENTS FROM ENDOGENOUS *XENOPUS* OOCYTE CURRENTS

An important consideration when using *Xenopus* oocytes to measure the hemichannel currents is the large number of endogenous channels whose activity can overlap with the hemichannel activity. The currents evoked by endogenous channels can mask the magnitude and the kinetics of heterologously expressed connexin hemichannels [24]. There are three main sources of current contamination: endogenous connexin (Cx38) hemichannels, Ca^{2+}-activated chloride channels, and voltage-activated sodium channels. When working with heterologously expressed connexin in oocytes, different scenarios present themselves depending on the stimuli and/or the recording protocols used to assay the hemichannel currents.

Here, we describe some approaches that help to isolate the connexin hemichannel currents from those that are endogenous.

Xenopus oocytes are metabolically coupled to follicular cells through gap junction channels. Cx38 is the major player forming both hemichannel and gap junction channels in these cells [25,26]. The intrinsic expression of Cx38 significantly varies among different oocyte batches and among oocytes from the same batch. The positive depolarizing membrane potentials ($>+10$ mV) or the reduction of extracellular Ca^{2+} concentration (below ~0.15 mM) activates the nonjunctional Cx38 hemichannels in *Xenopus* oocytes. These same experimental conditions have been used to activate hemichannels

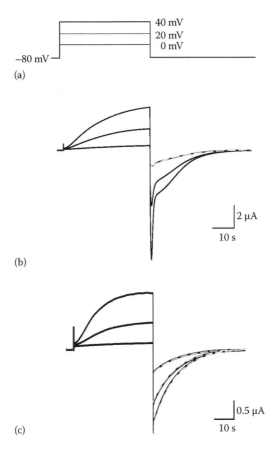

FIGURE 9.2 Ca^{2+}-activated chloride currents overlap with the connexin hemichannel currents. (a) Voltage steps protocols. (b) Depolarizing voltage pulses from −80 to 0, 20, and 40 mV elicited currents from oocytes expressing hCx26 hemichannels in the presence of 1.8 mM extracellular Ca^{2+}. The tail currents display multiple kinetic components following depolarizing pulses of 20 and 40 mV. (c) Same voltage steps in an oocyte expressing hCx26, but prior being injected with BAPTA, that displays tail currents with only monotonic deactivation kinetics. To get approximate to a final intracellular concentration of 120 μM BAPTA and 28 nL from 5 mM stock BAPTA solution are injected into the oocyte. The gray lines correspond to single-exponential fitting of the tail currents.

formed by heterologously expressed connexin (e.g., Cx26 or Cx46) in oocytes, so the activity of endogenous Cx38 hemichannels must be significantly attenuated to assay the activity of the heterologously expressed hemichannels. To achieve this, antisense oligonucleotide against Cx38 (1 mg/mL) is injected 4 hours after harvesting the oocytes. After 1 day, the same oocytes are coinjected with another dose of Cx38 antisense plus the mRNA encoding for the heterologous connexin. The levels of endogenous Cx38 hemichannel current should always be tested in the oocytes injected with Cx38 antisense alone; only the batches that show no or low endogenous Cx38 hemichannel currents should be used to assay heterologously expressed connexin hemichannels upon conditions that may activate Cx38 hemichannels (e.g., reduced extracellular Ca^{2+}).

A large number of Ca^{2+}-activated chloride channels (mostly formed by TMEM16A proteins) are present in the plasma membrane of *Xenopus* oocytes [27]. They generate the fertilization potential that provides a fast electrical block to polyspermy eggs [28]. These Ca^{2+}-activated chloride currents show slow activation/deactivation kinetics and voltage-dependent sensitivity to cytosolic Ca^{2+}. Large depolarizing voltages and Ca^{2+} influx through heterologous connexin hemichannels evoke these chloride currents [16,29]. Figure 9.2b shows the tail currents from oocytes expressing hCx26 hemichannels with multiphasic deactivation kinetics after the repolarization from voltages greater than 20 mV. Conversely, oocytes expressing hCx26 hemichannels that were previously injected with 120 μM BAPTA showed only monotonic deactivation kinetics of the tail currents with single-exponential fit of ~12 seconds (Figure 9.2c, dashed gray lines). Thus, the enhanced opening of the hemichannels promotes the activity of Ca^{2+}-activated Cl^- currents in the *Xenopus* oocytes; the BAPTA injections are necessary to eliminate these currents.

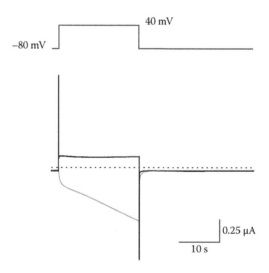

FIGURE 9.3 A long depolarizing pulse activates Na^+-dependent inward currents. Current traces in response to the depolarizing step from −80 to 40 mV elicited inward currents mediated by Na^+ permeant ions (gray trace). Replacing the bath solution with Na^+-free solution by using choline as a substitute prevents inward currents (black trace). Note that 5 μM atropine was added in the choline-containing solution.

Xenopus oocytes also show sodium-dependent inward currents that are activated by prolonged depolarization and are distinguished from voltage-activated sodium channels in excitable cells by slow activation kinetics and only modest sensitivity to tetrodotoxin [30–32]. Figure 9.3 shows the current traces in response to a depolarizing pulse of 40 mV elicited in a noninjected oocyte incubated in the presence of extracellular Na$^+$ (gray) or choline (black). These sodium inward currents have

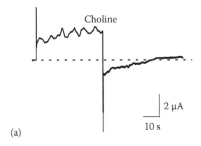

(a)

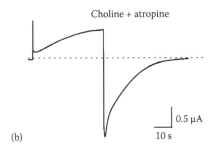

(b)

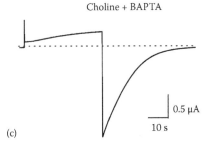

(c)

FIGURE 9.4 Choline substitution activates the endogenous muscarinic receptors in *Xenopus* oocytes. (a) Current trace in response to depolarizing pulse from −80 to 0 mV in the presence of a Na$^+$-free solution containing 114 mM choline as a substitute (choline–Ringer solution). Note the oscillatory currents along the trace. (b) Current trace elicited in the presence of a choline–Ringer solution containing 5 μM atropine. (c) Current trace elicited in an oocyte prior being injected with 120 μM BAPTA and recorded in the presence of a choline–Ringer solution. Note that in (b) and (c), the currents are similar to that displayed in normal Ringer solutions as seen in Figure 9.1a.

been shown to overlap with hemichannel currents, because they possess similar time and voltage dependence [24]. To avoid the contamination of the hemichannel currents by these channels, depolarizing step protocols could be restricted to voltages below <20 mV, unless the voltage sensitivity of connexin hemichannels is being examined. Alternatively, the extracellular replacement of sodium with large organic cations such as choline, TMA$^+$, or NMDG$^+$ can prevent the development of sodium inward currents due to the absence of permeant ions, as others have previously shown [24]. While, choline and its derivatives (TMA$^+$, TEA$^+$) have been used to study the permeability properties of nonselective cation channels heterologously expressed in oocytes [33–35], high concentrations of these cations can also activate endogenous muscarinic receptors due to their structural similarity with acetylcholine. Figure 9.4a shows that sodium ion replacement with choline generates oscillatory currents as the result of an increase in intracellular Ca^{2+} and, consequently, the activation of Ca^{2+}-activated chloride conductance. Atropine (5 μM), an inhibitor of muscarinic receptors, drastically reduces these oscillatory currents, but it does not affect the connexin hemichannel currents (Figure 9.4b). Similarly, intracellular injection with BAPTA (to a final intracellular concentration of 120 μM) completely abolishes the oscillatory currents (Figure 9.4c). Either treatment (atropine or intracellular BAPTA) can be effective when measuring the connexin hemichannel current in the presence of these organic cations.

Importantly, several human connexin mutations that alter the gating and/or the permeability of connexin hemichannels may indirectly affect the endogenous currents. For example, if a connexin mutation increases the permeability to Ca^{2+}, it may also promote higher levels of Ca^{2+}-activated chloride currents than would occur for wild-type connexin, which would consequently contaminate the kinetics of the connexin hemichannels currents from oocytes expressing these particular mutants. To unequivocally identify the electrophysiological properties of connexin mutants, the endogenous currents must be assessed and eliminated.

9.3 CONCLUDING REMARKS

Electrophysiological methods, including TEVC, reveal that connexin hemichannels have slow gating kinetics and low open probability at physiological recording conditions (negative potential and normal extracellular Ca^{2+}). Other large channels such as those formed by pannexin and the Ca^{2+} homeostasis modulator Ca^{2+} channel display similar properties. Such properties support the notion that major functions of these channels are not primarily due to rapid and transitory electrical signaling, as we recognize for more conventional ion channels. Although they may contribute in setting the resting membrane potential in some cell types [36], their major function seems to be associated with the release of small molecules (i.e., ATP, glutamate) that promote paracrine signaling or are important to maintain extracellular homeostasis (i.e., glutathione, NAD$^+$) [37]. Most cells express more than one connexin isoform, and when two compatible isoforms are expressed in the same cell, the hemichannels are likely to be heteromeric. Homomeric and heteromeric channels can have different properties of gating and ionic/molecular permeability. TEVC and patch clamp techniques applied to *Xenopus* oocytes are essential tools to characterize these properties among different homomeric and heteromeric hemichannels.

ACKNOWLEDGMENTS

This work was supported by National Institutes of Health/National Institute of General Medical Sciences (grant RO1-GM099490 to J. E. Contreras). We thank Dr. Andrew Harris for helpful comments on this chapter.

REFERENCES

1. Beyer EC, Ebihara L, Berthoud VM. Connexin mutants and cataracts. *Front Pharmacol.* 2013; 4:43.
2. Garcia IE, Maripillan J, Jara O, Ceriani R, Palacios-Munoz A, Ramachandran J et al. Keratitis-ichthyosis-deafness syndrome-associated Cx26 mutants produce nonfunctional gap junctions but hyperactive hemichannels when co-expressed with wild type Cx43. *J Invest Dermatol.* 2015; 135:1338–47.
3. Martinez AD, Acuna R, Figueroa V, Maripillan J, Nicholson B. Gap-junction channels dysfunction in deafness and hearing loss. *Antioxid Redox Signal.* 2009; 11:309–22.
4. Contreras JE, Sanchez HA, Veliz LP, Bukauskas FF, Bennett MV, Saez JC. Role of connexin-based gap junction channels and hemichannels in ischemia-induced cell death in nervous tissue. *Brain Res: Brain Res Rev.* 2004; 47:290–303.
5. Davidson JO, Green CR, Bennet L, Gunn AJ. Battle of the hemichannels—Connexins and pannexins in ischemic brain injury. *Int J Dev Neurosci.* 2015; 45:66–74.
6. Bennett MV, Garre JM, Orellana JA, Bukauskas FF, Nedergaard M, Saez JC. Connexin and pannexin hemichannels in inflammatory responses of glia and neurons. *Brain Res.* 2012; 1487:3–15.
7. Decrock E, De Bock M, Wang N, Bultynck G, Giaume C, Naus CC et al. Connexin and pannexin signaling pathways, an architectural blueprint for CNS physiology and pathology? *Cell Mol Life Sci.* 2015; 72:2823–51.
8. Contreras JE, Saez JC, Bukauskas FF, Bennett MV. Gating and regulation of connexin 43 (Cx43) hemichannels. *Proc Natl Acad Sci U S A.* 2003; 100:11388–93.
9. Wang N, De Bock M, Antoons G, Gadicherla AK, Bol M, Decrock E et al. Connexin mimetic peptides inhibit Cx43 hemichannel opening triggered by voltage and intracellular Ca2+ elevation. *Basic Res Cardiol.* 2012; 107:304.
10. Bennett MV, Contreras JE, Bukauskas FF, Saez JC. New roles for astrocytes: Gap junction hemichannels have something to communicate. *Trends Neurosci.* 2003; 26:610–7.
11. Pfahnl A, Dahl G. Gating of cx46 gap junction hemichannels by calcium and voltage. *Pflugers Arch.* 1999; 437:345–53.
12. Gomez-Hernandez JM, de Miguel M, Larrosa B, Gonzalez D, Barrio LC. Molecular basis of calcium regulation in connexin-32 hemichannels. *Proc Natl Acad Sci U S A.* 2003; 100:16030–5.
13. Trexler EB, Verselis VK. The study of connexin hemichannels (connexons) in *Xenopus* oocytes. *Method Mol Biol.* 2001; 154:341–55.
14. Schreibmayer W, Lester HA, Dascal N. Voltage clamping of *Xenopus* laevis oocytes utilizing agarose-cushion electrodes. *Pflugers Arch.* 1994; 426:453–8.
15. Goldin AL. *Expression of Ion Channels in Xenopus Oocytes: Expression and Analysis of Recombinant Ion Channels.* Wiley-VCH Verlag GmbH & Co. KGaA, Weinheim; 2006. p. 1–25.

16. Lopez W, Gonzalez J, Liu Y, Harris AL, Contreras JE. Insights on the mechanisms of Ca(2+) regulation of connexin26 hemichannels revealed by human pathogenic mutations (D50N/Y). *J Gen Physiol.* 2013; 142:23–35.

17. Harris AL, Contreras JE. Motifs in the permeation pathway of connexin channels mediate voltage and Ca (2+) sensing. *Frontiers Physiol.* 2014; 5:113.

18. Harris AL. Connexin channel permeability to cytoplasmic molecules. *Prog Biophys Mol Biol.* 2007; 94:120–43.

19. Talavera K, Nilius B. Electrophysiological methods for the study of TRP channels. In: Zhu MX (ed). *TRP Channels.* Boca Raton, FL; 2011.

20. Hille B. *Ion Channels of Excitable Membranes.* Sinauer Associates, Sunderland, MA; 2001.

21. Sakmann B. *Single-Channel Recording.* Springer Science & Business Media, Berlin; 2013.

22. Trexler EB, Bennett MV, Bargiello TA, Verselis VK. Voltage gating and permeation in a gap junction hemichannel. *Proc Natl Acad Sci U S A.* 1996; 93:5836–41.

23. Siebert AP, Ma Z, Grevet JD, Demuro A, Parker I, Foskett JK. Structural and functional similarities of calcium homeostasis modulator 1 (CALHM1) ion channel with connexins, pannexins, and innexins. *J Biol Chem.* 2013; 288:6140–53.

24. Ripps H, Qian H, Zakevicius J. Pharmacological enhancement of hemi-gap-junctional currents in *Xenopus* oocytes. *J Neurosci Methods.* 2002; 121:81–92.

25. Ebihara L, Beyer EC, Swenson KI, Paul DL, Goodenough DA. Cloning and expression of a *Xenopus* embryonic gap junction protein. *Science.* 1989; 243:1194–5.

26. Ebihara L. Xenopus connexin38 forms hemi-gap-junctional channels in the nonjunctional plasma membrane of *Xenopus* oocytes. *Biophys J.* 1996; 71:742–8.

27. Schroeder BC, Cheng T, Jan YN, Jan LY. Expression cloning of TMEM16A as a calcium-activated chloride channel subunit. *Cell.* 2008; 134:1019–29.

28. Cross NL, Elinson RP. A fast block to polyspermy in frogs mediated by changes in the membrane potential. *Dev Biol.* 1980; 75:187–98.

29. Sanchez HA, Mese G, Srinivas M, White TW, Verselis VK. Differentially altered Ca2+ regulation and Ca2+ permeability in Cx26 hemichannels formed by the A40V and G45E mutations that cause keratitis ichthyosis deafness syndrome. *J Gen Physiol.* 2010; 136:47–62.

30. Baud C, Kado RT, Marcher K. Sodium channels induced by depolarization of the *Xenopus* laevis oocyte. *Proc Natl Acad Sci U S A.* 1982; 79:3188–92.

31. Baud C, Kado RT. Induction and disappearance of excitability in the oocyte of *Xenopus* laevis: A voltage-clamp study. *J Physiol.* 1984; 356:275–89.

32. Parker I, Miledi R. Tetrodotoxin-sensitive sodium current in native *Xenopus* oocytes. *Proc Royal Soc B Biol Sci.* 1987; 232:289–96.

33. Villarroel A, Burnashev N, Sakmann B. Dimensions of the narrow portion of a recombinant NMDA receptor channel. *Biophys J.* 1995;68:866–75.

34. Yang XC, Sachs F. Characterization of stretch-activated ion channels in *Xenopus* oocytes. *J Physiol.* 1990;431.

35. Zhang Y, McBride DW, Jr., Hamill OP. The ion selectivity of a membrane conductance inactivated by extracellular calcium in *Xenopus* oocytes. *J Physiol.* 1998; 508 (Pt 3):763–76.

36. Bukauskas FF, Kreuzberg MM, Rackauskas M, Bukauskiene A, Bennett MV, Verselis VK et al. Properties of mouse connexin 30.2 and human connexin 31.9 hemichannels: Implications for atrioventricular conduction in the heart. *Proc Natl Acad Sci U S A.* 2006; 103:9726–31.

37. Wang N, De Bock M, Decrock E, Bol M, Gadicherla A, Vinken M et al. Paracrine signaling through plasma membrane hemichannels. *Biochim Biophys Acta.* 2013; 1828:35–50.

10 Functional Assays of Purified Connexin Hemichannels

Mohamed Kreir and Guillermo A. Altenberg

CONTENTS

Connexins assemble as hexamer complexes called hemichannels or connexons. Hemichannels traverse biological membranes and have a regulated central pore permeable to large hydrophilic solutes [1,2]. Two hemichannels, one from each of the neighboring cells, can dock head to head to form a gap junction channel, essential for cell-to-cell communication in animal eukaryotic cells [1,3,4]. There are 21 human connexin isoforms, with lengths between 226 and 543 amino acids, which form homomeric and heteromeric hemichannels of varying permeability properties, regulation, and associations with other proteins [1,3–5].

Whereas gap junction channels link two compartments of very similar composition (cytoplasms of adjacent cells), plasma membrane hemichannels connect the intracellular and extracellular fluids, which have very different compositions. Therefore, plasma membrane hemichannels are mostly closed, and their activity is tightly regulated. Connexin hemichannels at the plasma membrane display a low open probability (Po), but they have important roles in physiologic processes by mediating the transmembrane fluxes of signaling molecules/metabolites such as ATP, NAD^+, glutamate, glutathione, prostaglandin E(2), and glucose [6–13]. The abnormal opening of hemichannels can cause loss of metabolites, Ca^{2+} influx with alterations of signaling and protease activation, and equilibration of ionic gradients and cell swelling, which can result or contribute to cell damage and/or death. Normally, the combination of the cell membrane voltage and millimolar concentrations of extracellular divalent cations keeps the hemichannel Po low [14–18]. However, Cx43 hemichannels can be activated in the presence of extracellular divalent cations under a number of conditions including ischemia/hypoxia [19–25]. Studies in cardiomyocytes, astrocytes, and renal proximal tubule cells strongly suggest that Cx43 hemichannels are activated by ATP depletion and that the mechanism may involve dephosphorylation and oxidation [19–26]. In summary, connexin hemichannels are important in the normal cell physiology and in the pathophysiology of frequent disorders, and therefore understanding their basic properties and regulation is essential to explain normal and abnormal processes.

We have focused on Cx26 and Cx43 as models for studies of purified connexin hemichannels because of their medical importance and because these two isoforms are distinct in terms of their amino acid sequence and length of the C-terminal region (~20 amino acids in Cx26 and ~150 in Cx43). Success with the studies on purified functional Cx26 and Cx43 hemichannels will suggest that all isoforms are amenable to such approaches.

10.1 CONNEXIN 26 AND CONNEXIN 43 IN PHYSIOLOGY AND PATHOPHYSIOLOGY

Cx26 mutations are associated with deafness and skin diseases [27–33]. Genetic deafness occurs in ~1 in 2000 children, and the mutations of Cx26 are its major cause [27,30,34–41]. The cochlea in the inner ear contains the organ of Corti, which in turn contains the sensory cells (hair cells) that transduce sound waves into electrical impulses. The gap junctions in the cochlea are essential for hearing [30–32,42,43]. The absence of Cx26 causes death of the hair cells, but the mechanism is unknown. It has been speculated that deafness is a consequence of decreased K^+ recycling in the cochlea [43–45]. However, the notion of cochlear K^+ recycling has been questioned [46], and there are deafness-associated Cx26 mutants capable of K^+ transport, which present more subtle permeability changes that can affect the second messenger transport between cells or the efflux of signaling molecules through the hemichannels [18,47–56]. Since the ATP in the endolymph may be needed to maintain healthy hair cells, deafness could result from Cx26 mutations that decrease hemichannel-mediated ATP secretion [51,52]. A role of Ca^{2+} signaling in response to damaging stimuli has also been proposed, and therefore, an increase in Ca^{2+} permeability of some mutant Cx26 hemichannels (e.g., G45E mutant associated with keratitis ichthyosis deafness syndrome [15]) is also possible. Finally, it has also been proposed that leaky mutant hemichannels can lead to cell damage and deafness [57–60].

Cx43 is the most ubiquitous connexin isoform, and its C-terminal domain (CTD) includes sequences that mediate the interactions with tubulin, tyrosine kinases, ubiquitin ligase, zonula occludens 1 (ZO-1), and Na^+ channels [61–65]. The mutations of Cx43 are associated with a number of genetic disorders, including arrhythmias and oculodentodigital dysplasia, and alterations in Cx43 expression and function play important roles in pathophysiology of cardiac and brain ischemia and wound healing in diabetes [61,66–70]. The expression of Cx43 in the heart is high [71], normally at the intercalated disks, whereas its expression is low at the lateral membranes [71–75]. Cardiac remodeling occurs in response to a variety of cardiac disorders and is characterized by structural and electrical alterations that decrease heart electrical stability [71,72,75]. The changes in conduction are associated with lateralization (decreased Cx43 expression with a relative increase of Cx43 in lateral vs. intercalated disk), which is frequently associated with abnormal conduction and arrhythmias [71–75]. Lateralization is observed in a variety of acquired and inherited arrhythmic syndromes that include ischemic heart disease and several cardiomyopathies [71–73,75]. The mechanism of lateralization is not well understood, but it likely involves the activation of the tyrosine kinase Src with subsequent binding to ZO-1, competing with Cx43 for binding, which releases Cx43 from ZO-1, leaving ZO-1 at the intercalated disks, while the plaques form at the lateral membrane [72].

In summary, Cx26 and Cx43 are very different isoforms, and both are important in human physiology and pathophysiology.

10.2 EXPRESSION, PURIFICATION, AND CHARACTERIZATION OF PURIFIED CONNEXINS

Experiments in vivo are fundamental to understand biological processes, but in vitro studies using isolated systems under well-controlled conditions are equally important for a complete understanding of the normal function and the molecular mechanisms of diseases. In this context, studies of purified hemichannels provide direct functional and structural information that is an essential complement to the studies in more complex systems. In addition, although hemichannels have physiological and pathophysiological significances by themselves, studies of hemichannels contribute to our understanding of gap junction channels, since some basic properties and regulation of gap junction channel and hemichannel are very similar [19,76–85].

The procedures for the expression/purification of connexins include (1) the engineering of a DNA that codes for the desired connexin with the addition of affinity tags for purification and/or reporters such as a GFP and/or protease cleavage site(s) to remove tags/reporters for downstream applications (functional assays, crystallization); (2) the expression of the recombinant protein in an expression system suitable for the application. The insect cell/baculovirus expression system, *E. coli* or *Pichia pastoris* for application that need large amounts of purified protein, and mammalian cells for studies in more native environments; (3) the preparation of membranes by centrifugation and solubilization of the connexins with a detergent that does not irreversibly damage the proteins. In the case of connexins, because of their relative resistance to denaturing agents, it is possible to strip the membranes from proteins such as peripheral membrane proteins with a high salt concentration and/or alkali solutions. This results in a more reliable purity of the purified preparation; (4) the initial purification by liquid chromatography of the detergent-solubilized connexins. This can be accomplished by affinity chromatography. The most frequently used method is based on the affinity of a polyhistidine (His)-tag fused to the connexin for an immobilized divalent transition element (generally Ni^{2+}, Co^{2+}, Cu^{2+}, or Zn^{2+}). The connexin bound to the resin is then eluted with imidazole, which competes with the imidazole of the His. Connexins without tags can also be purified by ion exchange chromatography; (5) the final purification is generally performed by size-exclusion chromatography. This is important because it gives information on the purity, the oligomeric structure, and the monodispersity.

Recombinant connexins have been expressed in a variety of systems for cell biology, biochemical, and functional studies, including mammalian cell lines and frog oocytes. For biochemical and structural studies that require larger amounts of purified connexins, insect cells have been the preferred choice. The insect cell/baculovirus expression system yields milligram amounts of purified connexins, and it has been employed to express Cx26 and Cx43 [77,78,86–92]. Most frequently, we express connexins modified by the insertion of a protease cleavage site (thrombin or tobacco etch virus protease) and a poly-His tag at the C-terminal end [78,86,93]. These connexin DNA sequences cloned into the baculovirus transfer vectors are used to generate recombinant baculoviruses. We have used the Invitrogen Bac-to-Bac system to successfully generate viruses in Sf9 insect cells, which are then used for connexin production in either Sf9 or High Five cells [77,78,86]. In our hands, the Sf9 cells grown in suspension in serum-free medium are reliable. The cells grown at 26°C are infected (generally 2 viral particles/cell), and the cells are harvested ~2 days after infection, when the viability is ~40%.

Recently, we started experimenting with an *E. coli*-based connexin expression system that yields purified and functional Cx26 hemichannels in amounts equivalent to those obtained in insect cells [93]. Basically, we use XL10-Gold *E. coli* cells transformed with a partially optimized human Cx26 DNA into the plasmid pQE60. The cells are grown at 37°C in a modified M9 minimal medium supplemented with 10 mM $MgSO_4$, 1% glucose, and 0.4 mg/ml ampicillin, with induction at an OD_{600} of ~2 with 0.5 mM isopropyl-β-D-thiogalactopyranoside for 2 hours. The *E. coli* system has advantages over other available systems in terms of simplicity, cost, and turnover time, but one potential problem concerns posttranslational modifications. Connexins are not glycosylated, and there is no clear evidence of the direct regulation of Cx26 by posttranslational modifications, but this may be an issue for Cx43, which is regulated by phosphorylation [61,66]. However, many of the posttranslational modifications can be performed in vitro, as we have done on purified Cx43 [77,78]. The development of the *E. coli*-based expression system is significant because there are few if any examples of properly folded functional human recombinant membrane proteins in *E. coli* with milligram yields per liter culture. This new expression/purification system has the potential to increase the pace of structural and functional studies of connexins.

For purification, the insect cell pellets are resuspended in a 1 mM bicarbonate solution containing 1 mM protease inhibitors and then lysed. The bacteria are lysed on a microfluidizer, and crude membranes are harvested by centrifugation for subsequent processing. The insect cell and bacterial samples are then alkali extracted and solubilized with 2.5% *n*-dodecyl-β-D-maltoside or 1% Anzergent 3–12 in the presence of a high NaCl concentration (1–2 M), a chelator of divalent cations (e.g., 2 mM EGTA), a reducing agent, and 10% glycerol. The solubilized material is then purified based on the affinity of the His tag for Co^{2+}, followed by size-exclusion chromatography. If needed, the removal of the His tag is accomplished by site-specific proteolysis, which is followed by the isolation of the untagged connexin by size-exclusion chromatography. The insect cell/baculovirus and bacterial expression systems yield >0.2 mg/l culture of highly pure connexins [78,86,93].

The connexins expressed in insect cells have been employed in X-ray crystallography and cryoelectron microscopy studies [87–91,94–96], and we have recently characterized the biochemical and biophysical properties of Cx26 hemichannels purified from insect cells and bacteria [86,93]. The hemichannels were highly purified (Figure 10.1), and the apparent molecular weight of the protein–detergent complex was ~235 kDa, with an average hydrodynamic radius of 5.4 nm [86,93]. These data are consistent with a Cx26 hexamer–detergent complex. The purified Cx26 hemichannels were highly structured with a calculated α-helical content similar to that of the Cx26 crystal structure [86]. One important difference between Cx26 and Cx43 is on the stability of the hemichannels in the detergent. Cx26 is purified as hemichannels that are highly stable [86], whereas Cx43 hemichannels are stable in lipids but display an exchange of subunits in detergent [77]. The exchange of Cx43 hemichannel subunits in the solution allows for the generation of Cx43 hemichannels with a controlled subunit composition (see Section 10.4).

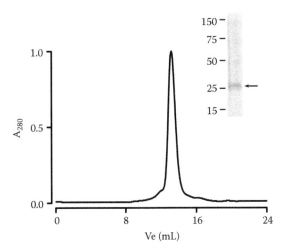

FIGURE 10.1 Purified recombinant human Cx26. Size-exclusion chromatogram of purified Cx26 expressed in *E. coli*. A_{280}: absorbance at 280 nm. Inset: Coomassie Blue-stained gel of the peak fraction. The positions of the molecular weight markers are indicated on the left. The identity of Cx26 was confirmed by Western blots with anti-Cx26 and anti-His antibodies. The figure shows a monodisperse and highly purified hemichannel preparation. (With kind permission from Springer Science+Business Media: *Bioscience Reports*, Functional hemichannels formed by human connexin 26 expressed in bacteria, 35(2), pii: e00177, 2015. Fiori MC, Krishnan S, Cortes DM, Retamal MA, Reuss L, Altenberg GA, Cuello LG.)

10.3 FUNCTIONAL ANALYSIS OF PURIFIED AND RECONSTITUTED CONNEXIN HEMICHANNELS

Recombinant purified gap junction channels and hemichannels expressed in insect cells have been extensively used for structural studies, but functional studies are very few. Recently, we developed a electrophysiological method to study the biophysical properties of purified Cx26 and Cx43 hemichannels reconstituted into giant unilamellar vesicles [89,97]. The purified proteins solubilized in 0.2% n-dodecyl-β-D-maltoside were incubated with giant unilamellar vesicles for 20–60 minutes. Then, the detergent was removed with polystyrene beads (Bio-Beads), and planar lipid bilayers were formed by bursting a giant unilamellar vesicle onto the glass surface containing a micrometer hole [89,97]. In single-channel studies, we found that the fusion of GFP to the C-terminal end of Cx43 influences the conductance levels and the Po of the hemichannels at low potentials [97], effects that were not observed when the Cx43 was studied in cells [14]. The results on the study of Cx26 showed unexpected low conductances compared to the values from the studies in cells [15,97], which might suggest that the lateral interaction between neighboring gap junction channels in junctional plaques plays an important role in determining hemichannel conductance. Even though single-channel analysis is ideal to study the purified channels, the experience with purified hemichannels is limited, with hemichannels often not showing the single-channel conductance and kinetics expected from the studies in cells [89,98,99]. This is clearly an area that needs additional experimental work.

Two commonly used assays are the transport-specific fractionation technique developed by Harris and collaborators [77,78,86,100,101], and the probe-permeation assays that use labeled solutes to determine whether they can enter into the liposomes containing hemichannels or be released from the liposomes preloaded with the probe [79,86,93,102]. The transport-specific fractionation technique is useful to assess whether the hemichannels are permeable or impermeable to sucrose and other hydrophilic solutes [77,78,86,100,101]. Although qualitative, the assay is very reliable because it is based on solute transport into the liposomes, as opposed to binding. Basically, the migration of liposomes seeded on top of a linear isoosmolar sucrose gradient in response to centrifugal force is followed. In the gradient, the increase in sucrose concentration from top to bottom is matched by a decrease in the urea concentration from top to bottom, with the osmolality remaining constant throughout the tube. The liposomes without hemichannels and the liposomes containing hemichannels impermeable to sucrose remain in the upper part of the tube, buoyed up by the entrapped urea solution of lower density. The heavier sucrose-loaded liposomes containing sucrose-permeable hemichannels migrate as a narrow band to a lower position in the tube (Figure 10.2a). Following the liposomes in the gradient is easier if they contain traces of fluorescent lipids (Figure 10.2b). The method can be combined with assays of permeability of other solutes. For such studies, the liposomes are preloaded with radiolabeled or fluorescent probes, and their permeation is determined from the retention of the probes in the liposomes (impermeable) or their loss (permeable). To increase the hemichannel Po, we routinely use a nominally divalent cation-free solution, and the fact that the Cx43 hemichannels are rendered

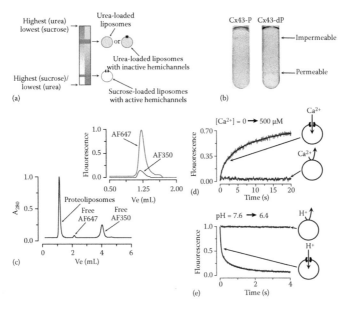

FIGURE 10.2 (See color insert.) Transport assays in purified and reconstituted hemichannels. (a) Principle of the transport-specific assay. (b) Migration of liposomes containing purified Cx43 hemichannels on an isoosmolar linear sucrose density gradient. The liposomes are colored by the presence of the trace lipids labeled with Rhodamine B. Cx43-P: liposomes containing Cx43 hemichannels with all six Ser368 residues phosphorylated by PKC, which are impermeable to sucrose; Cx43-dP: liposomes containing dephosphorylated Cx43 hemichannels, which are permeable to sucrose. (c) Permeability of reconstituted Cx26 hemichannels to Alexa Fluor probes. Liposomes preloaded with Alexa Fluor 350 (AF350) or 647 (AF647) were run on a size-exclusion column to separate free extraliposomal dyes from the dyes inside the liposomes. A_{280}: absorbance at 280 nm; top right: fluorescence associated with the liposomes containing Cx26. The data were normalized to the peak emission of the parallel experiments with liposomes without Cx26. In these experiments, if the dye is permeable through the hemichannels, it diffuses out because the extraliposomal dye concentration is essentially zero, and then, the dye retained inside the liposomes is very low. The impermeable dye (Alexa Fluor 647 is too large to permeate through the open Cx26 hemichannels) cannot diffuse out and is retained inside the liposomes. (d) Rate of Ca^{2+} influx into liposomes containing purified Cx26 hemichannels. Extraliposomal free $[Ca^{2+}]$ was rapidly increased from a few nanomolar to $500\,\mu M$ in a stopped-flow cell, and the rate of increase in the emission from Fluo-5N trapped into the liposomes was followed. The increase in fluorescence of Fluo-5N results from binding to Ca^{2+} in proportion to the intraliposomal $[Ca^{2+}]$ that increases as a result of the influx through the hemichannels. Black lines: multiexponential fit to the data, used to calculate the rate of Ca^{2+} influx. (e) Changes in the intraliposomal pH followed by the rate of fluorescein quenching. The fluorescein is attached to a phospholipid head group. The extraliposomal pH was rapidly reduced from 7.6 to 6.4 by mixing in a stopped-flow cell, and the rate of quenching of the fluorescein bound to the inner leaflet of the liposomes bilayer was used to calculate the rate of the decrease in intraliposomal pH. As the intracellular pH decreases because of the influx of H^+ and the protonated buffer through the hemichannels, the emission of fluorescein attached to the inner-leaflet head groups is quenched. Black line: multiexponential fit to the data, used to calculate the rate of liposome acidification. (Modified from Fiori MC, Reuss L, Cuello LG, Altenberg GA, *Frontiers in Physiology*, 5, 71, 2014. With permission.)

impermeable to sucrose after phosphorylation by protein kinase C (PKC) provides an excellent negative control for the assay. The transport-specific fractionation of the liposomes also helps in the determination of the percentage of the purified protein able to form functional hemichannels [77,78,98], a very important parameter to assess the quality of the preparation. If purified connexins are reconstituted in unilamellar liposomes of known size at a calculated hemichannel/liposome <1, the percentage of sucrose-permeable liposomes (containing one active hemichannel vs. those without hemichannels or with impermeable hemichannels) can be compared with the prediction based on the protein/lipid ratio. Our studies showed that essentially all purified Cx26 and Cx43 purified from insect cells and *E. coli* form functional hemichannels [77,86,93].

The uptake measurements assayed by rapid-filtration assays are relatively simple, and we have employed them to measure the influx of radiolabeled sucrose, maltose, and ethyleneglycol [78]. For these assays, hydrophilic solutes that can be detected by their radioactivity or other methods are added to the sample containing the hemichannel-containing liposomes and mixed rapidly. The sample is then placed on a filter membrane that binds the liposomes (nitrocellulose works well for liposomes made of phosphatidylcholine/phosphatidylserine), and the radioactivity of the filters is counted after a short but extensive washing at a low temperature in the presence of a hemichannel inhibitor such as Gd^{3+}; the low temperature and the inhibitor prevent the loss of the probe from the liposomes through the hemichannels during washing. The fluxes of hydrophilic solutes can be assessed by rapid filtration because their permeability through hemichannels is not high, and therefore the equilibration between intra- and extraliposomal spaces in liposomes containing ~1 hemichannel occurs in >10 seconds [86,103].

Another solute-permeability assay that we have used is based on the separation of extraliposomal probes that have been preloaded into the liposomes, from the probes retained inside the liposomes. We routinely use Alexa Fluor 350 (349 Da, permeable through hemichannels) and Alexa Fluor 647 (1300 Da, impermeable through hemichannels), but a variety of probes can be used, including signaling molecules such as ATP, inositol phosphates, and cyclic nucleotides [86,93,104,105]. Figure 10.2c shows an example were Alexa Fluor 350 and 647 were retained by the liposomes without the hemichannels, whereas only the latter was retained inside the liposomes with Cx26 hemichannels; the permeable probes come out of the liposomes by diffusion because the extraliposomal probe concentration is near zero, and they are therefore lost from inside the liposomes. From such studies, we can conclude that Alexa Fluor 350 permeates through the hemichannels, whereas Alexa Fluor 647 does not.

We have recently developed new fluorescence assays to assess the permeation of Ca^{2+}, Na^+, and H^+ through purified hemichannels. In spite of indirect evidence from studies in cells, the possibility of Ca^{2+} movement through hemichannels and gap junction channels had not been directly addressed until recently because of the sparsity of functional studies of purified hemichannels and the absence of a suitable assay. To fill this gap, we developed an assay to follow the time course of the changes in intraliposome $[Ca^{2+}]$ that result from Ca^{2+} influx through the hemichannels [86,93]. In liposomes preloaded with the low-affinity, hemichannel-impermeable, Ca^{2+}-sensitive fluorescent probe Fluo-5N (958 Da, −5 net charge), the elevation of free $[Ca^{2+}]$ from <10 nM to

500 μM increases the Fluo-5N emission if the liposomes contain hemichannels, but not in the absence of hemichannels. The permeation of Ca^{2+} through the hemichannels increases the intraliposomal $[Ca^{2+}]$ with the resulting binding of the cation to Fluo-5N and the increase in fluorescence; the Ca^{2+} influx and the permeability can then be assessed from the changes in fluorescence (Figure 10.2d). A similar approach can be used to estimate the hemichannel Na^+ permeability using the Na^+-sensitive fluorescent probe (1,3-benzenedicarboxylic acid, 4,4'-[1,4,10-trioxa-7,13-diazacyclopentadecane-7,13-diylbis(5-methoxy-6,2-benzofurandiyl)]bis (SBFI) instead of Fluo-5N. We found that purified Cx26 hemichannels have similar permeabilities to Na^+ and Ca^{2+}, pointing to a high Ca^{2+} permeability [86]. Since cell-to-cell cytosolic $[Ca^{2+}]$ gradients are very low, significant cell-to-cell Ca^{2+} fluxes through the gap junction channels will critically depend on the number of permeable channels and seem unlikely. In contrast, the large electrochemical driving force for the Ca^{2+} influx into the cells suggests a significant role of hemichannels in Ca^{2+} influx, at least under certain circumstances (e.g., ischemia), or in disease-causing mutants that display higher Ca^{2+} permeability (G45E mutant associated with keratitis ichthyosis deafness syndrome) [15,106].

We also developed a fluorescence-based assay to assess the permeation of H^+ equivalents through purified hemichannels (H^+/OH^-/buffer transport) [86]. Basically, we introduced traces of a phospholipid labeled with fluorescein in the liposome bilayer. Then, the influx of H^+ equivalents elicited by reducing the extraliposomal pH from 7.6 to 6.4 can be followed by the quenching of the fluorescein in the inner leaflet by acid. When the extraliposomal pH is reduced, the fluorescein emission from the outer leaflet of the liposome bilayer is rapidly quenched, whether the liposomes contain hemichannels or not; this quenching is too fast for detection in the stopped-flow setup, and the slower quenching of the inner-leaflet fluorescence, the result of intraliposomal acidification, is indicative of hemichannel-mediated decrease in intraliposomal pH (Figure 10.2e). In liposomes containing Cx26 hemichannels, the intraliposomal pH decreases, whereas it does not in the liposomes without hemichannels (Figure 10.2e). We estimated the permeability of Cx26 hemichannels to H^+ equivalents at ~10-fold that of Na^+ [86]. Since the sucrose permeability is low and the relative $K^+/cAMP$ and $K^+/Lucifer$ yellow permeability ratios are high [103], the Cx26 hemichannel permeability to organic buffers such as HEPES is lower than that to Na^+, and therefore, the data on H^+ equivalents/Na^+ permeability ratio are an underestimation of the true H^+/Na^+ permeability ratio.

10.4 GENERATION AND ANALYSIS OF PURIFIED CONNEXIN HEMICHANNELS OF CONTROLLED SUBUNIT COMPOSITION

For many studies, it will be desirable to correlate functional and structural observations in hemichannels with defined subunit composition. As mentioned earlier, purified Cx43 (but not Cx26) solubilized in detergent forms hemichannels where the connexins (subunits) exchange [77]. Basically, if the connexin hemichannels in the detergent are not very stable and their individual connexin subunits are in a dynamic equilibrium with the connexin monomers in the solution, it will be possible to obtain hemichannels of controlled subunit composition, e.g., hemichannels where a desired average subunit composition (percentage of phosphorylated or mutant subunits) can be obtained. We demonstrated the exchange of subunits and determined the subunit composition using size-exclusion chromatography and luminescence resonance energy transfer (LRET) [77]. For the former, we mixed Cx43 and Cx43-EGFP (Cx43 with an enhanced GFP fused to the C-terminus of the CTD). Each EGFP adds ~26 kDa, increasing the hydrodynamic size of the hemichannels in proportion to the number of Cx43-EGFP subunits in the hemichannel. Size-exclusion chromatography is sensitive but does not have the sufficient accuracy to reliably distinguish hemichannels containing, for example, three or five Cx43-EGFP subunits. To improve the accuracy, we used LRET, which measures the energy transfer from the lanthanides Tb^{3+} or Eu^{3+} to fluorescent acceptors [107]. LRET has atomic resolution, high sensitivity, and can provide information on the conformational changes in the millisecond to minute range [108–113]. The basic principles of LRET and its advantages have been described in detail [107] and are presented in Figure 10.3. The main advantages result from the long lifetime of the Tb^{3+} and Eu^{3+} excited states that makes possible the acquisition of long-lifetime emission in a gated mode (i.e., delaying acquisition for >30 microseconds). Gated acquisition minimizes the light scattering effects of structures such as detergent micelles and liposomes, and it results in minimal background with high signal-to-noise ratio. The dependence of the rate of energy transfer on donor–acceptor distance is more accurate than those assessed by Förster resonance energy transfer (FRET), because the donor and sensitized acceptor emissions in LRET are

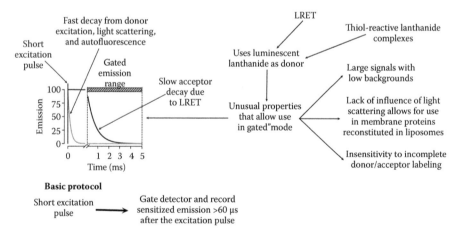

FIGURE 10.3 Basic principles and advantages of luminescence resonance energy transfer. The main advantages compared to traditional FRET are the long lifetime of the donor that allows for gated acquisition of the acceptor-sensitized emission, which essentially abolishes the contamination from light scattering of the excitation pulse, autofluorescence, and emission from the direct excitation of the acceptor, and it makes the measurements independent on the labeling stoichiometry. The donor sharp emission peaks with dark regions between the peaks allows for the isolation of the acceptor emission without contamination by the donor emission. (From Fiori MC, Reuss L, Cuello LG, Altenberg GA, *Frontiers in Physiology*, 5, 71, 2014. With permission.)

unpolarized, minimizing the uncertainty about κ (geometric factor related to the relative orientation of the donor and acceptor transition dipoles) [107,112]. Another major advantage is that the sensitized emission lifetime is independent of the labeling stoichiometry because long-lifetime acceptor emission can arise only from LRET [77,107]; i.e., the labeling stoichiometry affects the intensity of the signal but not its lifetime. Finally, the atomic-like lanthanide emission (sharp peaks with dark regions between them) allows for the measurements of acceptor emission without contamination from the donor emission [107,111,112].

For the analysis of Cx43 hemichannels by LRET, we labeled Cx43 with either donor (the Tb^{3+} chelator Tb^{3+}-DTPAcs124-EMCH) or acceptor (fluorescein maleimide) [77]. The stoichiometry of labeling was three probe molecules/hemichannel [77]. Since six of the nine Cx43 Cys are located in the extracellular loops, probably forming disulfide bonds [88,114], the labeling stoichiometry suggests that the remaining three CTD Cys are solvent accessible and reactive for labeling. After the labeling, Tb^{3+}-labeled, fluorescein-labeled, and unlabeled Cx43 are mixed in different proportions, using a low proportion of Tb^{3+}-labeled Cx43 (0.5 mol/hemichannel), and the mixed hemichannels are reconstituted. Under these conditions, the long lifetime-sensitized fluorescein emission that arises from the energy transfer from Tb^{3+} is proportional to the number of fluorescein-labeled subunits/hemichannel, and it can be used to determine the hemichannel subunit composition (Figure 10.4). The details and the validation have been published [77]. The LRET-based method can distinguish hemichannels containing ±1 acceptor-labeled subunits [77].

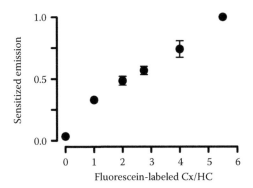

FIGURE 10.4 Use of luminescence resonance energy transfer to assay the hemichannel subunit composition. Sensitized fluorescence emission as a function of the average number of fluorescein-labeled subunits per hemichannel was determined. Under the conditions of the experiments, the measurements were performed in hemichannels containing one Tb^{3+}-labeled subunit (donor) and a variable average number of fluorescein-labeled (acceptor) subunits. The results show that the intensity of the fluorescein-sensitized emission is proportional to the number of donor subunits and can be used to determine the subunit composition of the hemichannels. Data were normalized to the peak value of one Tb/5.5 fluorescein-labeled preparation and are means ± SEM of seven to nine experiments. (Modified from Bao X, Lee SC, Reuss L, Altenberg GA, *Proceedings of the National Academy of Sciences of the United States of America*, 104, 4919–4924. 2007. With permission.)

10.5 USEFULNESS OF PURIFIED HEMICHANNELS TO ADDRESS MECHANISMS OF REGULATION

In this section, we present examples of the use of purified hemichannels to address the molecular mechanism of the regulation of Cx43 hemichannels by PKC-mediated phosphorylation. The phosphorylation of Cx43 hemichannels plays a critical role in gap junction remodeling, and the plasma membrane hemichannel opening in response to ischemic damage in the brain, heart, and kidney is linked to Cx43 dephosphorylation [20,23,72,115]. The activation of large nonselective Cx43 hemichannels during ischemia may overwhelm the normal membrane transport mechanisms and alter the intracellular composition, contributing to cell injury. The ATP depletion in ischemia can reduce phosphorylation and activate the Cx43 hemichannels, although other possibilities have been proposed [20,23,79–81,115–118].

The phosphorylation of Cx43 by PKC does not impede the electrical cell-to-cell coupling (fluxes of small inorganic univalent ions though gap junction channels), but it affects the chemical coupling (larger hydrophilic solutes) selectivity; the negatively charged solute permeation decreases and positively charged ones increases [76,78–81]. Mutagenesis experiments suggested that the phosphorylation of Ser368 is responsible for the functional effect [79,81], a conclusion corroborated in studies using purified Cx43 hemichannels reconstituted in liposomes [78]. Purified Cx43 dephosphorylated by alkaline phosphatase is permeable to organic hydrophilic probes such as sucrose; the phosphorylation by PKC in vitro yields hemichannels that are impermeable to the probes, and the effect of PKC is absent in the purified Cx43-Ser368A mutant, which is constitutively permeable to sucrose and carboxyfluorescein and does not respond to PKC [78,79]. These studies on purified Cx43 hemichannels show that the phosphorylation of Ser368 by PKC directly reduces solute permeability (not through regulatory intermediate steps).

The phosphorylation of Ser368 by PKC abolishes the sucrose permeability through the purified Cx43 hemichannels, but the phosphorylated hemichannels are still permeable to ethyleneglycol (62 vs. 342 Da for sucrose) (Figure 10.5). Therefore, PKC-phosphorylated Cx43 hemichannels are not fully closed, since they allow ethyleneglycol transport. Since the hydrodynamic radius of ethyleneglycol is larger than that of hydrated K^+ and Cl^-, these results can explain why the activation of PKC reduces dye transfer, but it has no major effect on cell-to-cell gap junction currents [76,80–84]. Since the stimulation of PKC decreases the frequency of the dominant ~100 pS conductance state of Cx43 gap junction channels in favor of a lower ~50 pS conductance state [76,81–83], it is tempting to speculate that the ~50 pS channels correspond to the phosphorylated hemichannels permeable to ethyleneglycol and impermeable to sucrose.

As described earlier, purified Cx43 hemichannels with defined average subunit composition can be generated by mixing subunits in detergent. This allowed us to determine the number of subunits that have to be phosphorylated by PKC to render the hemichannels impermeable to sucrose. Under conditions where the liposomes contain an average of 0.8 hemichannels the percentage of liposomes permeable to sucrose depends on whether the hemichannel in a liposome is permeable to sucrose or not. Assuming that dephosphorylated (Cx43-dP) and phosphorylated (Cx43-P) subunits randomly mix to form hemichannels, the subunit composition distribution

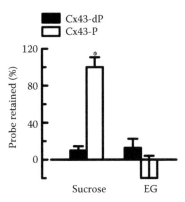

FIGURE 10.5 Partial closure of the Cx43 hemichannels by PKC-mediated phosphorylation. The percentage of radio-labeled probe retention after gel filtration was measured in liposomes preloaded with the probes. If the probe permeates through the hemichannels, it exits the liposomes, and therefore, 100% probe retained indicates impermeable hemichannels, whereas 0% retention indicates hemichannels permeable to the probe. EG: ethyleneglycol. *$p < 0.05$ versus liposomes containing Cx43-dP hemichannels. (Modified from Bao X, Lee SC, Reuss L, Altenberg GA, *Proceedings of the National Academy of Sciences of the United States of America*, 104, 4919–4924. 2007. With permission.)

can be calculated. For example, a 3/3 mixture of Cx43-dP/Cx43-P will contain ~31% of hemichannels with three Cx43-dP/3 Cx43-P (3/3) subunits, ~23% each with 2/4 and 4/2, ~9% each with 1/5 and 5/1, and ~2% each with all Cx43-dP and all Cx43-P. Essentially all liposomes containing hemichannels formed by Cx43-dP alone or Cx43-dP/Cx43-P mixtures of 5/1, 4/2, or 3/3 are permeable to sucrose [77]. Therefore, the presence of three phosphorylated subunits in a hemichannel does not produce sucrose-impermeable hemichannels. Increasing the average number of Cx43-P subunits/hemichannel decreases the percentage of sucrose-impermeable liposomes, but the percentage is only ~30% for an average of five Cx43-P/hemichannel [77]; if five Cx43-P/hemichannel were sufficient to abolish the sucrose permeability, the expected percentage of sucrose-impermeable liposomes is ~74%, the sum of the liposomes with hemichannels containing five (~40%) and six (~34%) Cx43-P subunits. This value is more than twice the value measured (~30%) but close to the percentage of liposomes containing all subunits phosphorylated (~34%). Therefore, since under our experimental conditions, the liposomes are either permeable to sucrose (contain at least one open hemichannel) or impermeable (no hemichannels open), the statistical analysis of the results indicates that all six hemichannel subunits have to be phosphorylated to abolish sucrose permeability; basically, with an average of five Cx43-P subunits per hemichannels, ~30% of the liposomes are sucrose permeable, instead of the ~75% predicted if five Cx43-P per hemichannel were sufficient to render the liposomes impermeable to the probe (Figure 10.6) [77]. Since the assays based on gel filtration only provide information on the permeability cutoff, our results cannot rule out that the partial phosphorylation of Cx43 hemichannels decreases the sucrose permeability, without abolishing it. Studies of the kinetics of transport such as those that we have developed recently [93,119] will be useful to address this issue.

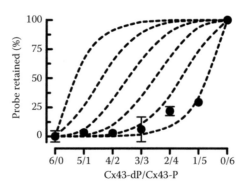

FIGURE 10.6 All Cx43 subunits have to be phosphorylated by PKC to abolish the hemichannel sucrose permeability. The dependence of the percentage of sucrose-permeable liposomes on the Cx34-dP/Cx43-P average ratios was determined in liposomes preloaded with radio-labeled sucrose. Dashed lines denote the expected percentage (from the binomial distribution) if the number of Cx43-P subunits necessary to render the hemichannel impermeable to sucrose are (from left to right) ≥ 1, ≥ 2, ≥ 3, ≥ 4, ≥ 5, or 6. Since the data fit best the dashed line on the far right, we concluded that all hemichannel subunits need to be phosphorylated to abolish the sucrose permeability. Data are means $\pm$ SEM of four to seven experiments. (Modified from Bao X, Lee SC, Reuss L, Altenberg GA, *Proceedings of the National Academy of Sciences of the United States of America*, 104, 4919–4924. 2007. With permission.)

Purified hemichannels are also useful to assess the conformational changes. In limited-proteolysis studies, we found trypsin-resistant CTDs in reconstituted hemichannels formed by PKC-phosphorylated Cx43 but not in those formed by dephosphorylated Cx43 or Cx43-S368A phosphorylated by PKC. We also found that the phosphorylation decreases the tryptophan fluorescence and produces a blueshift of the emission peak. These changes suggest that one or more tryptophan residues are in a more hydrophilic environment after phosphorylation [78]. Nuclear magnetic resonance spectroscopy studies of Ser368/Ser372 mutations to Asp showed a ~10% increase in the α-helical content structure of a Cx43 fragment consisting of transmembrane helix 4 and CTD, with changes in the chemical shifts of 14 residues [120]. Our partial-proteolysis studies suggest a more structured PKC-phosphorylated CTD in the purified Cx43 reconstituted in the liposomes compared to the largely unstructured isolated CTD [78,120]. Therefore, it is presently unknown whether the structure of the CTD fragments is physiologic or an artifact due to factors such as the absence of interactions with other regions and/or the abnormal folding. In summary, the phosphorylation of Ser368 produces conformational changes in Cx43, but the exact nature of the changes and how they relate to the alteration in permeability are unknown.

10.6 CONCLUSION

The methodologies for hemichannel expression/purification and the functional and structural analyses of connexin hemichannels are still underdeveloped. However, it seems likely that fundamental advances will take place in the near future because proven and reliable expression/purification systems and functional assays that constitute a solid base for detailed connexin hemichannel structure–function studies are currently available.

10.7 LIST OF REAGENTS AND EQUIPMENT

Here we list the more specialized reagents and equipment employed to express, purify, reconstitute, and characterize connexin hemichannels. Common reagents, standard laboratory equipment, and reagents specified in the main text are not listed here, but the details can be obtained by contacting us. Pictures and details of some of the equipment used for protein production and analysis can be found at http://www. ttuhsc.edu/som/physiology/mpcl.aspx.

1. Protein expression and purification
   a. Baculovirus/insect cell expression system: The pFastBac1 transfer plasmid is currently part of the ThermoFisher Bac-to-Bac Vector Kit (catalog #10360–014). The Bac-to-Bac Baculovirus Expression System for the generation of recombinant baculoviruses (catalog #10359–016), the Sf9 and other insect cells, as well as Grace's and serum-free HyClone CCM3 media were all obtained from ThermoFisher.
   b. Purification reagents: The detergent n-dodecyl-β-D-maltoside was from Inalco Pharmaceuticals (catalog #1758–1350), and Anzergent 3–12 was from Anatrace (catalog #AZ312). The immobilized Co^{2+} resin used for connexin purification was Clontech's Talon Superflow, and the high-resolution prepacked gel-filtration column to characterize the hydrodynamic properties of hemichannels was a Superdex 200 10/300 from GE Healthcare.
   c. Equipment: The medium-pressure chromatography systems used for connexin purification and hemichannel characterization were the AKTÄ Purifier from GE Healthcare and the LabAlliance APLC System from Agilent. The specialized detectors used to characterize the hemichannel size and permeability to fluorescent probes were an Agilent 1200 Series fluorescence detector, a Wyatt MiniDawn Treos multi-angle light scattering detector, and a Wyatt Optilab T-rEX refractive index detector.
2. Functional analysis of purified and reconstituted connexin hemichannels
   a. Reconstitution: All lipids used for reconstitution were purchased from Avanti Polar Lipids. For liposome extrusion to make unilamellar liposomes of uniform size, we used the Avanti Mini-Extruder from Avanti Polar Lipids (catalog #610000).
   b. Electrophysiology: For single-channel studies, we used an automated Nanion patch-clamp rig (Port-a-Patch, Nanion) with an EPC 10 HEKA amplifier. The data acquisition and analysis were done with the PatchMaster software from HEKA. Giant liposomes were made by electroswelling with the aid of the Nanion Vesicle Prep Pro.

   c. Stopped-flow H$^+$, Na$^+$, and Ca^{2+} influx: Fluo-5N, SBFI, and fluorescein-labeled phospholipid n-(fluorescein-5-thiocarbamoyl)-1,2-dihexadecanoyl-sn-glycero-3-phosphoethanolamine were purchased from ThermoFisher. The stopped-flow device was a SX20 from Applied Photophysics.

3. Spectroscopy: The thiol-reactive luminescence Tb^{3+} chelate can now be purchased from ThermoFisher (LanthaScreen Thiol reactive Tb Chelate). The luminescence resonance energy transfer data were acquired on a Photon Technology International spectrometer (QM3SS) or an Optical Building Blocks phosphorescence lifetime photometer (EasyLife L).

ACKNOWLEDGMENTS

This work was supported in part by grants from the National Institutes of Health (R01GM068586, R21DC007150, R01GM79629, and 3R01GM079629–03S1) and the American Heart Association (National Grant-in-Aid 0050353N and Texas Affiliate Grant-in-Aid 0455115Y).

REFERENCES

1. Nielsen MS, Nygaard Axelsen L, Sorgen PL, Verma V, Delmar M, Holstein-Rathlou NH: Gap junctions. *Comprehensive Physiology* 2012, 2(3):1981–2035.
2. Harris AL: Emerging issues of connexin channels: Biophysics fills the gap. *Quarterly Reviews of Biophysics* 2001, 34(3):325–472.
3. Mese G, Richard G, White TW: Gap junctions: Basic structure and function. *The Journal of Investigative Dermatology* 2007, 127(11):2516–2524.
4. Abascal F, Zardoya R: Evolutionary analyses of gap junction protein families. *Biochimica et Biophysica Acta* 2013, 1828(1):4–14.
5. Hua VB, Chang AB, Tchieu JH, Kumar NM, Nielsen PA, Saier MH, Jr.: Sequence and phylogenetic analyses of 4 TMS junctional proteins of animals: Connexins, innexins, claudins and occludins. *The Journal of Membrane Biology* 2003, 194(1):59–76.
6. Bruzzone S, Guida L, Zocchi E, Franco L, De Flora A: Connexin 43 hemi channels mediate Ca^{2+}-regulated transmembrane NAD$^+$ fluxes in intact cells. *FASEB Journal: Official Publication of the Federation of American Societies for Experimental Biology* 2001, 15(1):10–12.
7. Cherian PP, Siller-Jackson AJ, Gu S, Wang X, Bonewald LF, Sprague E, Jiang JX: Mechanical strain opens connexin 43 hemichannels in osteocytes: A novel mechanism for the release of prostaglandin. *Molecular Biology of the Cell* 2005, 16(7):3100–3106.
8. Kang J, Kang N, Lovatt D, Torres A, Zhao Z, Lin J, Nedergaard M: Connexin 43 hemichannels are permeable to ATP. *The Journal of Neuroscience: The Official Journal of the Society for Neuroscience* 2008, 28(18):4702–4711.
9. Rana S, Dringen R: Gap junction hemichannel-mediated release of glutathione from cultured rat astrocytes. *Neuroscience Letters* 2007, 415(1):45–48.
10. Retamal MA, Froger N, Palacios-Prado N, Ezan P, Saez PJ, Saez JC, Giaume C: Cx43 hemichannels and gap junction channels in astrocytes are regulated oppositely by proinflammatory cytokines released from activated microglia. *The Journal of Neuroscience: The Official Journal of the Society for Neuroscience* 2007, 27(50):13781–13792.
11. Ye ZC, Wyeth MS, Baltan-Tekkok S, Ransom BR: Functional hemichannels in astrocytes: A novel mechanism of glutamate release. *The Journal of Neuroscience: The Official Journal of the Society for Neuroscience* 2003, 23(9):3588–3596.

12. Wang N, De Bock M, Decrock E, Bol M, Gadicherla A, Vinken M, Rogiers V, Bukauskas FF, Bultynck G, Leybaert L: Paracrine signaling through plasma membrane hemichannels. *Biochimica et Biophysica Acta* 2013, 1828(1):35–50.

13. Orellana JA, Figueroa XF, Sanchez HA, Contreras-Duarte S, Velarde V, Saez JC: Hemichannels in the neurovascular unit and white matter under normal and inflamed conditions. *CNS & Neurological Disorders Drug Targets* 2011, 10(3):404–414.

14. Contreras JE, Saez JC, Bukauskas FF, Bennett MV: Gating and regulation of connexin 43 (Cx43) hemichannels. *Proceedings of the National Academy of Sciences of the United States of America* 2003, 100(20):11388–11393.

15. Sanchez HA, Mese G, Srinivas M, White TW, Verselis VK: Differentially altered Ca^{2+} regulation and Ca^{2+} permeability in Cx26 hemichannels formed by the A40V and G45E mutations that cause keratitis ichthyosis deafness syndrome. *The Journal of General Physiology* 2010, 136(1):47–62.

16. Bukauskas FF, Verselis VK: Gap junction channel gating. *Biochimica et Biophysica Acta* 2004, 1662(1–2):42–60.

17. Fasciani I, Temperan A, Perez-Atencio LF, Escudero A, Martinez-Montero P, Molano J, Gomez-Hernandez JM, Paino CL, Gonzalez-Nieto D, Barrio LC: Regulation of connexin hemichannel activity by membrane potential and the extracellular calcium in health and disease. *Neuropharmacology* 2013, 75:479–490.

18. Chen Y, Deng Y, Bao X, Reuss L, Altenberg GA: Mechanism of the defect in gap-junctional communication by expression of a connexin 26 mutant associated with dominant deafness. *FASEB Journal: Official Publication of the Federation of American Societies for Experimental Biology* 2005, 19(11):1516–1518.

19. Saez JC, Retamal MA, Basilio D, Bukauskas FF, Bennett MV: Connexin-based gap junction hemichannels: Gating mechanisms. *Biochimica et Biophysica Acta* 2005, 1711(2):215–224.

20. Contreras JE, Sanchez HA, Eugenin EA, Speidel D, Theis M, Willecke K, Bukauskas FF, Bennett MV, Saez JC: Metabolic inhibition induces opening of unapposed connexin 43 gap junction hemichannels and reduces gap junctional communication in cortical astrocytes in culture. *Proceedings of the National Academy of Sciences of the United States of America* 2002, 99(1):495–500.

21. John SA, Kondo R, Wang SY, Goldhaber JI, Weiss JN: Connexin-43 hemichannels opened by metabolic inhibition. *The Journal of Biological Chemistry* 1999, 274(1):236–240.

22. Li F, Sugishita K, Su Z, Ueda I, Barry WH: Activation of connexin-43 hemichannels can elevate [Ca(2+)]i and [Na(+)]i in rabbit ventricular myocytes during metabolic inhibition. *Journal of Molecular and Cellular Cardiology* 2001, 33(12):2145–2155.

23. Li WE, Nagy JI: Connexin43 phosphorylation state and intercellular communication in cultured astrocytes following hypoxia and protein phosphatase inhibition. *The European Journal of Neuroscience* 2000, 12(7):2644–2650.

24. Shintani-Ishida K, Uemura K, Yoshida K: Hemichannels in cardiomyocytes open transiently during ischemia and contribute to reperfusion injury following brief ischemia. *American Journal of Physiology Heart and Circulatory Physiology* 2007, 293(3):H1714–1720.

25. Vergara L, Bao X, Cooper M, Bello-Reuss E, Reuss L: Gap-junctional hemichannels are activated by ATP depletion in human renal proximal tubule cells. *The Journal of Membrane Biology* 2003, 196(3):173–184.

26. Retamal MA, Cortes CJ, Reuss L, Bennett MV, Saez JC: S-nitrosylation and permeation through connexin 43 hemichannels in astrocytes: Induction by oxidant stress and reversal by reducing agents. *Proceedings of the National Academy of Sciences of the United States of America* 2006, 103(12):4475–4480.

27. Lee JR, White TW: Connexin-26 mutations in deafness and skin disease. *Expert Reviews in Molecular Medicine* 2009, 11:e35.

28. Gerido DA, White TW: Connexin disorders of the ear, skin, and lens. *Biochimica et Biophysica Acta* 2004, 1662(1–2):159–170.

29. Zhao HB, Kikuchi T, Ngezahayo A, White TW: Gap junctions and cochlear homeostasis. *The Journal of Membrane Biology* 2006, 209(2–3):177–186.

30. Nickel R, Forge A: Gap junctions and connexins in the inner ear: Their roles in homeostasis and deafness. *Current Opinion in Otolaryngology & Head and Neck Surgery* 2008, 16(5):452–457.

31. Forge A, Becker D, Casalotti S, Edwards J, Marziano N, Nevill G: Gap junctions in the inner ear: Comparison of distribution patterns in different vertebrates and assessment of connexin composition in mammals. *The Journal of Comparative Neurology* 2003, 467(2):207–231.

32. Forge A, Wright T: The molecular architecture of the inner ear. *British Medical Bulletin* 2002, 63:5–24.

33. Liu W, Bostrom M, Kinnefors A, Rask-Andersen H: Unique expression of connexins in the human cochlea. *Hearing Research* 2009, 250(1–2):55–62.

34. Apps SA, Rankin WA, Kurmis AP: Connexin 26 mutations in autosomal recessive deafness disorders: A review. *International Journal of Audiology* 2007, 46(2):75–81.

35. Laird DW: The gap junction proteome and its relationship to disease. *Trends in Cell Biology* 2010, 20(2):92–101.

36. Martinez AD, Acuna R, Figueroa V, Maripillan J, Nicholson B: Gap-junction channels dysfunction in deafness and hearing loss. *Antioxidants & Redox Signaling* 2009, 11(2):309–322.

37. Ravecca F, Berrettini S, Forli F, Marcaccini M, Casani A, Baldinotti F, Fogli A, Siciliano G, Simi P: Cx26 gene mutations in idiopathic progressive hearing loss. *The Journal of Otolaryngology* 2005, 34(2):126–134.

38. Sabag AD, Dagan O, Avraham KB: Connexins in hearing loss: A comprehensive overview. *Journal of Basic and Clinical Physiology and Pharmacology* 2005, 16(2–3):101–116.

39. Steel KP: Science, medicine, and the future: New interventions in hearing impairment. *British Medical Journal Bmj* 2000, 320(7235):622–625.

40. Terrinoni A, Codispoti A, Serra V, Didona B, Bruno E, Nistico R, Giustizieri M, Alessandrini M, Campione E, Melino G: Connexin 26 (GJB2) mutations, causing KID Syndrome, are associated with cell death due to calcium gating deregulation. *Biochemical and Biophysical Research Communications* 2010, 394(4):909–914.

41. Zoidl G, Dermietzel R: Gap junctions in inherited human disease. *Pflugers Archiv: European Journal of Physiology* 2010, 460(2):451–466.

42. Steel KP: Perspectives: Biomedicine: The benefits of recycling. *Science* 1999, 285(5432):1363–1364.

43. Wangemann P: Supporting sensory transduction: Cochlear fluid homeostasis and the endocochlear potential. *The Journal of Physiology* 2006, 576(Pt 1):11–21.

44. Johnstone BM, Patuzzi R, Syka J, Sykova E: Stimulus-related potassium changes in the organ of Corti of guinea-pig. *The Journal of Physiology* 1989, 408:77–92.

45. Kudo T, Kure S, Ikeda K, Xia AP, Katori Y, Suzuki M, Kojima K, Ichinohe A, Suzuki Y, Aoki Y et al: Transgenic expression of a dominant-negative connexin26 causes degeneration of the organ of Corti and non-syndromic deafness. *Human Molecular Genetics* 2003, 12(9):995–1004.

46. Patuzzi R: Ion flow in stria vascularis and the production and regulation of cochlear endolymph and the endolymphatic potential. *Hearing Research* 2011, 277(1–2):4–19.

47. Beltramello M, Piazza V, Bukauskas FF, Pozzan T, Mammano F: Impaired permeability to Ins(1,4,5)P3 in a mutant connexin underlies recessive hereditary deafness. *Nature Cell Biology* 2005, 7(1):63–69.

48. Bruzzone R, Veronesi V, Gomes D, Bicego M, Duval N, Marlin S, Petit C, D'Andrea P, White TW: Loss-of-function and residual channel activity of connexin26 mutations associated with non-syndromic deafness. *FEBS Letters* 2003, 533(1–3):79–88.

49. Gossman DG, Zhao HB: Hemichannel-mediated inositol 1,4,5-trisphosphate (IP3) release in the cochlea: A novel mechanism of IP3 intercellular signaling. *Cell Communication & Adhesion* 2008, 15(4):305–315.

50. Zhao HB: Connexin26 is responsible for anionic molecule permeability in the cochlea for intercellular signalling and metabolic communications. *The European Journal of Neuroscience* 2005, 21(7):1859–1868.

51. Majumder P, Crispino G, Rodriguez L, Ciubotaru CD, Anselmi F, Piazza V, Bortolozzi M, Mammano F: ATP-mediated cell-cell signaling in the organ of Corti: The role of connexin channels. *Purinergic Signalling* 2010, 6(2):167–187.

52. Anselmi F, Hernandez VH, Crispino G, Seydel A, Ortolano S, Roper SD, Kessaris N, Richardson W, Rickheit G, Filippov MA et al: ATP release through connexin hemichannels and gap junction transfer of second messengers propagate Ca^{2+} signals across the inner ear. *Proceedings of the National Academy of Sciences of the United States of America* 2008, 105(48):18770–18775.

53. Goldberg GS, Moreno AP, Lampe PD: Gap junctions between cells expressing connexin 43 or 32 show inverse permselectivity to adenosine and ATP. *The Journal of Biological Chemistry* 2002, 277(39):36725–36730.

54. Goldberg GS, Lampe PD, Nicholson BJ: Selective transfer of endogenous metabolites through gap junctions composed of different connexins. *Nature Cell Biology* 1999, 1(7):457–459.

55. Deng Y, Chen Y, Reuss L, Altenberg GA: Mutations of connexin 26 at position 75 and dominant deafness: Essential role of arginine for the generation of functional gap-junctional channels. *Hearing Research* 2006, 220(1–2):87–94.

56. Mese G, Valiunas V, Brink PR, White TW: Connexin26 deafness associated mutations show altered permeability to large cationic molecules. *American Journal of Physiology Cell Physiology* 2008, 295(4):C966–974.

57. Gerido DA, DeRosa AM, Richard G, White TW: Aberrant hemichannel properties of Cx26 mutations causing skin disease and deafness. *American Journal of Physiology Cell Physiology* 2007, 293(1):C337–345.

58. Lee JR, Derosa AM, White TW: Connexin mutations causing skin disease and deafness increase hemichannel activity and cell death when expressed in Xenopus oocytes. *The Journal of Investigative Dermatology* 2009, 129(4):870–878.

59. Stong BC, Chang Q, Ahmad S, Lin X: A novel mechanism for connexin 26 mutation linked deafness: Cell death caused by leaky gap junction hemichannels. *The Laryngoscope* 2006, 116(12):2205–2210.

60. Retamal MA, Reyes EP, Garcia IE, Pinto B, Martinez AD, Gonzalez C: Diseases associated with leaky hemichannels. *Frontiers in Cellular Neuroscience* 2015, 9:267.

61. Solan JL, Lampe PD: Connexin43 phosphorylation: Structural changes and biological effects. *The Biochemical Journal* 2009, 419(2):261–272.

62. Agullo-Pascual E, Delmar M: The noncanonical functions of Cx43 in the heart. *The Journal of Membrane Biology* 2012, 245(8):477–482.

63. Agullo-Pascual E, Lin X, Pfenniger A, Lubkemeier I, Willecke K, Rothenberg E, Delmar M: A novel noncanonical role of cx43 in the heart: Ensuring the arrival of nav1.5 to the intercalated disk. *Heart Rhythm: The Official Journal of the Heart Rhythm Society* 2013, 10(11):1742.

64. Warn-Cramer BJ, Lau AF: Regulation of gap junctions by tyrosine protein kinases. *Biochimica et Biophysica Acta* 2004, 1662(1–2):81–95.

65. Herve JC, Derangeon M, Sarrouilhe D, Giepmans BN, Bourmeyster N: Gap junctional channels are parts of multiprotein complexes. *Biochimica et Biophysica Acta* 2012, 1818(8):1844–1865.

66. Marquez-Rosado L, Solan JL, Dunn CA, Norris RP, Lampe PD: Connexin43 phosphorylation in brain, cardiac, endothelial and epithelial tissues. *Biochimica et Biophysica Acta* 2012, 1818(8):1985–1992.

67. Churko JM, Laird DW: Gap junction remodeling in skin repair following wounding and disease. *Physiology* 2013, 28(3):190–198.

68. Giaume C, Leybaert L, Naus CC, Saez JC: Connexin and pannexin hemichannels in brain glial cells: Properties, pharmacology, and roles. *Frontiers in Pharmacology* 2013, 4:88.

69. Orellana JA, von Bernhardi R, Giaume C, Saez JC: Glial hemichannels and their involvement in aging and neurodegenerative diseases. *Reviews in the Neurosciences* 2012, 23(2):163–177.

70. Eugenin EA, Basilio D, Saez JC, Orellana JA, Raine CS, Bukauskas F, Bennett MV, Berman JW: The role of gap junction channels during physiologic and pathologic conditions of the human central nervous system. *Journal of Neuroimmune Pharmacology: The Official Journal of the Society on NeuroImmune Pharmacology* 2012, 7(3):499–518.

71. Fontes MS, van Veen TA, de Bakker JM, van Rijen HV: Functional consequences of abnormal Cx43 expression in the heart. *Biochimica et Biophysica Acta* 2012, 1818(8):2020–2029.

72. Duffy HS: The molecular mechanisms of gap junction remodeling. *Heart Rhythm: The Official Journal of the Heart Rhythm Society* 2012, 9(8):1331–1334.

73. Remo BF, Giovannone S, Fishman GI: Connexin43 cardiac gap junction remodeling: Lessons from genetically engineered murine models. *The Journal of Membrane Biology* 2012, 245(5–6):275–281.

74. Jeyaraman MM, Srisakuldee W, Nickel BE, Kardami E: Connexin43 phosphorylation and cytoprotection in the heart. *Biochimica et Biophysica Acta* 2012, 1818(8):2009–2013.

75. Miura T, Miki T, Yano T: Role of the gap junction in ischemic preconditioning in the heart. *American Journal of Physiology Heart and Circulatory Physiology* 2010, 298(4):H1115–1125.

76. Ek-Vitorin JF, King TJ, Heyman NS, Lampe PD, Burt JM: Selectivity of connexin 43 channels is regulated through protein kinase C-dependent phosphorylation. *Circulation Research* 2006, 98(12):1498–1505.

77. Bao X, Lee SC, Reuss L, Altenberg GA: Change in permeant size selectivity by phosphorylation of connexin 43 gap-junctional hemichannels by PKC. *Proceedings of the National Academy of Sciences of the United States of America* 2007, 104 (12):4919–4924.

78. Bao X, Reuss L, Altenberg GA: Regulation of purified and reconstituted connexin 43 hemichannels by protein kinase C-mediated phosphorylation of Serine 368. *The Journal of Biological Chemistry* 2004, 279(19):20058–20066.

79. Bao X, Altenberg GA, Reuss L: Mechanism of regulation of the gap junction protein connexin 43 by protein kinase C-mediated phosphorylation. *American Journal of Physiology Cell Physiology* 2004, 286(3):C647–654.

80. Kwak BR, van Veen TA, Analbers LJ, Jongsma HJ: TPA increases conductance but decreases permeability in neonatal rat cardiomyocyte gap junction channels. *Experimental Cell Research* 1995, 220(2):456–463.

81. Lampe PD, TenBroek EM, Burt JM, Kurata WE, Johnson RG, Lau AF: Phosphorylation of connexin43 on serine368 by protein kinase C regulates gap junctional communication. *The Journal of Cell Biology* 2000, 149(7):1503–1512.

82. Moreno AP, Fishman GI, Spray DC: Phosphorylation shifts unitary conductance and modifies voltage dependent kinetics of human connexin43 gap junction channels. *Biophysical Journal* 1992, 62(1):51–53.

83. Moreno AP, Saez JC, Fishman GI, Spray DC: Human connexin43 gap junction channels: Regulation of unitary conductances by phosphorylation. *Circulation Research* 1994, 74(6):1050–1057.

84. Takens-Kwak BR, Jongsma HJ: Cardiac gap junctions: Three distinct single channel conductances and their modulation by phosphorylating treatments. *Pflugers Archiv: European Journal of Physiology* 1992, 422(2):198–200.

85. Delmar M, Coombs W, Sorgen P, Duffy HS, Taffet SM: Structural bases for the chemical regulation of Connexin43 channels. *Cardiovascular Research* 2004, 62(2):268–275.

86. Fiori MC, Figueroa V, Zoghbi ME, Saez JC, Reuss L, Altenberg GA: Permeation of calcium through purified connexin 26 hemichannels. *The Journal of Biological Chemistry* 2012, 287(48):40826–40834.

87. Ambrosi C, Boassa D, Pranskevich J, Smock A, Oshima A, Xu J, Nicholson BJ, Sosinsky GE: Analysis of four connexin26 mutant gap junctions and hemichannels reveals variations in hexamer stability. *Biophysical Journal* 2010, 98(9):1809–1819.

88. Maeda S, Nakagawa S, Suga M, Yamashita E, Oshima A, Fujiyoshi Y, Tsukihara T: Structure of the connexin 26 gap junction channel at 3.5 A resolution. *Nature* 2009, 458(7238):597–602.

89. Gassmann O, Kreir M, Ambrosi C, Pranskevich J, Oshima A, Roling C, Sosinsky G, Fertig N, Steinem C: The M34A mutant of Connexin26 reveals active conductance states in pore-suspending membranes. *Journal of Structural Biology* 2009, 168(1):168–176.

90. Oshima A, Tani K, Hiroaki Y, Fujiyoshi Y, Sosinsky GE: Projection structure of a N-terminal deletion mutant of connexin 26 channel with decreased central pore density. *Cell Communication & Adhesion* 2008, 15(1):85–93.

91. Oshima A, Tani K, Hiroaki Y, Fujiyoshi Y, Sosinsky GE: Three-dimensional structure of a human connexin26 gap junction channel reveals a plug in the vestibule. *Proceedings of the National Academy of Sciences of the United States of America* 2007, 104(24):10034–10039.

92. Stauffer KA: The gap junction proteins beta 1-connexin (connexin-32) and beta 2-connexin (connexin-26) can form heteromeric hemichannels. *The Journal of Biological Chemistry* 1995, 270(12):6768–6772.

93. Fiori MC, Krishnan S, Cortes DM, Retamal MA, Reuss L, Altenberg GA, Cuello LG: Functional hemichannels formed by human connexin 26 expressed in bacteria. *Bioscience Reports* 2015, 35(2), pii: e00177.

94. Oshima A, Tani K, Toloue MM, Hiroaki Y, Smock A, Inukai S, Cone A, Nicholson BJ, Sosinsky GE, Fujiyoshi Y: Asymmetric configurations and N-terminal rearrangements in connexin26 gap junction channels. *Journal of Molecular Biology* 2011, 405(3):724–735.

95. Oshima A, Doi T, Mitsuoka K, Maeda S, Fujiyoshi Y: Roles of Met-34, Cys-64, and Arg-75 in the assembly of human connexin 26: Implication for key amino acid residues for channel formation and function. *The Journal of Biological Chemistry* 2003, 278(3):1807–1816.

96. Hoh JH, Sosinsky GE, Revel JP, Hansma PK: Structure of the extracellular surface of the gap junction by atomic force microscopy. *Biophysical Journal* 1993, 65(1):149–163.

97. Carnarius C, Kreir M, Krick M, Methfessel C, Moehrle V, Valerius O, Bruggemann A, Steinem C, Fertig N: Green fluorescent protein changes the conductance of connexin 43 (Cx43) hemichannels reconstituted in planar lipid bilayers. *The Journal of Biological Chemistry* 2012, 287(4):2877–2886.

98. Rhee SK, Bevans CG, Harris AL: Channel-forming activity of immunoaffinity-purified connexin32 in single phospholipid membranes. *Biochemistry* 1996, 35(28):9212–9223.

99. Buehler LK, Stauffer KA, Gilula NB, Kumar NM: Single channel behavior of recombinant beta 2 gap junction connexons reconstituted into planar lipid bilayers. *Biophysical Journal* 1995, 68(5):1767–1775.

100. Harris AL, Bevans CG: Exploring hemichannel permeability in vitro. *Methods in Molecular Biology* 2001, 154:357–377.

101. Harris AL, Walter A, Zimmerberg J: Transport-specific isolation of large channels reconstituted into lipid vesicles. *The Journal of Membrane Biology* 1989, 109(3):243–250.

102. Bevans CG, Kordel M, Rhee SK, Harris AL: Isoform composition of connexin channels determines selectivity among second messengers and uncharged molecules. *The Journal of Biological Chemistry* 1998, 273(5):2808–2816.

103. Kanaporis G, Mese G, Valiuniene L, White TW, Brink PR, Valiunas V: Gap junction channels exhibit connexin-specific permeability to cyclic nucleotides. *The Journal of General Physiology* 2008, 131(4):293–305.

104. Harris AL: Connexin channel permeability to cytoplasmic molecules. *Progress in Biophysics and Molecular Biology* 2007, 94(1–2):120–143.

105. Ayad WA, Locke D, Koreen IV, Harris AL: Heteromeric, but not homomeric, connexin channels are selectively permeable to inositol phosphates. *The Journal of Biological Chemistry* 2006, 281(24):16727–16739.

106. Schalper KA, Sanchez HA, Lee SC, Altenberg GA, Nathanson MH, Saez JC: Connexin 43 hemichannels mediate the Ca^{2+} influx induced by extracellular alkalinization. *American Journal of Physiology Cell Physiology* 2010, 299(6):C1504–1515.

107. Selvin PR: Principles and biophysical applications of lanthanide-based probes. *Annual Review of Biophysics and Biomolecular Structure* 2002, 31:275–302.

108. Posson DJ, Selvin PR: Extent of voltage sensor movement during gating of shaker K^+ channels. *Neuron* 2008, 59(1):98–109.

109. Posson DJ, Ge P, Miller C, Bezanilla F, Selvin PR: Small vertical movement of a K^+ channel voltage sensor measured with luminescence energy transfer. *Nature* 2005, 436(7052):848–851.

110. Zoghbi ME, Altenberg GA: Hydrolysis at one of the two nucleotide-binding sites drives the dissociation of ATP-binding cassette nucleotide-binding domain dimers. *The Journal of Biological Chemistry* 2013, 288(47):34259–34265.

111. Cooper RS, Altenberg GA: Association/dissociation of the nucleotide-binding domains of the ATP-binding cassette protein MsbA measured during continuous hydrolysis. *The Journal of Biological Chemistry* 2013, 288(29):20785–20796.

112. Zoghbi ME, Krishnan S, Altenberg GA: Dissociation of ATP-binding cassette nucleotide-binding domain dimers into monomers during the hydrolysis cycle. *The Journal of Biological Chemistry* 2012, 287(18):14994–15000.

113. Rambhadran A, Gonzalez J, Jayaraman V: Conformational changes at the agonist binding domain of the N-methyl-D-aspartic acid receptor. *The Journal of Biological Chemistry* 2011, 286(19):16953–16957.

114. Foote CI, Zhou L, Zhu X, Nicholson BJ: The pattern of disulfide linkages in the extracellular loop regions of connexin 32 suggests a model for the docking interface of gap junctions. *The Journal of Cell Biology* 1998, 140(5):1187–1197.

115. Hawat G, Benderdour M, Rousseau G, Baroudi G: Connexin 43 mimetic peptide Gap26 confers protection to intact heart against myocardial ischemia injury. *Pflugers Archiv: European Journal of Physiology* 2010, 460(3):583–592.

116. Lampe PD, Lau AF: Regulation of gap junctions by phosphorylation of connexins. *Archives of Biochemistry and Biophysics* 2000, 384(2):205–215.

117. Retamal MA, Schalper KA, Shoji KF, Bennett MV, Saez JC: Opening of connexin 43 hemichannels is increased by lowering intracellular redox potential. *Proceedings of the National Academy of Sciences of the United States of America* 2007, 104(20):8322–8327.
118. Retamal MA, Schalper KA, Shoji KF, Orellana JA, Bennett MV, Saez JC: Possible involvement of different connexin43 domains in plasma membrane permeabilization induced by ischemia-reperfusion. *The Journal of Membrane Biology* 2007, 218(1–3):49–63.
119. Fiori MC, Reuss L, Cuello LG, Altenberg GA: Functional analysis and regulation of purified connexin hemichannels. *Frontiers in Physiology* 2014, 5:71.
120. Grosely R, Kopanic JL, Nabors S, Kieken F, Spagnol G, Al-Mugotir M, Zach S, Sorgen PL: Effects of phosphorylation on the structure and backbone dynamics of the intrinsically disordered connexin43 C-terminal domain. *The Journal of Biological Chemistry* 2013, 288(34):24857–24870.

11 Methods to Determine Formation of Heteromeric Hemichannels

Agustín D. Martínez, Oscar Jara, Ricardo Ceriani, Jaime Maripillán, Paula Mujica, and Isaac E. García

CONTENTS

11.1 INTRODUCTION

The oligomerization of proteins involves the sequential recognition between multiple compatible monomeric protein subunits. In the case of ion channels, this complex multistep process ends with a protein organization that confers specific structural and physicochemical characteristics, which make the proteins suitable for ion conduction. In the case of a gap junction channel, the pore is wide, aqueous, and poorly selective, allowing the passage not only of ions but also of bigger molecules like ATP or glucose (Koval 2006).

Gap junction channel formation is a fine-tuning process that starts with the formation of a connexon or a hemichannel. Some connexins can oligomerize and form functional heteromeric channels (e.g., Cx43 and Cx45) (Martínez et al. 2002; Beyer et al. 2001; Koval, Molina, and Burt 2014), but others are incompatible (e.g., Cx43 and Cx26) (Koval, Molina, and Burt 2014; Gemel et al. 2004). Interestingly, heteromeric oligomerization increases the repertoire of channels with particular properties and thus increases the functional diversity.

In the next sections, we will review the most relevant aspect of connexin oligomerization and the mechanisms that allow the formation of heteromeric channels.

11.2 CONNEXIN OLIGOMERIZATION

The oligomerization of connexins takes place in the ER or the Golgi apparatus depending on the connexin type (Das Sarma, Wang, and Koval 2002; Maza, Das Sarma, and Koval 2005). First, it consists of the formation of connexons or hemichannels, which are hexameric subunit ensembles. After trafficking to the plasma membrane, the connexons can dock with complementary connexons provided by the neighboring cells to form gap junction channels. The hexameric arrangement may involve identical subunits (homomeric connexons) or various combinations of compatible connexins (heteromeric connexons). The connexin domains for oligomerization have been a matter of intensive studies. Transmembrane domains are involved in connexin oligomerization. Ahmad et al. (2001) have reported that a Cx32 chimera, in which transmembrane domain TM3 was replaced by a transmembrane domain from the cystic fibrosis transmembrane conductance regulator, a membrane protein, presents defective oligomerization and is retained in the cytoplasm. Using a chimeric construction, Maza, Das Sarma, and Koval (2005) have shown that the residues in the TM3 and E2 domains (R153 and Q173, respectively) regulate the Cx43 oligomerization by preventing the formation of Cx43 hexamers in the ER. In addition, by homology modeling, Fleishman et al. (2006) assigned the four transmembrane domains for Cx32 and showed by double complementarity experiments that the amino acid residues E146 and S139 in the TM3 interacted with the amino acid residues R32 in the TM1 and N206 in the TM4. According to the crystal structure of Cx26, the main interprotomer interactions reside in the extracellular half of transmembrane helices TM2 and TM4 and in the extracellular loops. Additionally, the structure suggests that TM3 could contribute to the interprotomer interactions through the formation of an aromatic cluster (Maeda et al. 2009).

One suitable methodology for the recognition of the protein motifs involved in oligomerization (e.g., dimerization) is TOXCAT (Duong et al. 2007; Mendrola et al. 2002; Polgar et al. 2010). The efficiency of TOXCAT as a dimerization predictor is equivalent to the calculations provided by using sedimentation equilibrium in detergents (Duong et al. 2007).

Using the TOXCAT strategy, our group recently described a VVAA motif (V37 to A40) located at the TM1 of the Cx26 to be critical for homodimerization and contributes to the hemichannel pore wall (Jara et al. 2012). Moreover, we propose that these residues form dimers as a first step for oligomerization. This hypothesis is supported by disulphide bond formation that stabilizes a TM1–TM1 dimer when a cysteine residue replaced the valine at position 37 (Jara et al. 2012). This finding was consistent with structural and functional measurements that suggested the TM1 as major contributor to the pore of gap junction channels and hemichannels (Kronengold et al. 2003; Maeda et al. 2009; Tang et al. 2009).

11.3 COMPATIBILITY BETWEEN CONNEXINS AND FORMATION OF HETEROMERIC HEMICHANNELS

The intrinsic properties, such as single-channel conductance, permeability, and regulatory properties, strictly depend on the nature of the connexon composition, i.e, homomeric or heteromeric gap junction channels and hemichannels. Due to the fact that most cells could express different connexin isoforms, it is possible that most channels may be heteromeric, which could confer major functional diversity (Ahmad et al. 1999; Berthoud et al. 2001; Brink et al. 1997; Das Sarma et al. 2001; Falk et al. 1997; Jiang and Goodenough 1996; Lagree et al. 2003; Martínez et al. 2002). We and other groups have shown that Cx43 forms heteromeric channels with Cx45, Cx40, and Cx37 but not with Cx26 or Cx32 (Gemel et al. 2004; Lagree et al. 2003; Martínez et al. 2002; Valiunas et al. 2001). Nonetheless, the motifs governing the compatibility among connexins are still not well defined.

The truncation of the C-terminal of Cx43 (Cx43tr251) allows the formation of functional homomeric channels but not gap junction plaques. The ability to form gap junctions was recovered when this truncated form was coexpressed with wild-type Cx43 or Cx45. These results demonstrate that the C-terminal of Cx43 is not critical for oligomerization, but it is required for proper gap junction plaque formation (Martínez et al. 2003).

In addition, specific physicochemical properties of the amino acids composing the N-terminal and the third transmembrane segments of Cx43 and Cx32 (two noncompatible connexins) are key players in determining their compatibility and function (Lagree et al. 2003). Using a Cx26/Cx43 chimera, in which Cx26 contains the TM3 of Cx43, we determined that the motifs located in the Cx43-TM3 define its compatibility to oligomerize (Martínez et al. 2011). However, for Cx26, the oligomerization compatibility motifs may be located in the N-terminal segment (Martínez et al. 2011). Consistently, syndromic Cx26 mutations located in this segment can oligomerize with Cx43 to form aberrant channels (García et al. 2015). In terms of heterotypic pairing, it has been shown that the amino acid sequences in the second extracellular loop regulate either the compatibility or the incompatibility among connexins (White et al. 1995).

11.4 CRITERIA AND METHODOLOGY FOR DETERMINATION OF HETEROMERIC HEMICHANNELS

The *cis* or *trans* dominant effect among connexin isoforms (mutant or wild type) can be interpreted as evidence for heterooligomerization, when (1) the wild-type connexins rescue the traffic of the mutant connexin to the plasma membrane, allowing the formation of gap junction plaques containing both proteins; (2) the mutant connexin might retain the wild-type connexin within the intracellular compartment; or (3) there is perfect colocalization of both connexin isoforms. (4) The conclusions drawn from these data should be supported by biochemical studies using pull-down strategies or sedimentation velocity through the sucrose gradients to separate different oligomers. (5) Finally, the interaction should produce measurable changes in the functional properties of the hemichannels and the gap junction channels. The approaches we have used are validated by previous work (Das Sarma et al. 2001; Lagree et al. 2003; Martinez et al. 2003; Maza et al. 2003; Wang et al. 2005).

11.5   IMMUNOFLUORESCENCE AND IMAGING METHODOLOGIES FOR MORPHOLOGICAL EVIDENCE OF HETEROMERIC INTERACTION

The methodology presented here is used to study the colocalization between Cx43, Cx26, or Cx45. This strategy can be adapted for any connexin combination simply by changing the antibodies and the expression systems. For in vitro studies, it is possible to use HeLa, N2A, or HEK 293 cell lines as cellular expression systems, since they are connexin free and can be easily transfected with the respective cDNAs (García et al. 2015). In case the antibodies fail to work, it is highly recommended to insert a C-terminal tag for specific immunological recognition, for instances, hemagglutinin (HA) tag or 6X-histidine tag (His-tag), which is also suitable for pull-down experiments (Martínez et al. 2002).

An example of using the immunofluorescence and imaging technique is shown in Figure 11.1. By confocal microscopy, we observed the colocalization between two different connexins. These images are obtained from the following protocol:

1. Cells grown on glass coverslips were fixed in methanol/acetone and permeabilized with 1% Triton X-100 prior to incubation with primary antibodies.
2. Single- and double-labeling immunofluorescence can be performed as previously described (Martínez et al. 2002) using rabbit polyclonal anti-HA antibodies and mouse monoclonal anti-Cx43 (Chemicon; MAB 3068), anti-Cx26 (Zymed 13-8100), or anti-Cx45 (Chemicon; MAB 3100) antibodies.
3. Cy3-conjugated goat anti-mouse and Cy2-conjugated goat anti-rabbit immunoglobulin G (IgG) antibodies can be obtained from Jackson ImmunoResearch (West Grove, Pennsylvania). The cells were examined using a Nikon confocal microscope. It should be considered that the emission wavelength of one of the fluorophores selected for immunodetection must not overlap the absorption wavelength of the other fluorophore; otherwise, a false positive may drive to misinterpretations.
4. The images are acquired using a Nikon Eclipse C1-Plus confocal microscope and processed with the EZ-C1 software (Nikon Instruments Inc). Briefly, the size and the fluorescence intensity are indicative of heterooligomerization results in gap junction plaque formation (full overlay of the

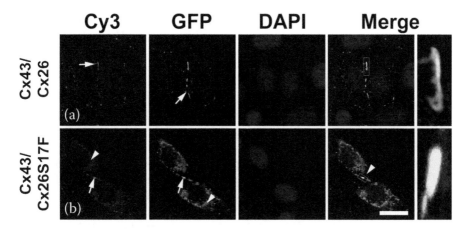

FIGURE 11.1 (**See color insert.**) Syndromic mutation Cx26S17F colocalizes with Cx43. Confocal images of HeLa cells c-expressing Cx43 (red labeling) with wild-type Cx26 or its deafness-associated syndromic mutant Cx26S17F. Arrows point to gap junction plaques. Arrowheads show perinuclear staining. The nuclei were stained with DAPI. Dashed rectangles in the merged panels show ROIs for 3D image projections. Bar = 10 μm. Although (a) Cx43 and Cx26 are located in the same gap junction plaque, they segregate in different regions, consistent with the lack of interaction between wild-type Cx26 and Cx43. However, Cx43 perfectly colocalizes with (b) mutant Cx26S17F in the gap junction plaque, suggesting that deafness syndromic mutation changes the compatibility of mutant Cx26S17F allowing it to interact with Cx43.

> fluorescent emission). The ROI should be taken with ×100 magnification and 0.5 AU pinhole, 0.5 μm in depth Z stacks. Then, 3D reconstruction of the gap junction plaque is performed with the NIS-Elements Viewer 4.0 software (Nikon Instruments Inc). This strategy allows a 360° view of a gap junction plaque, allows better interpretation of the results, and boosts the quality of the figures (García et al. 2015).

11.6 BIOCHEMICAL METHODOLOGIES TO STUDY HETEROMERIC INTERACTIONS

There are several methodologies that have been used to study connexin–connexin interactions such as sedimentationvelocity, pull-down, coimmunoprecipitation, gel filtration, and native polyacrylamide gel electrophoresis/immunoblot (García et al. 2015; Gemel et al. 2004; Lagree et al. 2003). Here, we describe pull-down and velocity sedimentation analyses, two common and relatively simple methodologies that we have used to study the oligomerization between compatible connexins (García et al. 2015; Jara et al. 2012; Martínez et al. 2011).

11.6.1 Pull-Down Using His-tag

The pull-down methodology using His-tag requires that the connexin should be constructed with a His-tag (His$_6$; six histidine in tandem), usually appended to its CT.

This methodology is based in the high affinity of His_6 for metals, like nickel or cobalt. After transfections with the connexin cDNAs (with or without His-tag), the connexons should be solubilized from the cell extracts as explained in next Section 11.6.2 and then affinity purified by applying Ni^{2+}-nitrilotriacetic acid (NTA) column or an equivalent method, as follow:

1. Connexons are solubilized by the incubation of a cell lysate with 1% Triton X-100, followed by centrifugation at 100,000g (Berthoud et al. 2001; Koval et al. 1997; Valiunas et al. 2001).
2. Then the Triton X-100-soluble supernatants containing connexin monomers and hexamers (Berthoud et al. 2001; Koval et al. 1997; Musil and Goodenough 1993) are subjected to affinity purification by binding to the Ni^{2+}–NTA column following manufacturer instructions.
3. After the extensive washing of the column with a washing solution (three to five times), the connexin–His specifically bound to the column is eluted with 1 mol/L imidazole together with any proteins associated to the connexin–His (Valiunas et al. 2001). Each eluted fraction contains about 500 μL.
4. Finally, the samples from each eluting fraction are subjected to electrophoresis through 8–10% polyacrylamide gels containing sodium dodecyl sulfate (SDS) and immunoblotted for the respective connexin antibodies.

11.6.2 Velocity Sedimentation Analysis of Connexin Oligomers

The velocity sedimentation analysis of connexin oligomers consists of the separation of oligomers using ultracentrifugation in sucrose gradients from Triton-X100 soluble fractions of cell homogenates. It is important to load the gradient prior to the ultracentrifugation with fresh samples devoid of clots and precipitated cellular debris, which can contain big protein aggregates and gap junction plaques that may lead to misinterpretations. During the ultracentrifugation, the hexamers could be separated following their match density with the sucrose gradient; therefore, the monomers and the oligomers will be isolated from low and high sucrose density fractions, respectively. A mandatory requirement for study heterooligomerization between connexins is that the sedimentation pattern of the respective homomeric hexamers (from the cells that express only one connexin) should produce peaks of sedimentation in different sucrose concentration for homomeric connexons formed by the different connexins. Thereafter, in the samples from cells that coexpress both connexins, these different connexins cosediment in a new oligomeric peak (García et al. 2015). The heteromeric channels formed by a mutant associated to syndromic deafness and Cx43 are shown in Figure 11.2. The procedure is as follows.

1. Monomers and oligomers are extracted by Triton X-100. The connexin monomers and oligomers are prepared and purified as follows (Martínez et al. 2011). Transfected HeLa cell cultures at 70% confluence (growing in two 100 mm culture dish) are harvested in ice-cold PBS containing 2 mM phenylmethanesulfonyl fluoride.

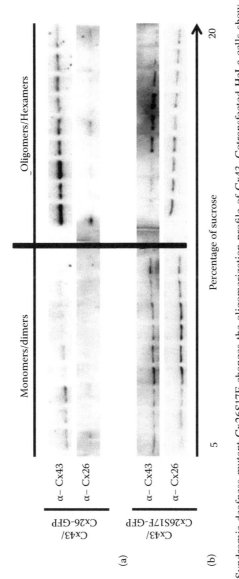

FIGURE 11.2 Syndromic deafness mutant Cx26S17F changes the oligomerization profile of Cx43. Cotransfected HeLa cells show that Cx43 and Cx26-GFP do not cosediment in (a) the same oligomeric fractions. Cotransfected HeLa cells with Cx43 and syndromic mutant Cx26S17F-GFP exhibit a change on the oligomeric peak and cosediment with Cx43 in (b) new oligomeric fractions.

2. The cell suspensions are centrifuged and the supernatants discarded. The pellet is resuspended in about 200 µL of incubation buffer (0.14 M NaCl, 5.3 mM KCl, 0.35 mM $Na_2HPO_4 \times 7H_2O$, 0.35 mM KH_2PO_4, 0.8 mM $MgSO_4$, 2.7 mM $CaCl_2 \times 2H_2O$, 20 mM HEPES, and pH 7.5) containing protease and phosphatase inhibitors.

3. Then, cell pellets are lysed in an incubation buffer by repeated passages (up and down) through a needle (20 passages through a 20-gauge needle followed by 10 passages through a 27-gauge needle), and the protein concentration is determined using the Qubit Protein Assay Kit (Thermo Fisher Scientific).

4. Cell lysates obtained in threes and are incubated on ice for 30 minutes in 1% Triton X-100 and mixed every 5 minutes by inversion. Do not vortex (this will destroy the hexamers).

5. Homogenates are span down for 30 minutes at 100,000g and 4°C. The resultant pellet contains the triton insoluble fraction, which includes gap junction plaques and many other protein aggregates. On the other hand, the supernatant contains the Triton X-100 soluble fraction, in which connexin monomers and hexamers are present.

6. Velocity sedimentation in the sucrose gradient. To resolve between connexin monomer and oligomers, the triton soluble fraction obtained in 5 is subjected to velocity sedimentation through 10 mL 5–20% (w/v) linear sucrose gradients containing 0.1% Triton as previously described (Berthoud et al. 2001).

7. Not more than 250 µL (300 µg) of soluble fractions are loaded on the top of the gradients and centrifuged for 18 hours at 28,600 rpm and 4°C, using a swinging bucket rotor (Sw 60 Ti, Beckman Optima XL-100K Ultracentrifuge). Thereafter, 250 µL fractions are collected from each gradient and analyzed by immunoblotting.

8. The monomeric fraction is determined. The fraction containing monomers are determined by treating the triton soluble fractions with SDS prior to separation in the sucrose gradients by ultracentrifugation.

9. Immunoblots for the respective connexins are performed to detect the connexin's levels in each sucrose fraction. The intensity of the bands in the different fractions is determined by densitometry. The values should be normalized with respect to the maximal densitometry signal.

10. Immunoblotting is done. The protein samples from each sucrose fraction (100 µg of the total homogenates or 50–100 µL of the Triton X-100-soluble cellular fractions) are resolved on 8% or 10% polyacrylamide gels.

11. The proteins are electrotransferred onto Immobilon-P membranes (Millipore, Bedford, Massachusetts).

12. The membranes are incubated in 5% nonfat milk in 0.1% Tween-20 in Tris-buffered saline (TBST) with pH 7.4 overnight at 4°C, and then incubated with the respective antibodies diluted in 5% nonfat milk in TBST for 3 hours at room temperature.

13. The membranes are repeatedly rinsed with TBST and then incubated for 30 minutes at room temperature with horseradish peroxidase-conjugated secondary goat antimouse or antirabbit IgG antibodies (Jackson ImmunoResearch).

14. After repeatedly rinsing in TBST, antibody binding is detected by chemiluminescence (Enhanced Chemiluminescence [ECL], Amersham, Arlington Heights, Illinois) using an imaging system (e.g., Epichemi3 Darkroom [Ultra-Violet Products Ltd. (UVP) BioImaging Systems]).

15. The intensity of the bands in different fractions is determined by densitometry using standard softwares such as PhotoShop or open source ImageJ.

11.7 FUNCTIONAL CHARACTERIZATION OF HETEROMERIC GAP JUNCTION CHANNELS AND HEMICHANNELS BY FLUORESCENT TECHNIQUES

For the undoubted demonstration of heteromeric gap junction channels and hemichannels, the supportive morphological and biochemical evidences should be accompanied by a functional characterization. The formation of heteromeric channels changes the permeability and the conductance properties of hemichannels and gap junction channels (García et al. 2015; Martínez et al. 2002). The latter could be demonstrated by fluorescent techniques, dye uptake (hemichannels), and dye coupling (gap junction channels) or by electrophysiological methods, single whole-cell (hemichannels) or double whole-cell voltage clamp (gap junction channels).

11.7.1 Dye Coupling

Intercellular coupling is examined after the microinjection of a permeability tracer of gap junction channel as previously described (Martínez et al. 2002, 2011). We can take advantage that the gap junction channel permeability to tracers that are of different size and charge, like neurobiotin (287, 3; +1) and Lucifer yellow (443, 4; −2), can change in heteromeric gap junction channels compared to their respective homomeric gap junction channels (Martínez et al. 2002). Therefore, we highly recommend performing simultaneous measurements of dye coupling using a combination of tracers. For initial screening, we recommend Lucifer yellow, a bulky molecule negatively charged, together with neurobiotin, a much more narrow and positively charged molecule, because it has been shown that some connexin gap junction channels are permeable to one but not the other permeability tracer (Martinez et al. 2002). Here, the procedure is as follows:

1. A cell grown on glass coverslips (around 80% confluent cultures) and mounted in the stage of a epifluorescence inverted microscope (e.g., NIKON TE-2000U) is impaled with a micropipette filled with a mix of 150 mM LiCl, 4% Lucifer yellow (Sigma Chemical Co.), and 4% neurobiotin (Vector Laboratories), both diluted in dimethyl sulfoxide. Optionally, other fluorescent nuclear acid tracers such as ethidium bromide (Sigma Chemical Co), YOPRO-1 (Invitrogen), or propidium iodide (Invitrogen) can be used.

2. Microinjection is accomplished with an Injectman-Femtoject system (Eppendorf) using 0.2–0.3-second pulses of 1–2 psi (1–2 minutes).

3. To label the neurobiotin, after injection, the cells are fixed with paraformaldehyde (4% in PBS) for 30 minutes and permeabilized with methanol/acetone (1:1) for 2 minutes at room temperature.
4. The neurobiotin tracer is detected after staining the cells with streptavidin-Cy3 conjugate (Sigma).
5. The extent of the intercellular transfer is determined by counting the number of cells containing the tracer excluding the impaled cell. Positive coupling is considered when the dye is diffused to two or more adjacent cells.

11.7.2   ASSESSMENT OF THE FUNCTIONAL STATE OF HEMICHANNELS BY DYE UPTAKE

The functional state of hemichannel is assessed by time-lapse imaging of the uptake of DNA tracers with different sizes and charges such as ethidium bromide (314, 4; +1), YOPRO-1 (375, 5; +2), and propidium iodide (424, 4; +2) as previously described (Contreras et al. 2002; Ebihara et al. 2011; Jara et al. 2012). In some experiments, the uptake of neurobiotin is also quantified (Jara et al. 2012). Follow the next procedure to perform the experiment.

1. Cells plated on glass coverslips are rinsed twice with Ca^{2+} free Hank's solution (Invitrogen) and incubated in this solution plus $5\,\mu M$ ethidium bromide or $20\,\mu M$ YOPRO-1 and $50\,\mu M$ propidium iodide or $100\,\mu M$ neurobiotin.
2. Fluorescent images of ROIs of different cells are observed using a 40× objective in an inverted microscope (e.g., NIKON TE-2000U).
3. For the DNA tracer uptake, the images are captured with a fast-cooled monochromatic digital camera (8-bit) every 30 seconds (exposure time = 30 milliseconds, gain = 0.5) (e.g., Nikon DS-2WBc).
4. To label the neurobiotin, the cells are fixed with paraformaldehyde (4% in PBS, for 30 minutes) after 10, 20, or 30 minutes of incubation in a free Ca^{2+} medium and permeabilized with 1% Triton-X100 in PBS for 2 minutes at room temperature. The neurobiotin is detected after staining the cells with streptavidin-Cy3 conjugate (Sigma). This experiment is done simultaneously for all conditions. ImageJ software can be used for image analysis and fluorescence intensity quantification over time.

11.7.3   CHARACTERIZATION OF HETEROMERIC HEMICHANNELS BY WHOLE-CELL RECORDINGS

To evaluate the functional activity of the hemichannels, electrophysiological methods are excellent approaches. Despite of several heterologous expression systems available, *Xenopus laevis* oocytes have been preferentially used for recording hemichannel currents (Ebihara 1996; García et al. 2015; López et al. 2013; Sánchez et al. 2010). As previously reported, hCx43 (Hansen et al. 2014; White et al. 1999) and deafness mutant Cx26S17F (Lee, Derosa, and White 2009; García et al. 2015) did not exhibit detectable hemichannel macroscopic currents (Figure 11.3a and 11.3b). The coinjection of

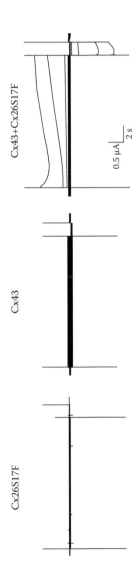

FIGURE 11.3 Large macroscopic currents generated by heteromeric hemichannels formed by Cx43 and Cx26S17F. Membrane currents from oocytes expressing Cx26S17F or Cx43 alone or coexpressing Cx43 with Cx26S17F were recorded using the TEVC technique. Depolarizing voltage steps were applied from a holding potential of −10 mV and stepped in 20 mV increments from −100 to +60 mV. Voltage steps over 0 mV induce hemichannels currents. Current voltage relationships can be obtained from the instantaneous tail currents and are sensitive to connexin channel blockers (García et al. 2015; López et al. 2013).

oocytes with the same amount of hCx43 and Cx26S17F cRNA (to obtain 1:1 protein expression ratio) promotes larger hemichannel currents (Figure 11.3c). For detailed protocols, please refer to Chapter 9 of this book.

11.7.4 CHARACTERIZATION OF HETEROMERIC GAP JUNCTION CHANNELS BY DUAL VOLTAGE CLAMP

Using a double patch clamp in a whole cell configuration is a powerful and versatile technique that allows performing electrophysiological recordings in pairs of cells with high input resistance and low junctional conductance. With this configuration, it is possible to study the electrical properties of gap junction channels using different heterologous expression systems (i.e., HeLa, N2A, etc.) (Martínez et al. 2002; Palacios-Prado et al. 2013; Xin, Gong, and Bai 2010). Single channel conductance of heteromeric gap junction channels can be distinguished from the homomeric channels conductances (Martínez et al. 2002). For detailed protocols, please refer to Chapter 4 of this book.

REFERENCES

Ahmad, S., J. A. Diez, C. H. George, and W. H. Evans. 1999. Synthesis and assembly of connexins in vitro into homomeric and heteromeric functional gap junction hemichannels. *Biochem J* 339 (Pt 2):247–53.

Ahmad, S., P. E. Martin, and W. H. Evans. 2001. Assembly of gap junction channels: Mechanism, effects of calmodulin antagonists and identification of connexin oligomerization determinants. *Eur J Biochem* 268 (16):4544–52.

Berthoud, V. M., E. A. Montegna, N. Atal, N. H. Aithal, P. R. Brink, and E. C. Beyer. 2001. Heteromeric connexons formed by the lens connexins, connexin43 and connexin56. *Eur J Cell Biol* 80 (1):11–9.

Beyer, E. C., J. Gemel, A. Martínez, V. M. Berthoud, V. Valiunas, A. P. Moreno, and P. R. Brink. 2001. Heteromeric mixing of connexins: Compatibility of partners and functional consequences. *Cell Commun Adhes* 8 (4–6):199–204.

Brink, P. R., K. Cronin, K. Banach, E. Peterson, E. M. Westphale, K. H. Seul, S. V. Ramanan, and E. C. Beyer. 1997. Evidence for heteromeric gap junction channels formed from rat connexin43 and human connexin37. *Am J Physiol* 273 (4 Pt 1):C1386–96.

Contreras, J. E., H. A. Sánchez, E. A. Eugenín, D. Speidel, M. Theis, K. Willecke, F. F. Bukauskas, M. V. Bennett, and J. C. Sáez. 2002. Metabolic inhibition induces opening of unapposed connexin 43 gap junction hemichannels and reduces gap junctional communication in cortical astrocytes in culture. *Proc Natl Acad Sci U S A* 99 (1):495–500.

Das Sarma, J., R. A. Meyer, F. Wang, V. Abraham, C. W. Lo, and M. Koval. 2001. Multimeric connexin interactions prior to the trans-Golgi network. *J Cell Sci* 114 (Pt 22):4013–24.

Das Sarma, J., F. Wang, and M. Koval. 2002. Targeted gap junction protein constructs reveal connexin-specific differences in oligomerization. *J Biol Chem* 277 (23):20911–8.

Duong, M. T., T. M. Jaszewski, K. G. Fleming, and K. R. MacKenzie. 2007. Changes in apparent free energy of helix-helix dimerization in a biological membrane due to point mutations. *J Mol Biol* 371 (2):422–34.

Ebihara, L. 1996. Xenopus connexin38 forms hemi-gap-junctional channels in the nonjunctional plasma membrane of Xenopus oocytes. *Biophys J* 71 (2):742–8.

Ebihara, L., J. J. Tong, B. Vertel, T. W. White, and T. L. Chen. 2011. Properties of connexin 46 hemichannels in dissociated lens fiber cells. *Invest Ophthalmol Vis Sci* 52 (2):882–9.

Falk, M. M., L. K. Buehler, N. M. Kumar, and N. B. Gilula. 1997. Cell-free synthesis and assembly of connexins into functional gap junction membrane channels. *EMBO J* 16 (10):2703–16.

Fleishman, S. J., A. D. Sabag, E. Ophir, K. B. Avraham, and N. Ben-Tal. 2006. The structural context of disease-causing mutations in gap junctions. *J Biol Chem* 281 (39):28958–63.

García, I. E., J. Maripillán, O. Jara, R. Ceriani, A. Palacios-Muñoz, J. Ramachandran, P. Olivero et al. 2015. Keratitis-ichthyosis-deafness syndrome-associated Cx26, utants produce nonfunctional gap junctions but hyperactive hemichannels when co-expressed with wild type Cx43. *J Invest Dermatol* 135 (5):1338–47.

Gemel, J., V. Valiunas, P. R. Brink, and E. C. Beyer. 2004. Connexin43 and connexin26 form gap junctions, but not heteromeric channels in co-expressing cells. *J Cell Sci* 117 (Pt 12):2469–80.

Hansen, D. B., T. H. Braunstein, M. S. Nielsen, and N. MacAulay. 2014. Distinct permeation profiles of the connexin 30 and 43 hemichannels. *FEBS Lett* 588 (8):1446–57.

Jara, O., R. Acuña, I. E. García, J. Maripillán, V. Figueroa, J. C. Sáez, R. Araya-Secchi et al. 2012. Critical role of the first transmembrane domain of Cx26 in regulating oligomerization and function. *Mol Biol Cell* 23 (17):3299–311.

Jiang, J. X., and D. A. Goodenough. 1996. Heteromeric connexons in lens gap junction channels. *Proc Natl Acad Sci U S A* 93 (3):1287–91.

Koval, M. 2006. Pathways and control of connexin oligomerization. *Trends Cell Biol* 16 (3):159–66.

Koval, M., J. E. Harley, E. Hick, and T. H. Steinberg. 1997. Connexin46 is retained as monomers in a trans-Golgi compartment of osteoblastic cells. *J Cell Biol* 137 (4):847–57.

Koval, M., S. A. Molina, and J. M. Burt. 2014. Mix and match: Investigating heteromeric and heterotypic gap junction channels in model systems and native tissues. *FEBS Lett* 588 (8):1193–204.

Kronengold, J., E. B. Trexler, F. F. Bukauskas, T. A. Bargiello, and V. K. Verselis. 2003. Single-channel SCAM identifies pore-lining residues in the first extracellular loop and first transmembrane domains of Cx46 hemichannels. *J Gen Physiol* 122 (4):389–405.

Lagree, V., K. Brunschwig, P. López, N. B. Gilula, G. Richard, and M. M. Falk. 2003. Specific amino-acid residues in the N-terminus and TM3 implicated in channel function and oligomerization compatibility of connexin43. *J Cell Sci* 116 (Pt 15):3189–201.

Lee, J. R., A. M. Derosa, and T. W. White. 2009. Connexin mutations causing skin disease and deafness increase hemichannel activity and cell death when expressed in Xenopus oocytes. *J Invest Dermatol* 129 (4):870–8.

López, W., Y. Liu, A. L. Harris, and J. E. Contreras. 2013. Divalent regulation and intersubunit interactions of human Connexin26 (Cx26) hemichannels. *Channels (Austin)* 8 (1).

Maeda, S., S. Nakagawa, M. Suga, E. Yamashita, A. Oshima, Y. Fujiyoshi, and T. Tsukihara. 2009. Structure of the connexin 26 gap junction channel at 3.5 A resolution. *Nature* 458 (7238):597–602.

Martínez, A. D., V. Hayrapetyan, A. P. Moreno, and E. C. Beyer. 2002. Connexin43 and connexin45 form heteromeric gap junction channels in which individual components determine permeability and regulation. *Circ Res* 90 (10):1100–7.

Martínez, A. D., V. Hayrapetyan, A. P. Moreno, and E. C. Beyer. 2003. A carboxyl terminal domain of connexin43 is critical for gap junction plaque formation but not for homo- or hetero-oligomerization. *Cell Commun Adhes* 10 (4–6):323–8.

Martínez, A. D., J., R. Acuña, P. J. Minogue, V. M. Berthoud, and E. C. Beyer. 2011. Different domains are critical for oligomerization compatibility of different connexins. *Biochem J* 436 (1):35–43.

Maza, J., J. Das Sarma, and M. Koval. 2005. Defining a minimal motif required to prevent connexin oligomerization in the endoplasmic reticulum. *J Biol Chem* 280 (22):21115–21.

Maza, J., M. Mateescu, J. D. Sarma, and M. Koval. 2003. Differential oligomerization of endoplasmic reticulum-retained connexin43/connexin32 chimeras. *Cell Commun Adhes* 10 (4–6):319–22.

Mendrola, J. M., M. B. Berger, M. C. King, and M. A. Lemmon. 2002. The single transmembrane domains of ErbB receptors self-associate in cell membranes. *J Biol Chem* 277 (7):4704–12.

Musil, L. S., and D. A. Goodenough. 1993. Multisubunit assembly of an integral plasma membrane channel protein, gap junction connexin43, occurs after exit from the ER. *Cell* 74 (6):1065–77.

Palacios-Prado, N., G. Hoge, A. Marandykina, L. Rimkute, S. Chapuis, N. Paulauskas, V. A. Skeberdis et al. 2013. Intracellular magnesium-dependent modulation of gap junction channels formed by neuronal connexin36. *J Neurosci* 33 (11):4741–53.

Polgar, O., C. Ierano, A. Tamaki, B. Stanley, Y. Ward, D. Xia, N. Tarasova, R. W. Robey, and S. E. Bates. 2010. Mutational analysis of threonine 402 adjacent to the GXXXG dimerization motif in transmembrane segment 1 of ABCG2. *Biochemistry* 49 (10):2235–45.

Sánchez, H. A., G. Mese, M. Srinivas, T. W. White, and V. K. Verselis. 2010. Differentially altered Ca^{2+} regulation and Ca^{2+} permeability in Cx26 hemichannels formed by the A40V and G45E mutations that cause keratitis ichthyosis deafness syndrome. *J Gen Physiol* 136 (1):47–62.

Tang, Q., T. L. Dowd, V. K. Verselis, and T. A. Bargiello. 2009. Conformational changes in a pore-forming region underlie voltage-dependent loop gating of an unapposed connexin hemichannel. *J Gen Physiol* 133 (6):555–70.

Valiunas, V., J. Gemel, P. R. Brink, and E. C. Beyer. 2001. Gap junction channels formed by coexpressed connexin40 and connexin43. *Am J Physiol Heart Circ Physiol* 281 (4):H1675–89.

Wang, M., A. D. Martínez, V. M. Berthoud, K. H. Seul, J. Gemel, V. Valiunas, S. Kumari, P. R. Brink, and E. C. Beyer. 2005. Connexin43 with a cytoplasmic loop deletion inhibits the function of several connexins. *Biochem Biophys Res Commun* 333 (4):1185–93.

White, T. W., M. R. Deans, J. O'Brien, M. R. Al-Ubaidi, D. A. Goodenough, H. Ripps, and R. Bruzzone. 1999. Functional characteristics of skate connexin35, a member of the gamma subfamily of connexins expressed in the vertebrate retina. *Eur J Neurosci* 11 (6):1883–90.

White, T. W., D. L. Paul, D. A. Goodenough, and R. Bruzzone. 1995. Functional analysis of selective interactions among rodent connexins. *Mol Biol Cell* 6 (4):459–70.

Xin, L., X. Q. Gong, and D. Bai. 2010. The role of amino terminus of mouse Cx50 in determining transjunctional voltage-dependent gating and unitary conductance. *Biophys J* 99 (7):2077–86.

12 Methods to Examine the Role of Gap Junction and Pannexin Channels in HIV Infection

Courtney A. Veilleux and Eliseo A. Eugenin

CONTENTS

12.1 INTRODUCTION

This chapter summarizes the mechanisms by which human immunodeficiency virus (HIV) regulates gap junctions and hemichannels and provides detailed methods uniquely optimized to investigate such conditions. In order to appreciate the various influences of the HIV life cycle on the regulation of gap junctions and hemichannels, a brief review of HIV replication is first presented. This is followed with a description of the techniques used to examine channel function and regulation in lymphocytes and astrocytes.

HIV entry into the immune cells requires binding of the envelop protein of the virus, glycoprotein-120 (gp120) to CD4 and coreceptor CCR5 or CXCR4. Upon binding, the fusion of the viral and the plasma membrane occurs resulting in the release of the capsid, which contains viral RNA and several viral proteins, including reverse transcriptase, into the cytoplasm. Although this process is well described and accepted, several events remain unclear. For example, it is known that actin rearrangement and actin polymerization mediate the virus–cell fusion events; however, many of the proteins activated by HIV to mediate these cellular changes are unknown [1]. HIV entry remains extensively studied, and new insights to viral entry will provide novel mechanisms of therapy in the future [2].

Following the plasma membrane and virus membrane fusion and the release of the capsid into the host cytoplasm, the viral protein reverse transcriptase facilitates the conversion of single-stranded RNA into double-stranded DNA (dsDNA). This dsDNA is then transported to the nucleus and inserted into the host DNA by the action of the viral enzyme integrase. Once inserted into the host DNA, the viral dsDNA is transcribed by the host polymerase to produce HIV structural proteins such as Gag, Pol, and Env, accessory proteins including Vif, Vpr, Vpu and Nef, and two regulatory proteins, Tat and Rev. Tat, or transactivator of transcription, enhances and stabilizes the replication of the virus [3]. Rev, or regulator of expression of virion proteins, assists in the export of transcribed viral RNA from the nucleus [4].

Tat and Rev are early expression viral proteins and are produced and released even in the absence of viral replication or in the presence of antiretroviral treatment [5]. These secreted proteins have profound toxic effects on the immune and nervous systems [6–8]. The efficient production of viral proteins and HIV RNA results in formation of new proviruses that subsequently mature upon activation of viral protease, which cleaves polypeptides into individual proteins. The viral maturation is completed after budding from the cell. HIV budding may occur at the plasma membrane, as in T cells, or from multivesicular bodies, as in macrophages [9]. Budded virions then continue a new cycle of infection.

12.2 HIV'S INFLUENCE ON GAP JUNCTIONS AND HEMICHANNELS

Despite the vast amounts of research elucidating HIV's cycle of infection, numerous steps of the viral cell cycle remain unknown. Recent data demonstrate that the opening of connexin and pannexin containing channels, in concert with ATP release and activation of purinergic receptors, is required for multiple events during the HIV viral cycle and HIV-associated toxicity, including viral entry and replication, virus-mediated migration and polarization, and amplification of toxicity from HIV-infected cells into uninfected cells [10–12]. An overview of the roles of gap junction channels and hemichannels in response to HIV is described along with the relevant methods to visualize and determine channel function and regulation.

12.2.1 An Introduction to Channel Dynamics during HIV Fusion in Leukocytes

During HIV infection, the circulating leukocytes are the major tropic cells, with the infection occurring in both macrophages and lymphocytes. The infection of leukocytes begins with cell–virus fusion and binding of the virus protein gp120 to the host CD4. Upon binding of the virus to CD4 and the chemokine receptors CCR5 or CXCR4, pannexin1 channels open and release ATP into the extracellular space through the channel pore [10]. This released ATP activates specific purinergic receptors involved in HIV entry and replication in $CD4^+$ T lymphocytes and macrophages [10,13,14]. The opening of pannexin1 channels in response to CD4 binding to gp120 occurs in a biphasic manner, with early opening during the interaction of the virus with CD4 and CCR5 or CXCR4 5–15 minutes after viral exposure and a late opening probably due to the new binding of newly released virions to CD4 and CCR5 or CXCR4, which occurs between 18 and 24 hours postinfection [10]. This biphasic opening of pannexin1 channels is crucial for the HIV viral cycle; blocking of pannexin1 channels totally abolishes HIV entry and viral replication [10].

In human macrophages, purinergic receptors, including $P2X_1$, $P2X_7$, and $P2Y_1$, participate in HIV replication [13,14]. It has also been demonstrated that $P2X_1$ receptors are essential for HIV entry into the human primary macrophages; however, the mechanism by which $P2X_7$ and $P2Y_1$ alter the replication is unknown [14]. Multiple groups have demonstrated that the activation of $P2Y_2$ is associated with HIV infection and replication [13,14]. The interactions between the receptors and these secondary messengers are illustrated in Figure 12.1. Notably, it has been demonstrated that extracellular ATP participates in the release of HIV from human macrophages [15]. Together, this demonstrates that pannexin1 channels, purinergic receptors, and extracellular ATP play a key role in HIV fusion and replication.

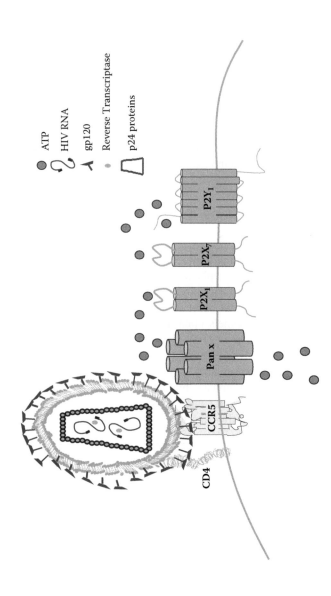

FIGURE 12.1 (See color insert.) Illustration of HIV binding to the cell surface of the host cells and facilitating pannexin1 channel (Panx) opening and purinergic receptor activation. The virion binds to the CD4 and the coreceptor CCR5 to initiate fusion, which results in ATP movement across the plasma membrane via pannexin1 channels. This ATP can activate purinergic receptors including P2X$_1$, P2X$_7$, and P2Y$_1$ that participate in HIV fusion and subsequent replication.

12.2.2 Methods to Assess Channel Dynamics during HIV Fusion in Leukocytes

Note on Safety and Precautions: For all experiments involving human cells, animals, and pathogens, documentation and approval must be met for all of the following: Institutional Review Board (IRB), Institutional Animal Care and Use Committee, Institutional Biosafety Committee, and Standard Operating Procedure. Additionally, all protocols involving fresh blood must be treated at least at biosafety level 2 according to the standard set by the Centers of Disease Control, due to the potential of infectious disease within the samples. Individuals completing experiments must be properly trained and approved by all necessary parties before beginning any experiments.

12.2.2.1 Cell Isolation and Separation

1. The leukopacks were obtained from the New York or the New Jersey blood center, and the peripheral blood mononuclear cells were isolated using Ficoll-Paque PLUS (GE Healthcare Life Sciences, Pittsburg, Pennsylvania) and differential centrifugation according to the manufacturer's guidelines found at https://www.gelifesciences.com.

2. The isolation of the leukocyte populations may be achieved from purified leukocytes using the following methods:

   a. CD14+ monocytes: CD14+ monocyte population was isolated using magnetic CD14+ specific beads (Stem Cell Technologies, Vancouver), according to the manufacturer's instructions.

   b. CD14+ CD16+ monocytes: CD14+ CD16+ monocytes represent only 1–5% of the circulating monocytes. To enrich this population, the monocytes were first isolated from leukocytes following the earlier described protocol. Once isolated, the monocytes were cultured with 10 ng/mL of macrophage colony-stimulating factor on Teflon flasks to prevent adherence. Further information regarding this isolation technique was previously published [16].

   c. CD4+ or CD8+ Lymphocytes: Monocytes were first excluded using the CD14 magnetic beads as described earlier in a. This is done because the monocytes express low levels of CD4. The remaining cells were isolated with CD4+-coupled magnetic beads from Stem Cell Technologies (Vancouver, British Columbia, Canada). CD8+ leukocytes may also be isolated using the same technique, but using CD8+ magnetic beads instead of CD4+ beads.

   d. Other circulating populations that may be isolated using magnetic separation include lymphocytes (CD3-conjugated magnetic beads), natural killer cells (CD56-conjugated magnetic beads), macrophages (CD11b-conjugated magnetic beads), dendritic cells (CD11c-conjugated magnetic beads), stem cells (CD34-conjugated magnetic beads), granulocytes (CD66-conjugated magnetic beads), endothelial cells (CD146-conjugated magnetic beads), epithelial cells (CD326-conjugated magnetic beads), or B Cells (CD19- or CD20-conjugated magnetic beads).

12.2.2.2   Infection of Cells with HIV

1. Immune cells were inoculated with HIV at a density of 10^3 cells per well in 48-well plates. The strains used for these experiments are HIV_{ADA}, HIV_{YU-2}, or HIV_{BAL}, which are common laboratory strains and obtained from the National Institutes of Health (NIH) repository (Germantown, Maryland) [17,18]. HIV_{BAL} was infected at a multiplicity of infection (MOI) of 0.001 or 0.001. HIV_{ADA} or HIV_{LAI} were infected at a concentration of 20 ng p24/mL/1×10^6 cells, which was calculated using a HIV p24 enzyme-linked immunosorbent assay (ELISA) kit (Perkin Elmer, Waltham, Massachusetts).

2. After incubating the cells with the virus for 2–24 hours at 37°C and 5% CO_2, the cells were washed with PBS and cultured with media to eliminate unbound virus.

3. The supernatants were collected, and the medium was changed every 24 hours until 7 days postinoculation to determine the viral production. A 24-hour cycle of collection is utilized because the full viral cycle to produce new progeny takes between 18–24 hours. Thus, the collection every 24 hours is representative of one cycle.

4. The viral replication was analyzed using an HIV p24 ELISA (Perkin Elmer System, Waltham, Massachusetts) according to the manufacturer's instructions.

12.2.2.3   HIV Entry Assay

1. Beta-lactamase containing HIV_{BaL} was generated as described previously [19]. Briefly, calcium phosphate was used to transfect the HEK-293T cells using calcium and two plasmids, one containing the HIV_{BaL} genome, identified as pWT/BaL (NIH repository, Germantown, Maryland) and the other containing β-lactamase fused to HIVvpr, identified as pMM310 (NIH repository, Germantown, Maryland). Sixteen hours after the transfection, the calcium phosphate was removed. The supernatant was collected at 24 and 48 hours posttransfection. This resulted in the production of complete viral particles containing active β-lactamase.

2. The enrichment of the virus can be completed using the sucrose gradient purification of the transfection supernatants [20]. Specifically, 500 μL of virus from the collected supernatant was floated on a 20% sucrose cushion diluted in PBS and ultracentrifuged for 10 minutes at 70,000g. The resulting pellet was then resuspended in 500 μL of PBS and floated over a discontinuous sucrose gradient ranging in concentration from 60% to 30%. The total gradient was then subjected to ultracentrifugation for 15 minutes at 100,000g. Viral concentration during enrichment can range depending on transfection success and is quantifiable using methods described in Section 12.2.2.2.

3. The fractions from the discontinuous gradient were stored for future use at −80°C and tested for reactivity against a p24 or other HIV specific antibody for the verification of HIV isolation.

4. Primary human macrophages were cultured in black 96-well plates with clear bottoms at a density of 4.3×10^4 cells per well.

5. Cell cultures were maintained for 7–8 days before the cells were inoculated with β-lactamase containing HIV_{BaL} at an MOI of 0.01, 0.005, or 0.001. During this step, the blockers of pannexin1 channel, and or connexin containing channels, or ATP receptors can be added to examine the effects on viral entry. Examples of blockers include mimetic peptides, oATP, and probenecid.

6. After the viral inoculation of primary human macrophages, the cells were washed with phenol red-free DMEM (Thermo Fisher Scientific, Waltham, Massachusetts) and loaded with CCF2-AM for 6 hours at room temperature using the GeneBLAzer in vivo Detection Kit (Thermo Fisher Scientific, Waltham, Massachusetss). It should be noted that CCF2-AM fluoresces green until cleaved by β-lactamase, which makes the molecule fluoresce blue.

7. The images were captured with a fluorescent microscope at an excitation wavelength of 405 nm and emission wavelengths at 460 nm or 530 nm to detect blue and green cells, respectively. Chroma Technologies Filter Set #71008 (Chroma Technologies, Bellows Falls, Vermont) or Omega Optical Filter Set #XF124 (Omega Optical Inc., Flushing, New York) are both appropriate for this procedure.

8. A total of eight fields from four separate wells per condition were captured. The percentage of positive cells for viral entry was determined by counting the number of cells that only emit blue compared to the total number of cells that emit both blue and green wavelengths.

12.2.2.4 Imaging of Pannexin or Connexin Channel Opening Using Ethidium Bromide

Note: Other dyes may be used in place of Ethidium bromide (Etd), such as TOPRO, YOPRO, Lucifer yellow and sulforhadamine; additionally, more complex protocols such as using neurobiotin tagged with avidin and stained with fluorescence can also be substituted for Etd. All samples are HIV infected and therefore require biosafety level 2+ conditions.

1. Control, HIV-infected, or gp120-treated cells were exposed to 5 µM Etd or another dye for 10 minutes at 37°C and 5% CO_2 in a culture medium.

2. The cells were washed for 2 minutes each, for a total of five times with Hank's balanced salt solution containing 137 mM NaCl, 5.4 mM KCl, 0.34 mM Na_2HPO_4, 0.44 mM KH_2PO_4, and 1.2 mM $CaCl_2$ all adjusted to pH 7.4.

3. The cells were fixed at room temperature for 30 minutes with 2% paraformaldehyde and mounted in an antifade reagent conjugated to DAPI (Thermo Fisher Scientific, Waltham, Massachusetts) prior to examination using a confocal laser-scanning microscope.

4. To accomplish the live cell imaging, the cells were plated in Locke's solution (154 mM NaCl, 5.4 mM KCl, 2.3 mM $CaCl_2$, 5 mM HEPES at pH 7.4) with 5 µM Etd and imaged using phase-contrast and fluorescence microscopy with time-lapse imaging. The live cell imaging was achieved using a

Zeis Axio Observer Z1 microscope (Carl Zeis AG, Oberkochen, Germany) every 30 seconds to 1 minute for a total of 1–72 hours.

5. The data were analyzed using ImageJ [21]. For live cell imaging, the average of four independent background fluorescence intensity measurements was subtracted from the fluorescence intensity of individual cells and averaged for 20 cells per treatment over time. The rate of the uptake is calculated from the slope curve of the fluorescence over time using the Microsoft Excel software.

12.2.2.5 Imaging of Gap Junction Channels Using Dye Coupling

1. To evaluate the function of gap junction channels, the intercellular transfer of Lucifer yellow (5% w/v in 150 mM LiCl) was evaluated by microinjecting the dye into a single cell and evaluating the diffusion of the dye into neighboring cells, as previously described [22].
2. The cells were microinjected with an Eppendorf microinjector (Eppendorf, Hamburg, Germany) until the injected cell brightly fluoresced. It is important to note that the pipettes must be immersed in bleach for at least 1 minute after every reloading or changing of pipettes, due to the sharpness and the biohazard of the pipettes in the context of HIV experiments.
3. The cells chosen to microinject were located either in pairs or in clusters in order to best facilitate the transfer of dye into the adjacent cells. The observation of dye coupling occurred using a microscope and an excitation wavelength at 450–490 nm and an emission wavelength at 520 nm. The recordings proceeded for up to 3 minutes.
4. The cells were scored as coupled if the dye transfer occurred to one or more adjacent cells. The dye transfer was evaluated using a fully motorized Zeiss Z microscope (Carl Zeiss AG, Oberkochen, Germany).
5. Four independent experiments were performed, in which a minimum of 20 cells were microinjected per experiment.
6. The incidence and index of the dye coupling was scored as the percentage of injections that resulted in dye transfer and the numbers of cells coupled to a single microinjected cell.

12.2.2.6 Evaluation of Channel Function Using Scrape Loading

Note: The following protocol is based on a previously published work [23]:

1. A confluent cell monolayer was obtained by growing cells on gelatin. Then, they were placed in a solution of 0.05% Lucifer yellow and dextran conjugated to Rhodamine. The dextran conjugated to Rhodamine (molecular weight ranging from 3000 to 70000 Da) is utilized as a marker of cell damage but not channel transport. This is because the dextran conjugated to Rhodamine is too large to pass through the channels.
2. After the initial incubation of the monolayer with the dyes, a scalpel or a needle was used to generate a scraped line on the monolayer.

3. The dye solution remained on the cells for 3 minutes, prior to briefly washing with PBS. It should be noted that all discarded solutions and instruments have to be mixed with bleach in case of HIV exposure.
4. As a control, 0.05% dextran conjugated to rhodamine without Lucifer yellow was used to scrape the cell monolayer. This dye was used to stain the damaged cells due to the scrape because of its high molecular weight.
5. Junctional communication is evaluated by the diffusion of Lucifer yellow versus the number of damaged cells stained by the dextran conjugated to rhodamine.

12.2.2.7 ATP Release Assay as a Readout of Pannexin1 Channel and Connexin Hemichannel Opening

Note: As mentioned in Section 12.1, in the event of the opening of pannexin1 channel and connexin hemichannels in response to HIV infection or gp120 treatment, ATP is released into the extracellular space. The release of ATP is therefore a dependable readout of the activity of these membrane channels. Conducting a readout of ATP release can be used to analyze multiple human fluids including serum and cerebral spinal fluid. ATP release into the extracellular milieu is achieved by three major mechanisms: cell death, vesicular release (e.g., neurotransmitters), and opening of pannexin1 channel and connexin hemichannels. Thus, in the absence of cell death and neurotransmission, ATP release is primarily mediated by the opening of the latter membrane channels.

1. Thirty-five thousand cells per well were cultured in 96-well plates. The medium was changed after 1 week to Rosewell Park Memorial Institute (RPMI) 1640 without phenol red and supplemented with 200 IU of penicillin and 200 µg/mL streptomycin and 10 mM HEPES.
2. The Cells were inoculated at 2, 5, 10, 15, 30, or 30 minutes with media, HIV, or gp120.
3. The supernatant was removed and analyzed for ATP concentration using the ATPlite Luminescence Assay System by Perkins Elmer (Waltham, Massachusetss). According to the manufacturer's instructions, 100 µL of the sample and an equal volume of the ATPlite reagent were combined, and the luminescence was measured using a Microplate Luminometer or other plate reader. The readings were then compared to a standard curve provided by the kit measuring from 0.39 nM ATP to 100 nM ATP.

Note: As a positive control, the cells are subjected to mechanical stress using a pipette and pressing on the cells. It is important that the plates containing experimental conditions are not agitated, as any form of mechanical stress or aggravation may cause ATP release and confound the results. Other controls for this experimental setup include HIV-related stimulating compounds, including $gp120_{SF162}$ and $gp120_{BAL}$ at 5, 50, or 500 nM. Furthermore, the actual virus may also be used at an MOI of 0.1–0.001 as an additional control (Please see "Note on Safety and Precautions," Section 12.2.2).

12.2.2.8   The Role of Pannexin or Connexin Hemichannels in Cell Migration

Note: For this protocol, the channel blockers may include probenecid, siRNA, peptides, connexin43 (Cx43) extracellular blocking antibody, or other blocking antibodies. Blocking should be applied during the initial stages of the migratory process.

1. A 5 µm pore polycarbonate filter was coated with sterile 0.2% gelatin, dried, and placed between the bottom and the top of Neuroprobe 48 Micro Chemotaxis Champers (Neuroprobe, Gaithersburd, Maryland).
2. Between 1×10^6 and 3×10^6 immune cells infected with the laboratory strains of HIV were added to the top chamber in media consisting of RPMI 1640 supplemented with 2% FBS. Chemotactic factors or channel blockers may be added. The calibration of chemotactic factors and channel blockers may be required.
3. A negative control used for this experiment is cells treated with RPMI 1640 with 2% FBS only. A positive control for this experimental setup is cells treated with RPMI 1640 with 2% FBS and chemokines or chemotactic peptides. Examples include N-formylmethionyl-leucyl-phenylalanine, a chemotactic peptide released by Gram-positive and Gram-negative bacteria.
4. The checkerboard analysis was also utilized to establish the effect of the gradients on cell migration. This was done by adding media with and without chemotactic factors or channel blockers to the bottom chamber and adding the cells treated with RPMI 1640 with 2% FBS to the top chamber above the membrane. Null gradients are used as a control by adding chemotactic media to both the upper and lower chambers.
5. Migration was allowed to proceed for 1–2 hours at 37 °C and 5% CO_2.
6. The membranes were removed, and the cells, which migrated through the membrane, were located on the underside of the insert facing the bottom of the well. These cells may be stained with Diff-Quick Stain Set (Siemens, Munich, Germany) or a similar Giemsa-based stain for further visualization.
7. The quantification of the migration was accomplished using a densitometry software.
8. The migration data were presented as the percentage of cells on the underside of the insert over the baseline of cells in nonchemotactic conditions.

12.2.3   The Role of Connexin and Pannexin Based Channels in HIV Neuropathy

HIV infection causes multiple disorders of the central nervous system (CNS), including dementia, encephalitis, and tumors; these conditions result from the direct infiltration of the virus into the brain parenchyma [24]. HIV infection of the brain, which causes neurocognitive impairment, is clinically defined as HIV-1-associated neurological disorders (HANDs), and it can be categorized as a minor cognitive motor disorder or the more severe form—HIV-1-associated dementia (HAD). Fortunately, the onset of highly active antiretroviral therapy has contributed to the decreased levels of HAD despite the viral success in the infiltration of the CNS within the initial weeks of infections and the establishment among cell populations of viral reservoirs within the first year [25].

The cell populations affected by HIV-1 in the CNS are predominately microglial cells and perivascular macrophages [26,27]. A small population (5–8%) of HIV-infected astrocytes has also been detected [12,24,27–33]. Despite the low infection rate in astrocytes, HIV-infected astrocytes successfully utilize gap junctions and hemichannels to increase the toxicity in the milieu and trigger inflammation in uninfected neighboring cells.

One mechanism of inflammation by HIV-infected astrocytes is via Cx43 opening on the surface of astrocytes; this mechanism remains incompletely understood, and it is triggered by virus interaction at unknown receptors. Downstream signal transduction cascades result in astrocyte-mediated disruption of the neuronal processes in adjacent neurons by a Cx43-dependent alteration in dickkopf-1 protein (DKK1) [34]. DKK1 is not only a known soluble Wnt pathway inhibitor, but it has also been identified as a key regulator of interferon gamma-mediated enhancement of HIV replication and HIV reactivation in astrocytes [35]. The Cx43-dependent DKK1 induction in HIV-infected astrocytes clearly represents a mechanism occurring during HIV replication, and it is a new avenue of investigation to study the mechanisms by which HIV hijacks the host cell communication systems, particularly gap junctions and hemichannels, to spread toxicity and inflammation in the brain milieu, which can lead to HAND.

HIV viral replication and production of new virions in the CNS and in astrocytes, is not a major contributor to HAND; however, the secondary messengers produced by the small population of infected astrocytes can spread to other cells via gap junctions and hemichannels and initiate multiple signal transduction pathways, resulting in bystander apoptosis of uninfected neighboring cells [34,36].

This spread of toxicity and inflammation from infected astrocytes can pass to other nearby cells including endothelial cells, neurons, and other astrocytes [37]. These toxic and inflammatory signals are mediated by many intracellular factors such as cytochrome C-dependent molecules, including IP_3 and intracellular free Ca^{2+} [36]. Infected astrocytes have also altered the regulation of connexin gap junction channel opening, which affects glutamate metabolism and chemokine (C-C motif) ligand 2 secretion; this supports that HIV infection of astrocytes is essential for neuronal compromise and leukocyte transmigration, respectively [12]. Despite the large damage facilitated by the small population of HIV-infected astrocytes, these HIV infected cells remain protected from apoptosis by an unknown mechanism [36,38,39]. The prolonged survival of infected cells, and apoptosis of nearby uninfected cells, may represent a unique viral reservoir within the CNS and be a key aspect to the neuropathogenesis of HIV, as shown in Figure 12.2.

The apoptosis of nearby uninfected cells in the brain milieu has been demonstrated among microglia [40], astrocytes [37,39], between astrocytes and neurons [41], and also between astrocytes and endothelial cells [37]. Although the vast majority of recent evidence suggests that this process is mediated by gap junctions [37,39], it remains controversial due to the differences in the models used, the intensity of the injury, and the methods of analysis. Additional roles of gap junctions during HIV infection include the amplification of ischemic damage [42] and the apoptosis during

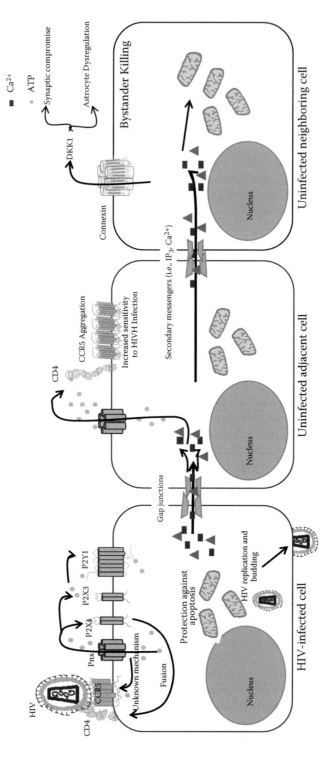

FIGURE 12.2 **(See color insert.)** HIV infection and mechanism of bystander apoptosis in the host cells involving pannexin1 channels and connexin containing channels, as well as secondary messengers including IP$_3$, calcium ions, and ATP. The leftmost cell depicts HIV infection and resultant protection from apoptosis. The gap junction channels allow the movement of the secondary messengers to pass into the uninfected cells connected by the gap junctions with the HIV-infected cells. This movement increased the cells susceptibility to HIV infection, as shown in the center cell with CCR5 aggregation, or bystander killing, as depicted in the far right cell. HIV infection directly affects the cellular function to promote viral survival and indirectly mediates apoptosis in uninfected nearby cells by a gap junction-mediated mechanism.

HIV infection [12]. Interestingly, this contrasts the studies showing that the blockage of gap junctional proteins increases the neuronal vulnerability to oxidative stress or ischemic insult [43–47].

Cell death facilitated by gap junctions has been described in other diseases [11]. Apoptosis induced by metabolic stress is directly dependent on the gap junction expression and the coupling capability in neighboring glioma cells, even when the antiapoptotic mitochondrial protein Bcl-2 is transfected for protection [42]. The ability to resist apoptosis with heightened expression of Bcl-2 does not protect the cells from death if a weaker neighbor is compromised, as long as there is active gap junction coupling [42].

HIV alteration of the gap junction channel function may allow the development of apoptosis resistance in infected human astrocytes, at the same time inducing death of uninfected bystanders by a gap junction-dependent mechanism [37,39]. This bystander cell death can be reduced by directly blocking gap junctions. Therefore, HIV-infected astrocytes may be protected from cell death while killing uninfected neighboring cells by a gap junction-dependent mechanism and with the use of secondary messengers, as depicted in Figure 12.2.

12.2.4 METHODS TO STUDY CONNEXIN AND PANNEXIN DYNAMICS DURING HIV INFECTION IN THE CENTRAL NERVOUS SYSTEM

12.2.4.1 Human Astrocyte Isolation

Note: The explicit approval of the use of human fetal tissue is needed through the IRB. A complete procedure for the generation of human cortical fetal astrocytes can be attained in research described previously [48]. Specifically, the following:

1. Fetal tissue of 16–20 weeks was obtained and used in accordance with institutional approval.
2. The meninges from the cortical hemisphere was minced and shaken in HEPES-buffered saline with 1× trypsin-ethylenediaminetetraacetic acid (EDTA) and 1× DNAse I for 1 hour at 37 °C.
3. The tissue was filtered using a 250 μm and a 150 μm filter, sequentially. These cells were then plated onto 150 cm^2 tissue flasks at a concentration of 9×10^7 cells per flask in DMEM containing 25 mM HEPES, 10% FBS, 200 IU penicillin and 200 μg/mL streptomycin, and 1% nonessential amino acids.
4. The cells were enriched for astrocytes by splitting at low confluence (approximately 20%) a total of three times over the course of 40 days. Low confluence splitting limits neuronal support from the astrocytes and leads to their degeneration, resulting in cultures enriched to contain 97–99% astrocytes.

12.2.4.2 Experiments Using Astrocytes Obtained from Glial Fibrillary Acidic Protein–Connexin 43-Deleted Mice

Note: This technique is especially important because of the large inability to examine the gap junction channel function in an HIV-infected human brain tissue during the pathogenesis of the disease. For this reason, the majority of literature is centered on postmortem human tissue. This mouse model used to examine Cx43 in astrocytes is heavily supported [49,50], although a similar model with knockout of Cx30 has also been used to examine CNS gap junction channels [51,52]. These animals can be used for in vivo experiments or for the isolation of astrocytes for in vitro experiments.

1. Cre-recombinase strain mice under the transcriptional control of glial fibrillary acidic protein (GFAP) were crossed with Cx43 deletion mouse to produce Cx43fl/+ mice [53].
2. These mice presented a lacZ expression profile depicting astrocyte-specific cell-restricted expression of endogenous Cx43.
3. To confirm the mice strains, comparative X-gal staining of the Cx43 deletion and the Cx43fl/+ mice was completed.

12.2.4.3 An In Vivo Dual-Species Model of HIV Regulation of Gap Junction Communication and Bystander Apoptosis

1. Human astrocytes were infected with HIV at an MOI of 0.01–0.1 for at least 24 hours.
2. The cells were lifted from their culture flasks and pipetted into a single-cell suspension using 0.05% trypsin-EDTA. The cells were plated onto new flasks and maintained in a medium without FBS to avoid immune activation in the brains of the animals.
3. A 0.1 mL Hamilton syringe was loaded with a 2–5 µL solution containing 500 astrocytes and kept on ice. The loading of 500 cells allows for only approximately 25 cells to be infected due to the low infectivity of human astrocytes (around 5%).
4. Using the same Hamilton syringe, 2 µL of cell solution described in step 3 were injected into the cortex of the mouse through an 0.1 mm opening in the skull.
5. In addition, or alternatively, the mouse was injected in the tail with 1 mg/mL albumin labeled with Evans Blue to determine the extent, if any, of blood brain barrier (BBB) compromise. Normally, albumin conjugated to Evans Blue is not permeable to the brain parenchyma. The BBB disruption was visible as the blue dye leaked into the tissue upon harvest.
6. Experiments in vitro have shown that the gap junction communication is restored after 2 hours of contact between human and mouse cells. Therefore, apoptosis is detected in a mouse as early 12 or 24 hours postinjection. Bystander apoptosis results from the intracellular messengers generated in the HIV-infected human astrocytes, which are then transmitted to the mouse cells. The mouse cells are resistant to HIV and remain resistant despite the interaction with human infected astrocytes. This is because the mouse cells lack cyclin T1, which is crucial for HIV replication.
7. The cognitive status was determined 24 hours after the injection of the HIV-infected human cells into the mouse cortex. Examples of standardized testing of cognitive behavior that may be administered include the open field test [54].

8. After the selected time points, the animals were sacrificed, and the brains were fixed in formalin, sectioned into 10–20 μm slices, and analyzed for bystander CNS damage, apoptosis, and synaptic compromise using staining and confocal microscopy. Multiple factors were analyzed including bystander apoptosis of mouse cells, survival of injected HIV-infected human astrocytes, and diffusion of secondary messengers into adjacent cells.

Note: Because the mouse cells are not infected by HIV, this procedure optimally allows for the analysis of the gap junction formation by human cells infected with HIV versus the gap junction formation in uninfected human cells. Additionally, because only 5% of the human astrocytes are infected with HIV, it is possible to assess the importance of the passage of the secondary messengers between the gap junctions during infection. An illustration of this methodology is represented in Figure 12.3.

12.2.4.4 An In Vitro Dual-Species Model of HIV Regulation of Gap Junction Formation

1. The mouse astrocytes were purchased from any commercial vendor.
2. A 1:10 or 1:50 mixed culture of human astrocytes to mouse cell culture (astrocytes, neurons, or endothelial cells) were plated in DMEM containing 10% FBS.
3. The cells were incubated at 37 °C and 5% CO_2 for 1–2 hours to allow for optimal gap junction channel contact and formation.
4. Mixed cells were infected with HIV at an MOI of 0.001 for the HIV_{BAL} strain or 20 ng p24/mL/1 × 10^6 cells for HIV_{ADA} or HIV_{LAI} strains for 24 hours. HIV-p24 can be determined by ELISA using a Perkin Elmer commercial kit (Waltham, Massachusetts).
5. Bystander apoptosis was determined using terminal deoxynucleotidyl transferase dUTP nick end labeling (TUNEL) or annexin-5 staining. The mixed cultures were fixed and permeabilized in 70% ethanol for 20 minutes at −20°C before the TUNEL reaction mixture kit (Roche Holding AG, Basel, Switzerland) was added for 1 hour at 37 °C and 5% CO_2. Cyclin type I staining can be used to differentiate human and mouse cells, as it is not present in mice.
6. After incubation, the cells were washed three times in PBS and blocked in a blocking solution for at least 4 hours.
   a. The blocking solution consists of 1 mL of 0.5 M EDTA, 100 μL of fish gelatin, 0.1 g BSA, and 1% horse serum in a total of 10 mL.
7. The primary antibodies, based on the experimental setup, were incubated overnight at 4°C. Potential primary antibodies may include anti-p24 at 1:50 concentration (Genentech, San Francisco, California), anti-cyclin T1 to identify the human cells at 1:300 (Abcam, Cambridge, UK), anti-Cx43 at 1:2000 concentrations, and anti-GFAP at a concentration of 1:800 (Sigma,

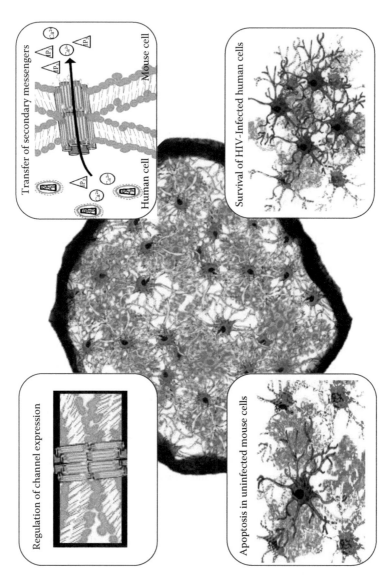

FIGURE 12.3 **(See color insert.)** A dual species model of HIV infection in the CNS allows for the characterization and the analysis of HIV gap junctions and bystander apoptosis. Human HIV-infected astrocytes are injected into a 0.1 mm opening into the mouse cortex. As the mice cells are resistant to HIV, this method allows for the analysis of channel expression, transfer of secondary messengers between human and mouse cells including IP₃, calcium, and ATP, apoptosis in mouse cells, and survival of HIV-infected human astrocytes.

St. Louis, Missouri). In addition, the cells can be stained by TUNEL or with annexin-5 staining to examine the apoptosis.

8. The fixed cells were brought to room temperature in PBS with light shaking for 1 hour before being washed with PBS four times for 5 minutes each.

9. Secondary antibody conjugates were then applied for 1 hour at room temperature.

10. The cells were washed for 1 hour in PBS to remove excess secondary antibody.

11. The cells were mounted using an antifade reagent with DAPI (Thermo Fisher Scientific, Waltham, Massachusetts) and examined using confocal microscopy.

12. The readout was calculated by determining the apoptosis or the spread of secondary messengers from human cells to mouse cells.

Note: The antibody specificity was affirmed by staining the control cells with a nonspecific myeloma protein matching the same isotype of the primary antibodies used. Nonimmune sera is an alternative control for antibody specificity validation.

12.2.4.5 Evaluation of the Spread of Toxic Signals Using Laser Microdissection

1. The cells and the tissues were prepped as 8–20 μm sections using cryostat (for tissue sections) or fixated in 70% ethanol and kept on ice. Several options are available for microdissection mounting:

   a. Glass-foiled polyphenylene napthalate (2 μm) or polyphenylene napthalate is excellent for protein RNA and DNA isolation, but not DIC imaging mode.

   b. Frame-foiled polysthylene terethalate (1.4 μm) or polyester (0.9 μm) is great for downstream matrix-assisted laser desorption/ionization time-of-flight mass spectrometry and high-performance liquid chromatography or chromosome and gene microanalysis, respectively.

   c. Fluo membranes are also available for fluorescent and DIC imaging modes.

   d. Glass Director® slides (Expression Pathology, Rockville, Maryland) are excellent for proteomic applications.

2. After the selection of the appropriate slide, the cells or the tissues were stained for HIV and cellular markers (such as those for neurons, astrocytes, glial, and endothelial cells or cyclin T1 to identify human cells from mouse cells).

3. By using the labeled cells and the laser capture microdissection system, the HIV-infected cells were isolated, in addition to the surrounding uninfected cells, using Leica LMC 6500 equipment (Leica Microsystems, Wetzlar, Germany). This was done to precisely compare their phenotype and identify the second messengers transmitted from the HIV-infected cells into mouse cells.

4. After the collection of the desired material, several applications can be performed as described in Figure 12.4.

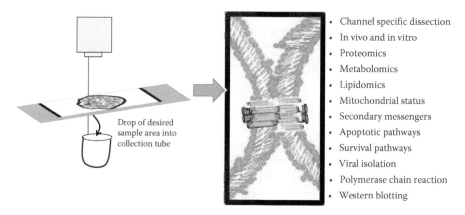

Drop of desired
sample area into
collection tube

- Channel specific dissection
- In vivo and in vitro
- Proteomics
- Metabolomics
- Lipidomics
- Mitochondrial status
- Secondary messengers
- Apoptotic pathways
- Survival pathways
- Viral isolation
- Polymerase chain reaction
- Western blotting

FIGURE 12.4 **(See color insert.)** A schematic of microdissection as a method for isolating the channels for precise downstream analysis. The channel-specific dissection of gap junctions using microdissection allows for numerous downstream applications including proteomics, lipidomics, metabolomics, and others. Additionally, this may be utilized in vivo or in vitro and is a unique methodology for analyzing pathways including those of mitochondrial stress, apoptosis, and survival. PCR, polymerase chain reaction.

12.3 CONCLUSION

The methodology to examine the HIV regulation of gap junctions and hemichannels is both feasible and highly adaptable. The methods described herein provide a basic framework to examine the channels during the infection using both in vivo and in vitro models, human and mouse cell systems, and a variety of potential staining targets for microscopy. Importantly, many of the described methods can be used regardless of an in vivo or an in vitro starting point. The examination of the interaction that HIV has on connexin and pannexin containing channels will elucidate the key steps into the pathology of the infection and the ability of the virus to hijack channel mechanisms in host cells.

REFERENCES

1. Pontow, S. E. et al. Actin cytoskeletal reorganizations and coreceptor-mediated activation of Rac during human immunodeficiency virus-induced cell fusion. *J Virol*, 2004. **78**(13): p. 7138–7147.
2. Melikyan, G. B. HIV entry: A game of hide-and-fuse? *Curr Opin Virol*, 2014. **4**: p. 1–7.
3. Karn, J. Tackling Tat. *J Mol Biol*, 1999. **293**(2): p. 235–54.
4. Fischer, U. et al. The HIV-1 Rev activation domain is a nuclear export signal that accesses an export pathway used by specific cellular RNAs. *Cell*, 1995. **82**(3): p. 475–483.
5. Kline, E. R., and R. L. Sutliff. The roles of HIV-1 proteins and antiretroviral drug therapy in HIV-1-associated endothelial dysfunction. *J Investig Med Off Publ Am Fed Clin Res*, 2008. **56**(5): p. 752–769.
6. Neri, E., V. Musante, and A. Pittaluga. Effects of the HIV-1 viral protein TAT on central neurotransmission: Role of group I metabotropic glutamate receptors. *Int Rev Neurobiol*, 2007. **82**: p. 339–56.

7. Pollard, V. W., and M. H. Malim. The HIV-1 Rev protein. *Annu Rev Microbiol*, 1998. **52**: p. 491–532.

8. King, J. E. et al. HIV tat and neurotoxicity. *Microbes Infect*, 2006. **8**(5): p. 1347–57.

9. Nguyen, D. G. et al. Evidence that HIV budding in primary macrophages occurs through the exosome release pathway. *J Biol Chem*, 2003. **278**(52): p. 52347–54.

10. Orellana, J. A. et al. Pannexin1 hemichannels are critical for HIV infection of human primary CD4+ T lymphocytes. *J Leukoc Biol*, 2013. **94**(3): p. 399–407.

11. Velasquez, S., and E. A. Eugenin. Role of Pannexin-1 hemichannels and purinergic receptors in the pathogenesis of human diseases. *Front Physiol*, 2014. **5**: p. 96.

12. Eugenin, E. A., and J. W. Berman. Gap junctions mediate human immunodeficiency virus-bystander killing in astrocytes. *J Neurosci*, 2007. **27**(47): p. 12844–50.

13. Seror, C. et al. Extracellular ATP acts on P2Y2 purinergic receptors to facilitate HIV-1 infection. *J Exp Med*, 2011. **208**(9): p. 1823–34.

14. Hazleton, J. E., J. W. Berman, and E. A. Eugenin. Purinergic receptors are required for HIV-1 infection of primary human macrophages. *J Immunol*, 2012. **188**(9): p. 4488–95.

15. Graziano, F. et al. Extracellular ATP induces the rapid release of HIV-1 from virus containing compartments of human macrophages. *Proc Natl Acad Sci*, 2015. **112**(25): p. E3265–E3273.

16. Williams, D. W. et al. Mechanisms of HIV entry into the CNS: Increased sensitivity of HIV infected CD14+CD16+ monocytes to CCL2 and key roles of CCR2, JAM-A, and ALCAM in diapedesis. *PLoS One*, 2013. **8**(7): p. e69270.

17. Gendelman, H. E. et al. Efficient isolation and propagation of human immunodeficiency virus on recombinant colony-stimulating factor 1-treated monocytes. *J Exp Med*, 1988. **167**(4): p. 1428–1441.

18. Li, Y. et al. Characterization of antibody responses elicited by human immunodeficiency virus type 1 primary isolate trimeric and monomeric envelope glycoproteins in selected adjuvants. *J Virol*, 2006. **80**(3): p. 1414–26.

19. Cavrois, M., C. De Noronha, and W. C. Greene. A sensitive and specific enzyme-based assay detecting HIV-1 virion fusion in primary T lymphocytes. *Nat Biotechnol*, 2002. **20**(11): p. 1151–4.

20. Gluschankof, P. et al. Cell membrane vesicles are a major contaminant of gradient-enriched human immunodeficiency virus type-1 preparations. *Virology*, 1997. **230**(1): p. 125–33.

21. Schneider, C. A., W. S. Rasband, and K. W. Eliceiri. NIH image to ImageJ: 25 years of image analysis. *Nat Meth*, 2012. **9**(7): p. 671–675.

22. Eugenin, E. A. et al. Gap junctional communication coordinates vasopressin-induced glycogenolysis in rat hepatocytes. *Am J Physiol*, 1998. **274**(6 Pt 1): p. G1109–16.

23. el-Fouly, M. H., J. E. Trosko, and C. C. Chang, Scrape-loading and dye transfer: A rapid and simple technique to study gap junctional intercellular communication. *Exp Cell Res*, 1987. **168**(2): p. 422–30.

24. Bagasra, O. et al. Cellular reservoirs of HIV-1 in the central nervous system of infected individuals: Identification by the combination of in situ polymerase chain reaction and immunohistochemistry. *AIDS*, 1996. **10**(6): p. 573–85.

25. Dahiya, S. et al. Genetic Variation and HIV-Associated Neurologic Disease. *Adv Virus Res*, 2013. **87**: p. 183–240.

26. Cosenza, M. A. et al. Human brain parenchymal microglia express CD14 and CD45 and are productively infected by HIV-1 in HIV-1 encephalitis. *Brain Pathol*, 2002. **12**(4): p. 442–55.

27. Wiley, C. A. et al. Cellular localization of human immunodeficiency virus infection within the brains of acquired immune deficiency syndrome patients. *Proc Natl Acad Sci U S A*, 1986. **83**(18): p. 7089–93.

28. Conant, K. et al. In vivo and in vitro infection of the astrocyte by HIV-1. *Adv Neuroimmunol*, 1994. **4**(3): p. 287–9.

29. Ohagen, A. et al. Apoptosis induced by infection of primary brain cultures with diverse human immunodeficiency virus type 1 isolates: Evidence for a role of the envelope. *J Virol*, 1999. **73**(2): p. 897–906.

30. Tornatore, C. et al. HIV-1 infection of subcortical astrocytes in the pediatric central nervous system. *Neurology*, 1994. **44**(3 Pt 1): p. 481–7.

31. Nuovo, G. J. et al. In situ detection of polymerase chain reaction-amplified HIV-1 nucleic acids and tumor necrosis factor-alpha RNA in the central nervous system. *Am J Pathol*, 1994. **144**(4): p. 659–66.

32. Takahashi, K. et al. Localization of HIV-1 in human brain using polymerase chain reaction/in situ hybridization and immunocytochemistry. *Ann Neurol*, 1996. **39**(6): p. 705–11.

33. Churchill, M. J. et al. Extensive astrocyte infection is prominent in human immunodeficiency virus-associated dementia. *Ann Neurol*, 2009. **66**(2): p. 253–8.

34. Orellana, J. A. et al. HIV increases the release of dickkopf-1 protein from human astrocytes by a Cx43 hemichannel-dependent mechanism. *J Neurochem*, 2013. **128**(5): p. 752–63.

35. Li, W. et al. IFN-gamma mediates enhancement of HIV replication in astrocytes by inducing an antagonist of the beta-catenin pathway (DKK1) in a STAT 3-dependent manner. *J Immunol*, 2011. **186**(12): p. 6771–8.

36. Eugenin, E. A., and J. W. Berman. Cytochrome c dysregulation induced by HIV infection of astrocytes results in bystander apoptosis of uninfected astrocytes by an IP and calcium-dependent mechanism. *J Neurochem*, 2013. **127**(8): p. 644–54.

37. Eugenin, E. A. et al. Human immunodeficiency virus infection of human astrocytes disrupts blood-brain barrier integrity by a gap junction-dependent mechanism. *J Neurosci*, 2011. **31**(26): p. 9456–65.

38. Eugenin, E. A. et al. Differences in NMDA receptor expression during human development determine the response of neurons to HIV-tat-mediated neurotoxicity. *Neurotox Res*, 2011. **19**(1): p. 138–48.

39. Eugenin, E. A., and J. W. Berman. Gap junctions mediate human immunodeficiency virus-bystander killing in astrocytes. *J Neurosci Off J Soc Neurosci*, 2007. **27**(47): p. 12844–50.

40. Ribot, E. et al. Microglia used as vehicles for both inducible thymidine kinase gene therapy and MRI contrast agents for glioma therapy. *Cancer Gene Ther*, 2007. **14**(8): p. 724–37.

41. Loov, C. et al. Engulfing astrocytes protect neurons from contact-induced apoptosis following injury. *PLoS One*, 2012. **7**(3): p. e33090.

42. Lin, J. H. et al. Gap-junction-mediated propagation and amplification of cell injury. *Nat Neurosci*, 1998. **1**(6): p. 494–500.

43. Blanc, E. M., A. J. Bruce-Keller, and M. P. Mattson. Astrocytic gap junctional communication decreases neuronal vulnerability to oxidative stress-induced disruption of Ca2+ homeostasis and cell death. *J Neurochem*, 1998. **70**(3): p. 958–70.

44. Nakase, T., S. Fushiki, and C. C. Naus. Astrocytic gap junctions composed of connexin 43 reduce apoptotic neuronal damage in cerebral ischemia. *Stroke*, 2003. **34**(8): p. 1987–93.

45. Nakase, T., and C. C. Naus. Gap junctions and neurological disorders of the central nervous system. *Biochim Biophys Acta*, 2004. **1662**(1–2): p. 149–58.

46. Nakase, T., Y. Yoshida, and K. Nagata. Enhanced connexin 43 immunoreactivity in penumbral areas in the human brain following ischemia. *Glia*, 2006. **54**(5): p. 369–75.

47. Siushansian, R. et al. Connexin43 null mutation increases infarct size after stroke. *J Comp Neurol*, 2001. **440**(4): p. 387–94.

48. Eugenin, E. A. et al. MCP-1 (CCL2) protects human neurons and astrocytes from NMDA or HIV-tat-induced apoptosis. *J Neurochem*, 2003. **85**(5): p. 1299–1311.

49. Theis, M. et al. Accelerated hippocampal spreading depression and enhanced locomotory activity in mice with astrocyte-directed inactivation of connexin43. *J Neurosci*, 2003. **23**(3): p. 766–76.

50. Zhuo, L. et al. hGFAP-cre transgenic mice for manipulation of glial and neuronal function in vivo. *Genesis*, 2001. **31**(2): p. 85–94.

51. Cohen-Salmon, M. et al. Connexin30 deficiency causes instrastrial fluid-blood barrier disruption within the cochlear stria vascularis. *Proc Natl Acad Sci U S A*, 2007. **104**(15): p. 6229–34.

52. Teubner, B. et al. Connexin30 (Gjb6)-deficiency causes severe hearing impairment and lack of endocochlear potential. *Hum Mol Genet*, 2003. **12**(1): p. 13–21.

53. Schlaeger, T. M. et al. Uniform vascular-endothelial-cell-specific gene expression in both embryonic and adult transgenic mice. *Proc Natl Acad Sci U S A*, 1997. **94**(7): p. 3058–63.

54. Walsh, R. N., and R. A. Cummins. The open-field test: A critical review. *Psychol Bull*, 1976. **83**(3): p. 482–504.

Index

Page numbers followed by f and t indicate figures and tables, respectively.

T - #0041 - 171024 - C320 - 234/156/15 - PB - 9780367658441 - Gloss Lamination